AF344411

PROCESSING OF LONG LENGTHS OF Superconductors

PROCESSING OF LONG LENGTHS OF Superconductors

Proceedings of the symposium
on Processing of Long Lengths of Superconductors,
held during Materials Week '93 in Pittsburgh, Pennsylvania,
October 17 - 21, 1993

EDITED  BY

U. Balachandran

E.W. Collings

A. Goyal

a Publication of

A Publication of The Minerals, Metals & Materials Society
420 Commonwealth Drive
Warrendale, Pennsylvania 15086
(412) 776-9000

Printed in the United States of America
Library of Congress Catalog Number 94-75873
ISBN Number 0-87339-271-X

If you are interested in purchasing a copy of this book, or if you would like to receive the latest TMS publications catalog, please telephone 1-800-759-4867.

Preface

In this proceedings of the symposium on processing of long lengths of superconductors, held during Materials Week '93 (Pittsburgh, October 17-21, 1993), 34 of the 38 papers accepted for presentation at the five-session symposium are included. All manuscripts reproduced here have been reviewed by experts in the field. Sponsored by The Minerals, Metals & Materials Society, the symposium focused on progress, issues, and prospects in the processing of long lengths of both niobium-based and oxide-based superconductors. Considerable progress has been made in the oxide superconductors, and magnets made with high-T_c oxide superconductors have been demonstrated with fields of $\approx$2.6 Tesla. The principal topics of the symposium included processing for high critical current densities, flux-pinning mechanisms, processing for the development of flux-pinning centers, film deposition on flexible substrates, chemistry, phase relations, stoichiometry, effects of strain, processing/microstructure relationships, and prototype magnets for practical applications.

On the first day of the symposium (Tuesday, October 19), speakers elaborated on present and future wire-fabrication techniques for high-T_c superconductors, as well as on methods for characterization. Important subjects discussed on the second day were fabrication of thick films, powder-in-tube processing, fabrication of prototype magnets from long-length conductors, synthesis, flux pinning and texture. In the final session on Thursday, October 21, speakers concentrated on conventional low-temperature superconductors.

We, the editors, express our appreciation to the speakers, attendees, and session chairs for making this endeavor successful. We are indebted to the manuscript reviewers for their very timely reviews. We acknowledge the valuable assistance of the TMS staff in facilitating the development of the program and the publication of this proceedings volume. Finally, we thank Charles Malefyt for editing several of the manuscripts, and Sylvia Hagamann for preparing many of the manuscripts in camera-ready format.

U. (Balu) Balachandran
Argonne National Laboratory
Argonne, IL

E.W. Collings
Battelle Columbus Division
Columbus, OH

Amit Goyal
Oak Ridge National Laboratory
Oak Ridge, TN

November 1993

Table of Contents

TAPE/WIRE FABRICATION

SYNTHESIS AND CHARACTERIZATION

LOW-TEMPERATURE SUPERCONDUCTORS

TAPE/WIRE FABRICATION

PRESENT AND FUTURE WIRE FABRICATION TECHNIQUES FOR HIGH T_c SUPERCONDUCTORS

S. Jin

AT&T Bell Laboratories, Murray Hill, NJ 07974

ABSTRACT

Fabrication of high-J_c, long-length conductors is essential for successful applications of the high T_c superconductors. Several different approaches for high-J_c wire fabrications have been demonstrated, however, none of them meets all the requirements in terms of desirable length, critical currents, and operating temperature ranges. High-J_c YBCO wires have been made only in length less than a meter. The BSCCO wires/ribbons exhibit high J_c and can be made long but they are not particularly useful for major applications above ~ 30K. For TBCCO, neither high-J_c long wire comparable to BSCCO material nor sufficient flux pinning comparable to YBCO has been demonstrated, even with the promising single Tl-O layered compound. This paper describes various microstructural and processing problems limiting the development of desirable long-length superconductors, and discusses some alternative approaches that may be utilized to overcome the problems.

Processing of Long Lengths of Superconductors
Edited by U. Balachandran, E.W. Collings and A. Goyal
The Minerals, Metals & Materials Society, 1994

I. INTRODUCTION

For successful applications of bulk high T_c superconductors (HTSC), the fabrication of long-length wires is essential. While significant progress has been made, high-J_c, long-length wires with all the desirable characteristics at high temperatures ar 1 high fields are yet to be produced. The materials issues such as how to minimize the grain boundary weak link and flux creep problems, or how to fabricate long lengths (e.g., in excess of 1000 meters) with uniform, reproducible properties, still need to be addressed. Also, at some point in the future, the high T_c superconductor community will have to decide which of the three types of materials, YBCO, BSCCO, or TBCCO, is the most suitable system for desirable long-length wires so that the limited resources can be concentrated on engineering development for large-scale wire applications. This paper reviews various microstructural and processing problems limiting the development of long-length superconductors, and discusses some alternative processing approaches that may be utilized to overcome the problems.

II. VARIOUS WIRE FABRICATION TECHNIQUES

For the high T_c superconductor materials to be commercially useful in high-field magnets, energy storage devices, motors/generators, or power transmission cables, they need to be shaped into composite multifilamentary wires or ribbons containing both the superconductor and a normal metal. Such a composite geometry is desirable for several important reasons: (1) parallel electrical conduction paths to serve as a bypass in case of local failure, (2) heat sinking, (3) minimal loss of energy by magnetic flux motion, (4) ac loss reduction, (5) mechanical protection of the brittle superconductor core against Lorentz force and stresses during handling, and (6) protection from atmospheric environment.

As the high T_c superconductors are mechanically hard and brittle ceramic materials, they can not be drawn into fine wires. However, several different approaches to the fabrication of YBCO, BSCCO, and TBCCO wires have been reported. The wires or continuous conductor ribbons can be prepared by either bulk processing or thin film/thick film processing.

The wires prepared by bulk processing include metal-clad composite wires or ribbons (by powder-in-tube technique using oxide[1,2] or metallic powder[3]), metal-core composites[4] (by surface coating technique), bare ceramic wires[5] (by slurry-extrusion method), jelly roll wires,[6] melt-spun ribbons of oxide[7] or metallic precursor[8] materials, and zone-melted wires[9] (by directional melt-texturing). These wire fabrication techniques are schematically illustrated in Fig. 1. These techniques have been utilized to prepare wires from all three types of high T_c superconductors, Y-Ba-Cu-O, Bi-Sr-Ca-Cu-O, and Tl-Ba-Ca-Cu-O with varying degrees of success.

The wires/ribbons of high T_c superconductors can also be fabricated using the deposition of thin or thick films as shown schematically in Fig. 2. Physical deposition,[10,11] chemical vapor deposition, dip coating or spray coating of a slurry containing the superconductor powder,[12,13] electrodeposition,[14] tape casting[15] and other techniques such as plasma spray coating have been used.

Because of the desired stabilization by normal metal, most of the wires or ribbons have metal substrate or sheath in contact with the superconductor material. Silver has been the most frequently used metal due to its inertness. Bare ceramic superconductor wires would have to be somehow coated with normal metal, e.g., by plating, plasma spray coating, etc.

Not all the wires/ribbons shown in Figs. 1 and 2 exhibit desirable high J_c properties. The BSCCO material fabricated into metal-clad wire, jelly roll wire, and dip-coated wire, the YBCO material fabricated into zone-melted wire and physically deposited ribbon, and the TBCCO material processed by metal clad approach, thick film coating, and electrodeposition exhibit high current carrying capability. One common problem encountered in some of the other wires/ribbons is the grain boundary weak-link problem and resultant low J_c with its severe field dependent deterioration.

III. GRAIN BOUNDARY WEAK LINKS

The grain boundary weak link problem in bulk polycrystalline high T_c materials may be overcome by employing the following fundamentally different processing approaches.

(i)   Decoration of grain boundaries with a metallic coating thinner than the normal metal coherence length--Normal metals such as silver, when they become superconducting through proximity effect, have a coherence length on the order of hundred angstroms, much larger than 4-30 Å in HTSC.

(ii)   Modification or correction of grain boundary oxygen stoichiometry --- It has been observed than an ozone gas heat treatment of YBCO films[16] suppresses the weak link behavior and significantly increases the transport J_c.

(iii)   Introduction of a large number of specially oriented non-weak-link grain boundaries[17] (near-coincident-site lattice orientations). --- Suitable processing conditions (perhaps involving high temperatures or long heat treatment times) may be employed in order to enhance the formation of these boundaries and allow high transport J_c in polycrystalline materials.

(iv)   Formation of crystallographic texture --- In the highly textured HTSC, the presence of undesirable, high-angle grain boundaries in the paths of transport currents can be avoided.

The first three approaches have not yet been fully proven by actual experiments; clever processing details will have to be incorporated in order to overcome the anticipated experimental difficulties in these approaches. The last approach utilizing the crystallographic texture has been amply demonstrated in terms of achieving high transport J_c in polycrystalline HTSC.

IV. BSCCO WIRES

High J_c BSCCO wires/ribbons have been demonstrated by many research groups.[18-22] The J_c's in short wires are in excess of 10^5 A/cm^2 at 4.2K and H = 25T field, and those for the long wires (e.g., ~ 100 meters) about 10^4 A/cm^2. Demonstration solenoids have been constructed with generated magnetic fields as high as 2.6 tesla.[22] These high J_c wires are typically produced by either metal-clad approach using Ag-sheath or by doctor blade approach using Ag foil substrate. There are also several variations of these basic approaches that have been employed for BSCCO wire fabrication.

The Ag-clad approach is easier for producing multi-filamentary wire using re-bundling/wire drawing method. Because of the relatively extensive deformation employed for this ductile metal/brittle ceramic composite (often greater than 99% deformation), the possibility of "sausage effect" and reduced J_c in long wires is of concern. Further R&D is needed to find out if such an effect is a major problem or not.

While the doctor-blade or slurry coating approach is not the best for fabricating multifilamentary wire geometry, the process makes it possible to produce long wires without (or with very small amount of) deformation. In this case, no sausage effect is anticipated, and the J_c of a long wire is likely to be essentially the same as a short wire.

The high J_c values in BSCCO wires/ribbons are attributed to the strong c-axis texture. The principal mechanism of the texture formation appears to be related to the interface-induced texturing during partial melt processing of the BSCCO wires/ribbons. The layer-like growth of BSCCO crystals aided by two-dimensionality at the Ag-superconductor interface leads to the strong c-axis texture, and it is perhaps no coincidence that high J_c BSCCO wires are obtained only when Ag is involved either as a clad or a substrate.

One drawback in these high J_c BSCCO wires is the flux creep problem at temperatures higher than ~ 30K, and H $\geq$ 1T (especially for the field direction of H $\perp$ ab) as shown in Fig. 3. Thus the usefulness of the present BSCCO wires is confined to the operating temperature T < 30K (for magnetic fields much greater than ~ 1 tesla) unless flux pinning is enhanced by some new, industrially viable methods in the future.

V. YBCO WIRES

The YBCO superconductor exhibits the least flux creep problem and intergrowth problem among the three major HTSC materials (YBCO, BSCCO and TBCCO), and hence is the most desirable material if long wires can be made. The development of long YBCO wires has been somewhat slower than that of BSCCO mainly for the following reasons.

1. The nature of the grain boundary weak links in YBCO is different from that in BSCCO, i.e., they are a lot less forgiving and thus most of the tilt or twist boundaries (except for some special orientations) are severely weak-linked. In YBCO wires, therefore, an essentially three dimensional (tri-axial) grain alignment is necessary to achieve high J_c (e.g. by melt-texturing) as compared to the only two dimensional (one orientation) grain alignment needed for high J_c BSCCO wires.

2. The kinetics of peritectic phase transformation needed for directional solidification during melt texture processing is too slow to allow long wire fabrication in an industrially desirable time scale.

3. Unlike BSCCO or TBCCO the melting temperature of the Y-123 superconductor is higher than that of the Ag-clad material, so that the partial melt processing of the superconductor can not be carried out easily without damaging the Ag-cladding in the wire.

Some possible new approaches yet to be explored for fabrication of long YBCO wires include a modification of phase diagram to allow congruent melting (rather than peritectic melting) for high speed zone melting, a use of ultrafine 211 phase inclusions as local chemical reservoirs to reduce the diffusion distance for peritectic reaction, a use of thin and narrow dimension to exploit the natural crystal growth direction from YBCO melt for desired texturing, and a use of mechanically self-aligning particle shape such as elongated-plate shape YBCO crystal particles.

The use of epitaxial thin film deposition technique for high J_c YBCO wire fabrication (such as shown in Fig. 2(a)) has recently been demonstrated.[11] The YBCO film was deposited on a yttrium-stabilized-zirconia (YSZ) buffer layer on a metal substrate such as Hastelloy. The YSZ buffer layer has been biaxially aligned through ion-beam-assisted deposition. During the subsequent deposition of $YBa_2Cu_3O_{7-\delta}$ film (~ 1 μm thick) by laser ablation, an epitaxial growth occurs with the resultant YBCO film textured not only in the c-axis but in the a- and b-axes as well. Significantly improved transport J_c values of 2.5×10^5 A/cm^2 at 77K, H = 0, and 2.2×10^4 A/cm^2 at H = 8T (H // ab) have been obtained.

For the thin film approach to be commercially viable for long wire fabrications, the deposition speed will have to be significantly increased. Another problem that needs to be addressed is the very small volume fraction of the superconductor as compared to the substrate material, and consequently the much reduced overall current carrying capacity per unit cross-sectional area of the wire. It would be desirable to increase the thickness of the superconductor and reduce that of the substrate alloy sheet. This may be partly achieved by multilayer deposition of epitaxial films.

VI. TBCCO WIRES

The TBCCO superconductor appears to have materials characteristics somewhat in-between YBCO and BSCCO. The texturing and J_c values in TBCCO wires (e.g. prepared by electrodeposition technique shown schematically in Fig. 2(d)) are better than in YBCO but not as good as in BSCCO wires. The flux creep resistance of the Tl-1223 material is better than that of BSCCO but is worse than that of YBCO. Further studies are required to understand whether two or three dimensional texture is needed for high J_c in TBCCO wires.

While the thallium toxicity and the availability issues need to be addressed before TBCCO wires are developed and widely used, well-textured, high-J_c TBCCO wires (e.g. 1223 type) with higher operating temperatures than those of BSCCO could become valuable if the use of simpler or cheaper cryocooling technique is desired.

VII. SUMMARY

Various wire fabrication techniques for YBCO, BSCCO and TBCCO superconductors have been reviewed, and the advantages/disadvantages of each technique have been compared. Some possible new approaches for overcoming various microstructural and processing problems and for long wire fabrications have also been discussed.

REFERENCES

1. S. Jin, R. C. Sherwood, T. H. Tiefel, R. B. van Dover and D. W. Johnson, Jr., Appl. Phys. Lett. **51**, 203, (1987).

2. K. Sato, T. Hikata, H. Mukai, M. Ueyama, N. Shibuta, T. Kato, T. Masuda, M. Nagata, K. Iwata, and T. Mitsui, IEEE Trans. Magn. **27**, 1231 (1991).

3. A. Otto, L. J. Masur, J. Gannon, E. Podtburg, D. Daly, G. J. Yurek, and A. P. Malozemoff, "Multifilamentary Bi-2223 Composite Tapes Made by a Metallic Precursor Route", in *IEEE Transactions on Applied Superconductivity* to be published.

4. T. H. Tiefel, S. Jin, R. C. Sherwood, R. B. van Dover, R. A. Fastnacht, M. E. David, D. W. Johnson, Jr., and W. W. Rhodes, J. Appl. Phys. **64**, 5896 (1988).

5. R. B. Poeppel, B. K. Flandermeyer, J. T. Dusek, and I. D. Bloom, Chemistry of High Temperature Superconductors, (ACS, Washington, DC) ACS Series **351**, 261 (1987).

6. C. C. Tsuei, C. C. Chi, T. Frei, D. B. Mitzi, T. Kazyaka, T. Haugan, J. Ye, S. Patel, D. T. Shaw, and M. K. Wu, Materials Chemistry and Physics **32**, 95 (1992).

7. S. Jin, R. C. Sherwood, T. H. Tiefel, G. W. Kammlott, R. A. Fastnacht, M. E. Davis and S. M. Zahurak, Appl. Phys. Lett. *52*, 1628 (1988).

8. P. Haldar, Y. Z. Lu, and B. C. Giessen, Appl. Phys. Lett. **51**, 538 (1987).

9. M. J. Neal, D. B. Chandler, L. J. Klemptner, and M. V. Parish, IEEE Trans. Appl. Supercond. **1**, 175 (1991).

10. E. Narumi, L. W. Song, F. Yang, S. Patel, Y. H. Kao, and D. T. Shaw, Appl. Phys. Lett. **58**, 1202 (1991).

11. Y. Iijima, N. Tanabe, O. Kohno, and Y. Ikeno, Appl. Phys. Lett. **60**, 769 (1992).

12. J. Shimoyama, T. Morimoto, H. Kitaguchi, H. Kumakura, K. Togano, H. Maeda, K. Nomura, and M. Seido, Jpn. J. Appl. Phys. **31**, L163 (1992).

13. S. Jin, J. E. Graebner, T. H. Tiefel, R. B. van Dover, A. E. White, and G. W. Kammlott, Physica **C177**, 1917 (1991).

14. R. N. Bhattacharya, P. A. Parilla, R. Noufi, P. Arendt, and N. Elliott, J. Electrochem. Soc. **139**, 67 (1992).

15. D. W. Johnson, Jr., E. M. Gyorgy, W. W. Rhodes, L. C. Feldman, and R. B. van Dover, Adv. Ceram. Mater. **2**, 364 (1987).

16. M. Kawasaki, P. Chaudhri, and A. Gupta, Phys. Rev. Lett. **68**, 1065 (1992).

17. S. E. Babcock, X. Y. Cai, D. L. Kaiser, and D. L. Larbalestier, Nature **347**, 167 (1990).

18. K. Heine, J. Tenbrink, and M. Thoener, Appl. Phys. Lett. **55**, 2441 (1989).

19. K. Sato, N. Shibuta, H. Mukai, T. Hikata, M. Ueyama, and T. Kato, J. Appl. Phys. **70**, 6484 (1991).

20. K. Togano, H. Kumakura, K. Kadowaki, H. Kitaguchi, H. Maeda, J. Kase, J. Shimoyama, and K. Nomura, Cryogenic Eng. **38**, 1081 (1992).

21. S. Jin, R. B. van Dover, T. H. Tiefel, J. E. Graebner, and N. D. Spencer, Appl. Phys. Lett. **58**, 868 (1991).

22. P. Haldar, J. G. Hoehn, Jr., U. Balachandran, and L. R. Motowidlo, Proc. Symp. on ''Processing of Long Lengths of Superconductors'', 1993 TMS-AIME Fall Meeting, Pittsburgh, October 17-21, 1993.

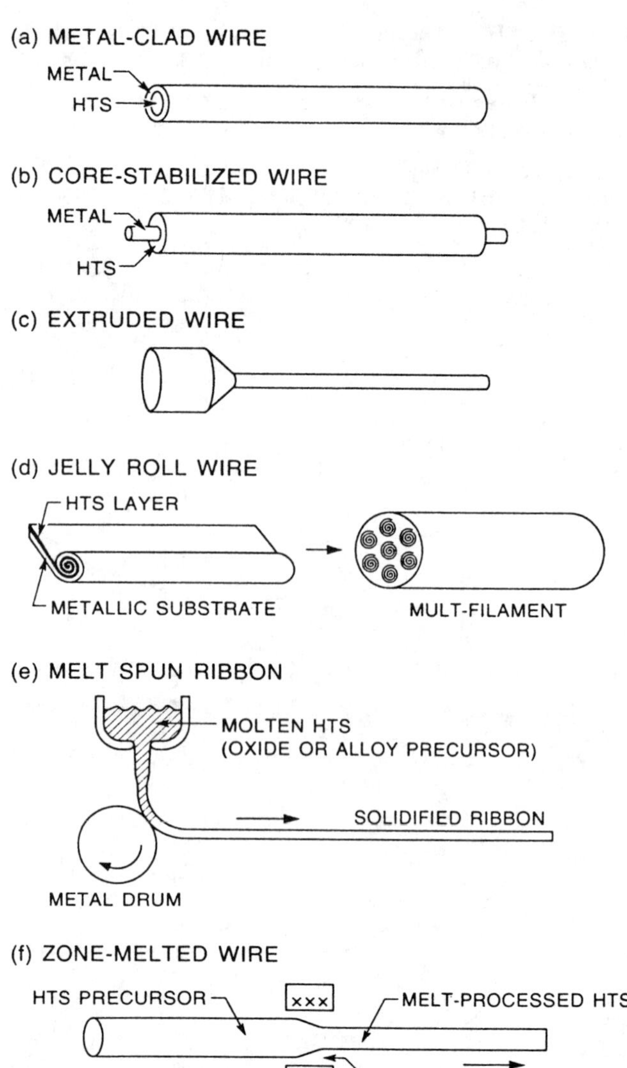

Fig. 1 Schematic illustration of various wire fabrication techniques for high T_c superconductors.

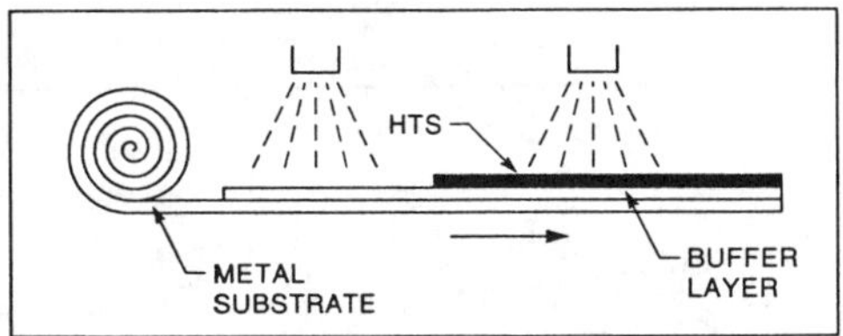

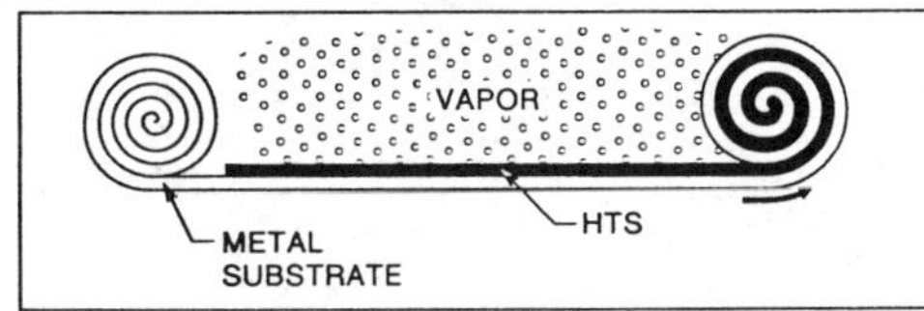

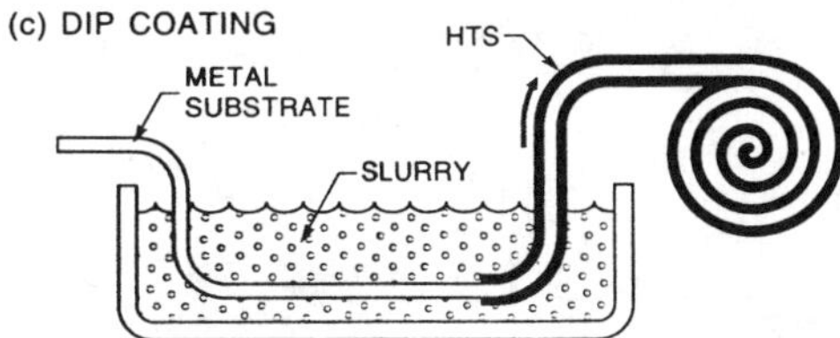

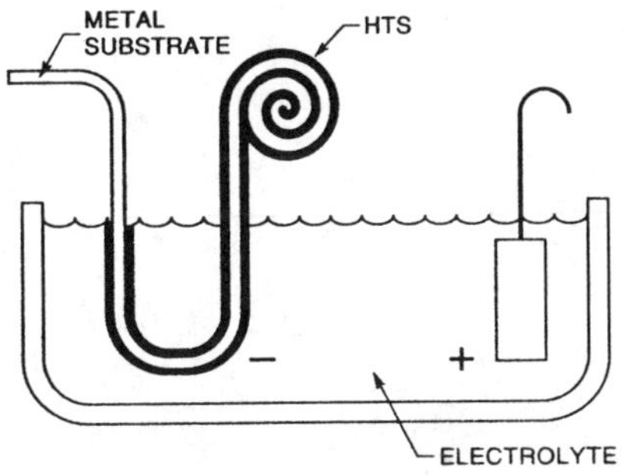

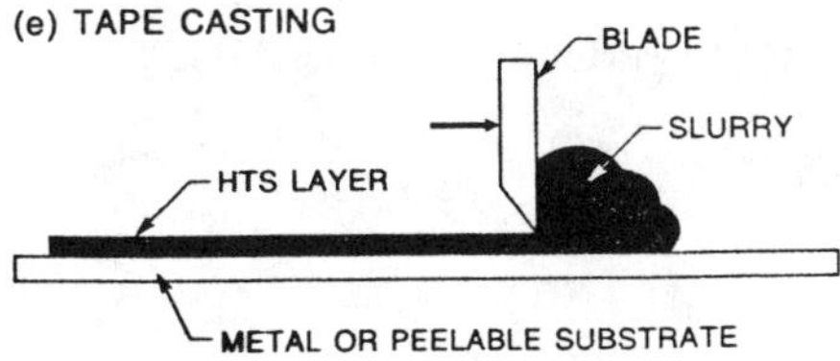

Fig. 2   Fabrication of HTSC ribbons by various deposition techniques.

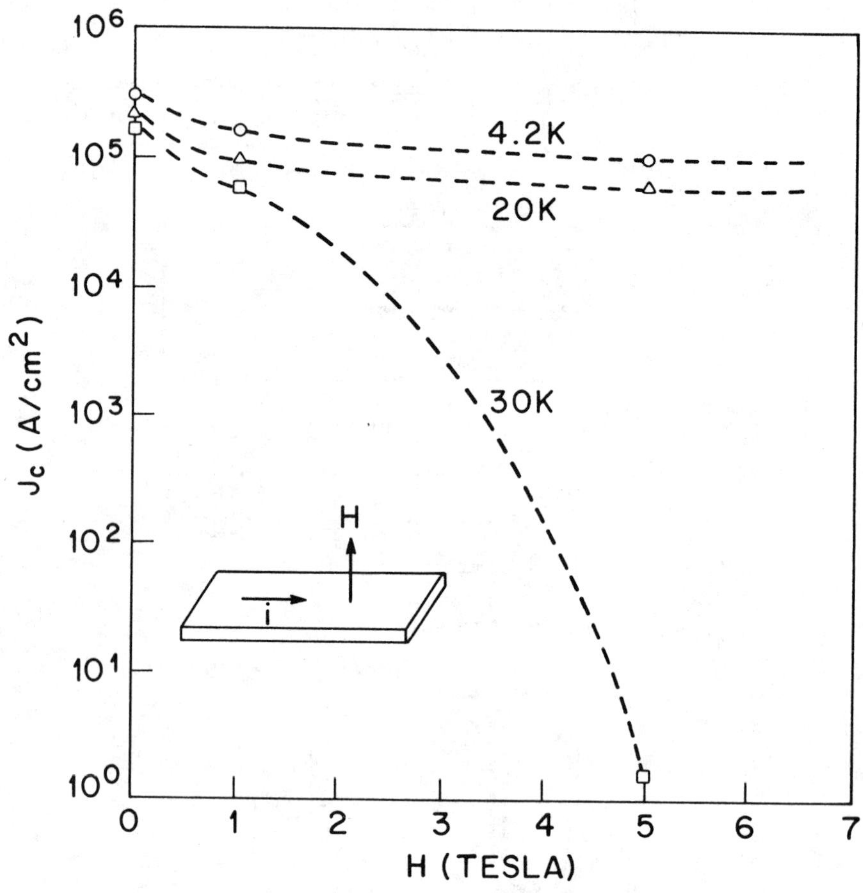

Fig. 3   Transport J_c vs H in c-axis textured BSCCO ribbons.[21]

POWDER SYNTHESIS, FABRICATION, AND TRANSPORT

PROPERTIES OF LONG-LENGTH Ag-CLAD Bi-2223 CONDUCTORS

A. N. Iyer[1], and U. Balachandran[1]
P. Haldar[2], J.G. Hoehn, Jr.[2], and L.R. Motowidlo[3]

[1]Energy Technology Division, Argonne National Laboratory, Argonne, IL 60439
[2]Intermagnetics General Corporation, Guilderland, NY 12084
[3]IGC Advanced Superconductors, Inc., Waterbury, CT 06704

Abstract

Prereacted polyphase powders of Pb-doped Bi-Sr-Ca-Cu-O (2223) were prepared by the solid-state reaction of appropriate amounts of Bi_2O_3, PbO, $SrCO_3$, $CaCO_3$, and CuO. The calcined powders were characterized by thermal analysis, x-ray diffraction and scanning electron microscopy. The powders were then used to fabricate long lengths of flexible Ag-clad 2223 superconductor tapes via the powder-in-tube technique. Growth of the 2223 phase in the tapes after each heat-treatment step was then studied. At 4.2 and 27 K, improved process conditions yielded transport critical current density (J_c) values greater than 10^5 A/cm^2 at zero field; at 77 K, J_c exceeded 4×10^4 A/cm^2. Long tapes (up to 70 m in length) were fabricated and cowound into superconducting pancake coils, and their transport properties were characterized at 4.2, 27, 64, and 77 K.

Processing of Long Lengths of Superconductors
Edited by U. Balachandran, E.W. Collings and A. Goyal
The Minerals, Metals & Materials Society, 1994

Introduction

Since the discovery by Maeda et al. [1] of the 110 K 2223 phase in the Bi-Sr-Ca-Cu-O system, considerable effort has been expended to obtain single phase 2223. Yet two major issues, namely the kinetics and the stability of the phase, seem to complicate its development. The 2223 phase is stable only in a very narrow temperature range and the kinetics of its formation is so slow that it is almost impossible to obtain phase pure 2223 [2-5]. Because the characteristics of the high T_c superconductors developed with BSCCO powder are dictated by the amount and alignment of the 2223 phase, overcoming the above mentioned problems is of utmost importance.

Over the past few years, several synthesizing techniques have been developed so that powders with the desired properties can be obtained. These powders have been used to fabricate superconductors for practical application, using techniques such as powder-in-tube (PIT), sinter forging, and hot isostatic pressing [4,6]. At 77 K, the transport properties of the conventionally consolidated and sintered 2223 samples are very low [7]. On the other hand, tapes fabricated by the PIT process seem to exhibit very good transport properties. Using a series of thermomechanical treatments, the superconducting compound can be textured to obtain very high critical current densities [2,4,6,8-10]. The metallic sheath used in fabrication, protects the superconductor core from chemical and mechanical abrasion [6]. Also, if the superconductor goes normal for any reason, the metallic sheath acts as a shunt. Additionally, the main advantage of the PIT process is that for high energy and power transmission applications, the superconductor should be fabricated in the form of long flexible wires, for which the PIT process seems to be a very convenient technique.

Considerable progress has been made in the development of the PIT process, and several groups have successfully demonstrated high current density (J_c) in short-tape samples of silver clad Bi-2223 [8-12]. While, at 77 K, J_c exceeds 4×10^4 A/cm^2, at 4.2 K and magnetic fields up to 20 T, the J_c values are significantly higher than those of the currently available low-temperature NbTi and Nb$_3$Sn superconductors [2,7]. Long lengths of the tapes have been fabricated with the PIT process. These tapes have been cowound into prototype pancake-shaped coils by the "wind and react" approach. In this paper, we discuss not only some of the latest results of short tape samples but also the fabrication procedures and transport properties of long conductors.

Experimental Procedure

The precursor powder for the PIT process was obtained by the standard solid-state reaction of high-purity oxides and carbonates of Bi, Pb, Sr, Ca, and Cu. The polyphase powders were mixed and calcined at 800-850°C for ≈50 h to obtain a partially reacted powder. This powder consisted of a mixture of Bi-2212, calcium cuprates, and other secondary phases. In order to obtain a more homo-geneous powder, intermittent grinding was employed during the calcination stage. For some of the best results, the cation ratios used were Bi:Pb:Sr:Ca:Cu = 1.8:0.4:2:2.2:3. Addition of lead not only changes the physical properties of the different members of the Bi-Sr-Ca-Cu-O system but also increases their stability. Moreover lead enhances the formation of the 110 K 2223 phase [12,14]. The

14

precursor powders were characterized by differential thermal analysis (DTA), X-ray diffraction (XRD), scanning electron microscopy (SEM) and wet chemical analysis. They were then packed by mechanical agitation into high-purity silver tubes. (Ag which has very good thermal and electrical properties, is normally used as the sheath material). The packed tubes were swaged and drawn to a final diameter of ≈ 2 mm, then rolled at the rate of about 10% reduction per pass to a final thickness of 0.1 mm. To obtain optimal properties while avoiding sausaging of the tapes, the reduction during mechanical working must be kept below 15%. After every 30% reduction the tubes were given an intermittent annealing treatment for about 5 min at approximately 450°C. This is done primarily to relieve the strain associated with cold working of the Ag sheath.

The tapes were also characterized by XRD and SEM. Transport properties were measured by the standard four-point probe dc measurement with 1.0 µV/cm used as the criteria for these J_c measurements. J_c values were determined over the cross section of the superconducting core only. The ratio of the Ag to the superconductor is 3:1.

Short tapes were cut and subjected to a series of uniaxial pressing and heat treatment schedules. Heat treatment was carried out at 820-850°C for ≈ 50-250 h. Long lengths of tapes were fabricated by a two-step rolling method. These tapes are characterized by winding them into pancake-shaped coils. One to six tapes are cowound in parallel on an alumina mandrel with an insulating layer separating each layer of tape. The tapes were then heat treated by the "wind and react" approach. Small test magnets were fabricated by stacking the pancake coils on top of each other.

<u>Results and Discussion</u>

Previous studies on powder synthesis have indicated that nearly phase-pure 2223 can be obtained by calcining at 850°C for ≈ 100 h. Phase purity of the powder was determined by XRD. The powders were then compacted into pellets and sintered in air at 840-850°C; the sintered pellets were found to be only 68% dense and exhibited low J_c values. Calcining the powder at temperatures less than 850°C and for shorter time resulted in a mixed-phase powder consisting of 2223, 2212 and secondary phases such as alkaline earth cuprates. DTA analysis of the powders was used as the basis for selecting sintering temperatures. The DTA trace exhibited two endotherms with onset temperatures of ≈ 835°C and ≈ 850°C. Pellets made from the mixed-phase powder showed not only enhanced densification (≈ 80%) but also better transport properties. This implies that a transient liquid phase, which is consumed by an in-situ reaction during sintering, is important to achieve increased densification and improved transport properties.

As mentioned earlier, some of the best results were achieved with a cation ratio of Bi:Pb:Sr:Ca:Cu = 1.8:0.4:2:2.2:3. Investigation by Oota et al. [15] indicated that the presence of excess Ca and Cu enhances the formation of the 2223 phase. Figure 1 shows the effect of Ca/Cu contents on the J_c of Ag-clad 2223 tapes at 77 K and with total heat treatment times at 830-850°C in air. Although both samples were processed similarly, sample A (with relatively low Ca/Cu content) took much longer time to achieve high J_c , as compared to sample B (with relatively high Ca/Cu content). This implies that the kinetics of the 2223 phase formation in sample A, is relatively slow as compared to sample B [10,16].

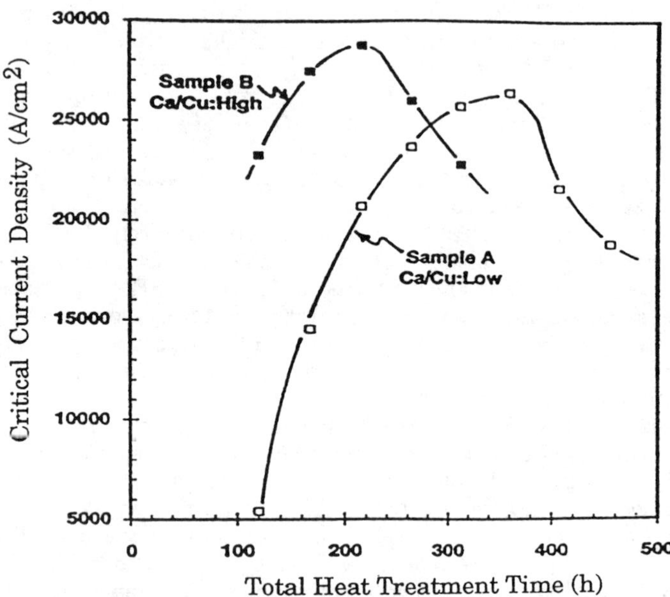

Figure 1 - Effect of Ca/Cu ratio on critical current density (Jc) of Ag-clad 2223 tapes at 77 K and total heat treatment time at 830-850°C.

Short tape samples were cut from the long tapes and subjected to a series of thermomechanical treatments. The sequence consisted of first heat treating at 820-850°C for ≈50 h, followed by uniaxial pressing the tapes and then heat treatment for an additional 100 h. These steps were repeated at least four or five times. After each step, the tapes were characterized by XRD, SEM, and critical current measurements. Qualitative analyses of J_c, as a function of number of thermomechanical treatments, indicate that J_c initially increases, reaches a peak, and then finally decreases. This variation has been reported by several groups and it may be due to the variations in degree of texturing and local inhomogenities of the tapes [17-19].

Plots A, B, and C in Fig. 2 correspond to XRD patterns of the short tape samples, after the first, third, and fifth steps, respectively, of the thermomechanical treatment. The XRD patterns were obtained by peeling off the silver layer while keeping the bulk of the superconductor core intact. The XRD pattern of the tape after the first heat treatment shows the presence of a large amount of 2212 phase and some minor secondary phases. Subsequent pressing and heat treatment of the tape resulted in increased texturing of the grains and also in the amount of the principal current-carrying 2223 phase. This is evident from XRD pattern B. As illustrated in Fig. 2, further treatment resulted in the degradation of the properties because the samples began to decompose. The XRD pattern shows that after ≈450 h of heat treatment, the samples decompose to form 2212 and 2201 phases.

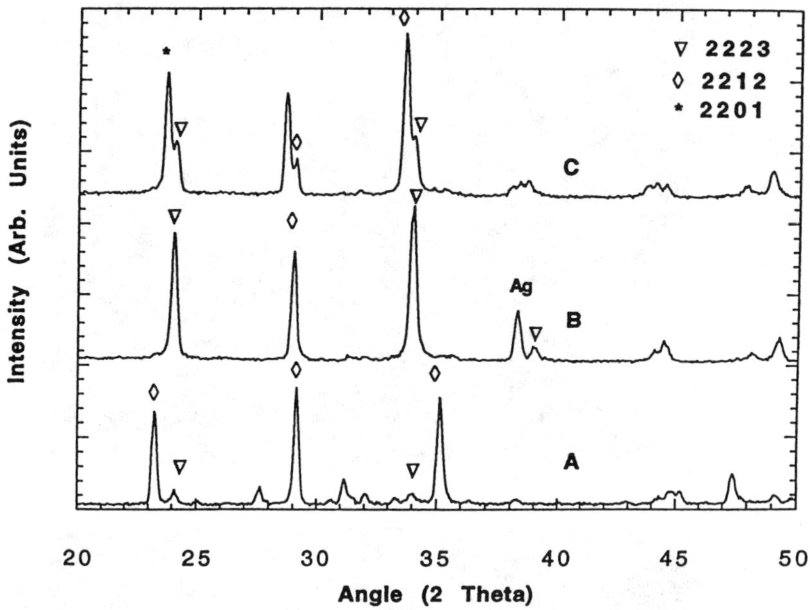

Figure 2 - X-ray diffraction of 2223 tapes after first (Plot A), third (Plot B) and fifth (Plot C) steps of thermomechanical treatments.

SEM photomicrographs of the as-fabricated (untreated) Ag-clad 2223 tape, shown in Fig. 3(a), indicate no texturing. The matrix shows large amounts of secondary phases. On the other hand, the SEM micrograph of the tapes after ≈250 h of heat treatment and two rounds of uniaxial pressing shows large plates of the well textured 2223 phase. Mechanical working resulted in the fragmentation and more even distribution of the secondary phases. The increase in grain alignment and amount of 2223 phase caused an increase in J_c of the tapes. The tape shown in Fig. 3(b) carried almost 30 A of current at 77 K and zero applied field. The transient liquid phase that is formed in-situ helps to heal cracks induced during mechanical working.

Previous studies have shown that the transport properties of the uniaxially pressed samples are much better than those obtained by rolling. This difference can be attributed to the more uniform deformation achieved during pressing than during rolling. During rolling, the pressure varies along the arc of contact between the roll and the sample; this not only results in inhomogeneous deformation of the tapes, but also introduces cracks. In turn, this deteriorates the transport properties of tapes [2,20,21]. But because uniaxial pressing is not suitable for production of long lengths of tape, effort is currently being made to optimize the rolling parameters. Thorough study of the rolling process and careful optimization of its parameters should enable us to fabricate long lengths of Ag-clad 2223 tape with transport properties similar to those obtained using uniaxial pressing.

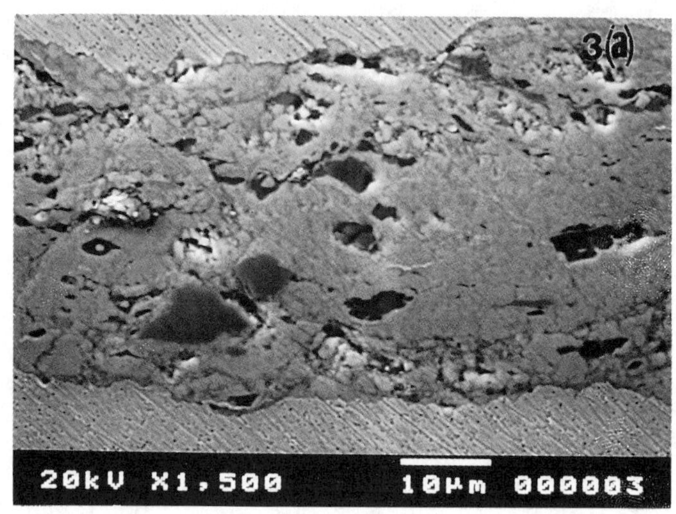

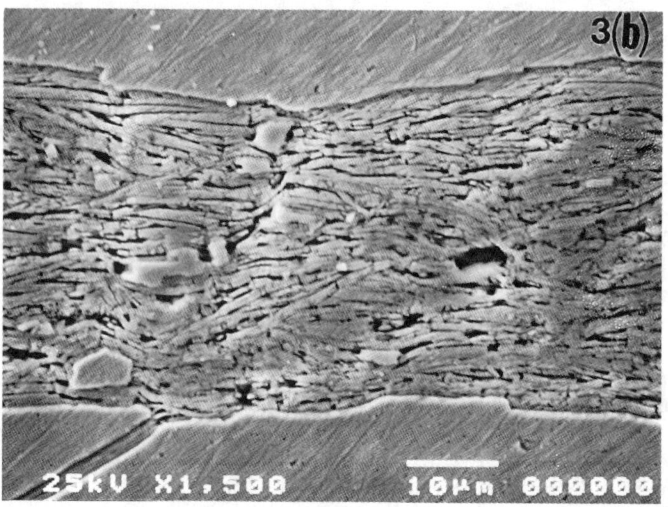

Figure 3 - SEM photomicrographs of Ag-clad 2223 tape (a) as fabricated and (b) after 250 h at ≈850°C and two rounds of uniaxial pressing.

We studied the behavior of short tape samples in the presence of magnetic fields up to 20 T at 4.2 and 27 K. The field was applied both parallel and perpendicular to the flat direction of the tape. At 4.2 K, J_c was found to be independent of applied field, but at 27 K, the applied field in the perpendicular direction tends to drive the superconductor to the normal state in field > 15 T [2,7]. At 4.2 K and 27 K, J_c of the samples was almost 10^5 A/cm^2.

Several parameters are involved in the processing of the Ag-clad 2223 tapes using the PIT method. These parameters should be carefully tailored to obtain the desired properties. Additionally, study by Flukiger et al. [22] have shown that carbon content of the powder could be a limiting factor for the transport properties of the tapes. The carbon could be segregating to the grain boundaries, thus causing poor connectivity between the grains [4,22]. Carbonates used in the precursor powder could be one source of the carbon. In addition, handling and proper storage of the powder also seems to play an important role. Currently, we are monitoring the carbon contents of our powder. The powders are stored in a standard dessicator. Preliminary results indicate that the powder tends to pick up a great deal of carbon ; exposure for ≈30 days almost doubled the carbon content of the powder from 0.038 wt% to 0.066 wt%. To study the effect of carbon on transport properties, grain alignment, and 2223 phase formation, tapes were prepared from powders containing both low and high carbon content. These tapes are currently being processed.

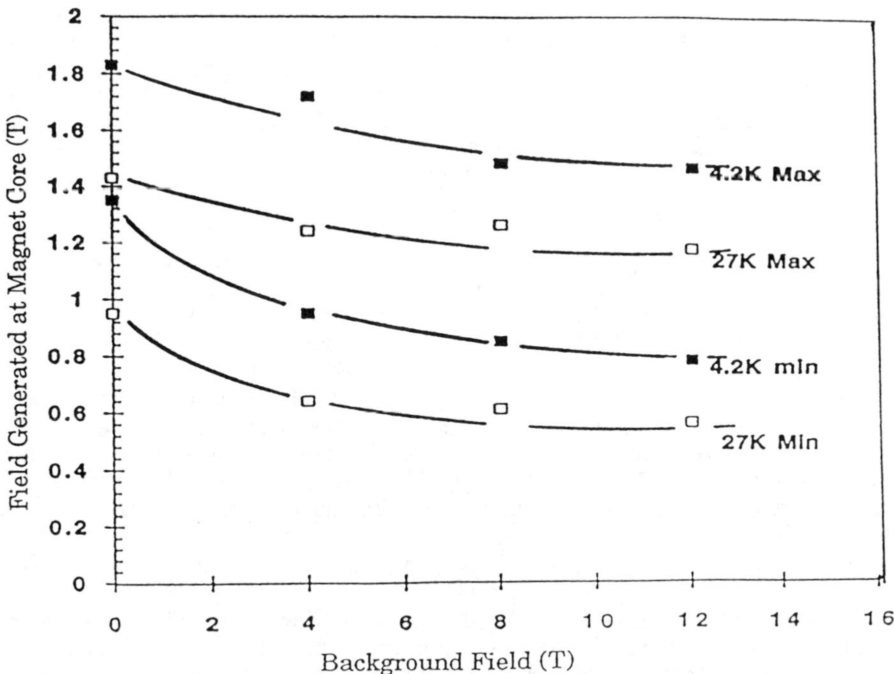

Figure 4 - Performance of a test magnet fabricated from 10 pancake coils
(≈480 m of tape) as a function of background field at 4.2 and 27 K.

Long lengths of Ag-clad 2223 tapes ($\approx$70 m) have been fabricated using a two-step rolling method. During rolling, the tapes were given an intermediate heat treatment. The long conductors are tested by winding them onto an alumina mandrel. 70 m long conductor carried a critical current (I_c) of 23 A at 77 K in self field, which translates to a J_c of $\approx$2x10^4 A/cm^2.

The behavior of these long conductors in a magnetic field was tested by winding them into pancake-shaped coils, which were fabricated by the "wind and react" approach. Three lengths of 16 m long tapes were cowound onto a 2.5 cm diameter mandrel. The resulting coil had an outer diameter of $\approx$11.30 cm and a total of 70 turns. Transport properties of the coil were determined using the 1.0 μV/cm criteria. The critical current at 4.2 , 27, and 77 K were 167,130, and 22 A respectively. Ten pancake coils, each containing three 16 m lengths of Ag-clad 2223 tapes, have been fabricated. Total length of tapes used in this coil is $\approx$480m. The coils were stacked together, connected in series to form a test magnet, and tested at 4.2 and 27 K as a function of applied magnetic field up to 12 T. Figure 4 shows the performance of this test magnet as a function of background field at 4.2 and 27 K. Because each pancake coil in the test magnet performed differently, data for central field corresponding to the minimum and maximum currents are plotted in Fig. 4. At 4.2 K, the stack carried a maximum critical current of 167 A and a minimum of 124 A in self field. The maximum magnetic field generated was $\approx$1.83 T in zero applied field. At 27 K, maximum critical current was 130 A and field generated was 1.43 T. Recently, we achieved a milestone when a stack of 10 pancake coils carried a current of $\approx$8 A at 64 K and in a field of 6 T. This coil generated a field of $\approx$0.53 T and $\approx$0.1 T at 64 K in zero and 6 T applied fields respectively [23].

Summary

The advantage of the powder-in-tube process over the conventional techniques of fabricating high-temperature oxide superconductors has been outlined in this paper. Reiterating, several parameters are involved in the processing of Ag-clad 2223 tapes. These parameters should be carefully controlled in order to obtain the desired properties. Because uniaxial pressing is not suitable for fabrication of long lengths of conductors, a two-step rolling technique has been employed. Long lengths of tapes were formed into pancake shaped coils. A test magnet that generated a field of $\approx$1.83 T at 4.2 K was then produced by stacking the pancake coils. The characteristics of the coils and the test magnet in the presence of magnetic fields at various temperatures have been studied.

Acknowledgments

Work at Argonne National Laboratory and part of the work at Intermagnetics General Corporation is supported by the U.S. Department of Energy (DOE), Energy Efficiency and Renewable Energy, as part of a DOE program to develop electric power technology, under Contract W-31-109-Eng-38. The authors thank Dr. M. Suenaga of Brookhaven National Laboratory, and Dr. Y. Iwasa of the Francis Bitter National Magnet Laboratory at the Massachusetts Institute of Technology, for low-temperature and high-field measurements.

References

1.   H. Maeda, Y. Tanaka, M. Fukutomi, and T. Asano, "A New High-T_C Oxide Superconductor without a Rare Earth Element," <u>Jpn. J. Appl. Phys</u>, **27** (1988), L209.

2.   U. Balachandran, A. N. Iyer, P. Haldar, and L. R. Motowidlo, "Processing and Properties of Commercially Viable High-Tc Bi-2223 Tapes and Coils Fabricated by the Powder-In-Tube Technique," <u>J. Metals</u>, **45**, No. 9 (1993), 54-57.

3.   D. Y. Kaufman et al., "Thermomechanical Processing of Reactively Sintered Ag-Clad $(Bi,Pb)_2Sr_2Ca_2Cu_3O_x$ Tapes," <u>Appl. Supercond.</u>, **1**, No. 1/2 (1993), 81-91.

4.   S. X. Dou and H. K. Liu, "Ag-Sheathed Bi(Pb)SrCaCuO Superconducting Tapes," <u>Supercond. Sci. Technol.</u>, **6** (1993), 297-314.

5.   P. Majewsky, B. Hettich, K. Schulze, and G. Petzow, "Preparation of Unleaded $Bi_2Sr_2Ca_2Cu_3O_{10}$," <u>Adv. Mater</u>., **3**, No. 10 (1991), 488-491.

6.   K. H. Sandhage, G. N. Riley, Jr., and W. L. Carter, "Critical Issues in the OPIT Processing of High-J_C BSCCO Superconductors," <u>J. Metals</u>, **43** (1991) 21-25.

7.   U. Balachandran, A. N. Iyer, C. A. Youngdahl, L. R. Motowidlo, J. G. Hoehn, Jr., and P. Haldar, "Fabrication, Properties, and Microstructures of High-T_C Tapes and Coils Made From Ag-Clad Bi-2223 Superconductors," <u>Adv. in Cryogenic Engineering</u>, **40**, (in press 1993).

8.   K. Sato et al., "High-J_C Silver-Sheathed Bi-Based Superconducting Wires," <u>IEEE Trans. Mag.</u>, **27** (1991), 1231 -1238.

9.   Y. Yamada, B. Obst, and R. Flukiger, "Microstructural Study of Bi(2223)/Ag Tapes with J_c(77K, 0 T) Values up to 3.3 x 10^4 A cm^{-2}," <u>Supercond. Sci. Technol.</u>, **4** (1991), 165-171.

10.   P. Haldar and L. Motowidlo, "Processing High Critical Current Density Bi-2223 Wires and Tapes," <u>J. Metals</u>, **44**, No. 10 (1992), 54-58.

11.   P. Haldar, J. G. Hoehn, Jr., U. Balachandran, and L. R. Motowidlo, "Processing and Transport Properties of High J_C Silver Clad Bi-2223 Tapes and Coils," (Paper presented at TMS Annual Meeting in Denver, CO, February 22-25, 1993).

12.   R. Flukiger et al., "High Critical Current Densities in Bi(2223)/Ag Tapes," <u>Supercond. Sci. Technol.</u>, **5** (1992), S61-68.

13.   S. X. Dou, H. K. Liu, A. J. Bourdillon, M. Kviz, N. X. Tan, and C. C. Sorell, "Stability of Superconducting Phases in Bi-Sr-Ca-Cu-O and the Role of Pb Doping," <u>Physica Review B.</u> **40**, No. 7 (1989), 5266

14. A. Ehmann et. al., "High T_c Superconductors in the System Bi-Pb-Sr-Ca-Cu-O," <u>J. Less Common Metals,</u> **151** (1989), 55-61.

15. A. Oota et al., "Growth Process of the (2223) Phase in Pb-Added Bi-Sr-Ca-Cu-0," <u>Jpn. J. Appl. Phys.</u>, **28** (1989), L1171.

16. P. Haldar et al., "Characteristics of Small Superconducting Coils Fabricated From Long Lengths of Silver-Clad Bi-2223 Tapes," (Paper presented at International ISTEC/MRS Workshop on Superconductivity, Honolulu, HI, June 23-26, 1992).

17. R. Flukiger, T. Graf, M. Decroux, C. Groth, and Y. Yamada, "Critical Currents in Ag-Sheathed Tapes of the 2223-Phase in (Bi, Pb)-Sr-Ca-Cu-O," <u>IEEE Trans. Mag.</u>, **27**, No. 2 (1991), 1258-1263.

18. R. Flukiger, B. Hensel, A. Jeremie, A. Perin, and J. C. Grivel, "Phase Formation and Inhomogenities in Bi(2223)/Ag Tapes at 77 K," (Paper presented at International ISTEC/MRS Workshop on Superconductivity, Honolulu, HI, June 23-26, 1992.)

19. J. Joo, J. P. Singh, and R. B. Poeppel, "Effects of Thermomechanical Treatment on Phase Development and Properties of Ag-Sheathed Bi(Pb)-Sr-Ca-Cu-O Superconducting Tapes," <u>Supercond. Sci. Technol.</u>, **6** (1993), 421-428.

20. P. Haldar, J. G. Hoehn, Jr., J. A. Rice, and L. R. Motowidlo, "Enhancement in Critical Current Density of Bi-Pb-Sr-Ca-Cu-O Tapes by Thermomechanical Processing: Cold Rolling versus Uniaxial Pressing," <u>Appl. Phys Lett.</u>, **60** (1992), 495-497.

21. G. E. Dieter, <u>Mechanical Metallurgy</u>, (New York: McGraw-Hill 1976), 608

22. R. Flukiger, A. Jeremie, B. Hensel, E. Seibt, J. Q. Xu, and Y. Yamada, "The Influence of Carbon Impurities on Jc in Ag/Bi(2223) Tapes," <u>Adv. in Cryogenic Engineering</u>, **38B** (1991), 1073-1080.

23. P. Haldar, J. G. Hoehn, Jr., U. Balachandran and L. R. Motowidlo, "High-Temperature Superconducting Coils from Silver-Sheathed Bi-2223 Tapes," Submitted to Proc. of the 1993 TMS Fall Meeting, Pittsburgh, PA, October 17-21, 1993.

HIGH-TEMPERATURE SUPERCONDUCTING MAGNETS AND COILS

FROM SILVER-SHEATHED Bi-2223 TAPES

P. Haldar,* J.G. Hoehn, Jr.,* L.R. Motowidlo,** and U. Balachandran***

*Intermagnetics General Corporation, Guilderland, NY 12084
**IGC Advanced Superconductors , Inc., Waterbury, CT 06704
***Energy Technology Division, Argonne National Laboratory,
Argonne, IL 60439

ABSTRACT

Long lengths (30 to 100 m) of silver-sheathed Bi-2223 tapes with high critical current density(J_c) were fabricated by the conventional powder-in-tube technique and used to make compact and robust coils by the wind-and-react approach. The tapes were insulated, co-wound in parallel, and heat treated to produce a fused conductor of larger cross section in the form of pancake coils that were stacked in series to make test magnets. The coils and test magnets were characterized at liquid nitrogen (77 K) and liquid helium (4.2 K) temperatures. The highest field generated by the test magnets was 0.36 T at 77 K and 2.6 T at 4.2 K. Progress in the prototype manufacturing of long lengths of HTS tape conductor is also discussed. Lengths of 30 to 100 m are now being made, and 70 m lengths were observed to carry 23 A ($J_c \approx 15{,}000$ A/cm^2) when immersed in liquid nitrogen.

Processing of Long Lengths of Superconductors
Edited by U. Balachandran, E.W. Collings and A. Goyal
The Minerals, Metals & Materials Society, 1994

Introduction

The powder-in-tube (PIT) process has proved to offer the greatest potential for manufacturing long lengths of high-temperature superconductor (HTS) wire manufacturing potential, together with the greatest improvement in current-carrying ability [1-4]. The $(Bi,Pb)_2Sr_2Ca_2Cu_3O_{10}$ compound (Bi-2223) is particularly suitable for the PIT approach because it can be readily deformed, textured, and densified by a sequence of thermomechanical operations. Another advantage of the PIT technique is its similarity to the industrial processes used in the manufacture of low-temperature superconductors such as NbTi and Nb_3Sn.

In addition to low-field applications at liquid nitrogen temperature, the high-temperature superconductors provide the potential of operating electric power devices and high-field magnets at temperatures (above 4.2 K) and fields (above 20 T) beyond the current capability of low-temperature superconductors. This is possible if long lengths of HTS conductor with high current densities can be fabricated economically on an industrial scale. In this paper, we report critical current data for long-length ($\approx$70 m) tapes, as well as progress made in improving the manufacturability of these brittle materials. Long lengths of tapes were also processed, wound into small pancake coils and used to construct test magnets. Parameters of these test magnets and their critical current data at 77 and 4.2 K are presented.

Experimental Procedure

Long lengths of silver-clad Bi-2223 HTS tapes were fabricated by the PIT process as described earlier [3-4]. Prereacted precursor powders were initially prepared from a mixture of high-purity (>99.99%) Bi, Pb, Sr, Ca, and Cu oxides or carbonates. These were carefully mixed, heat treated, ground, and packed into high-purity silver tubes. The packed billets were lightly swaged, drawn through a series of dies, and then rolled to final size. A series of intermediate heat treatments was performed between 800 to 840°C to enable growth and alignment of the Bi-2223 phase in the tape core. Long (30 to 100 m) monofilament tapes have been processed this way.

Test magnets were made by stacking and connecting in series a set of double pancake coils. Each double-pancake coil set was made by the "wind-and-react" approach. Moreover, each pancake coil comprised three to five lengths of monofilament tape, co-wound in parallel to form a larger monolith conductor with ceramic insulation separating each turn. Two coils wound on a single form were separated by ceramic insulation and heat treated together to make a set of double-pancake coils. Epoxy resin provided the necessary structural strength and rigidity. The double-pancakes were then stacked to form the test magnet.

Short tape samples were measured for their critical current by the four-probe technique at 77 K. Longer tapes ($\approx$70 m) were also characterized by the four-probe technique by spirally winding them on large mandrels and immersing them in liquid nitrogen .

The test magnets were measured by placing them in a cryostat; critical currents were determined in liquid helium and liquid nitrogen. Voltage taps were placed at the ends of each pancake coil to determine their performance individually.

The criterion for critical current was 1 μV/cm, and J_c values were determined for the superconducting and overall cross-sectional areas. Note that the high aspect ratio of the tape, combined with the irregularities of the superconducting core, makes it extremely difficult to obtain an accurate value for the core cross sections [4]. For example, techniques such as weighing the sample before and after etching the core, weighing of cut micrographs, measuring of enlarged optical micrographs, and measurements from an image analyzer provide a wide range of values, from 8×10^{-4} cm^2 to 1.7×10^{-3} cm^2. Thus, there can be a large range in core J_c, depending on measurement technique used. Although it is useful to determine the value of core J_c, it is more useful to know the overall or engineering J_c, because this value provides a design base line for magnets and other devices. In this paper, we report overall J_c of the composite to elucidate the problems associated with enhancing the core J_c.

Results and Discussion

Short tape samples (3.5 to 4.0 cm in length) that had been subjected to several cycles of uniaxial pressing and heat treatment always carried the highest currents. Values above 40 A were typically attained at liquid nitrogen temperature, with the highest being 51 A. Data for short samples that were uniaxially pressed and heat treateo, as well as those that were cold-rolled and heat treated, are presented in Table 1 and compared with those of long (34 m and 70 m) tapes. The long monofilament tapes were processed by a two-cycle cold-rolling and heat treatment operation. Overall J_c, determined from the cross-sectional area given by thickness times width (not considering taper of the tape edges), is also indicated in Table 1 for all samples. Also provided for comparison are the approximate superconductor fractions for each sample type. It can be noted from this table that short samples that were subjected to rolling still perform considerably below those that were pressed. The highest superconductor J_c level in rolled short samples have more recently begun to approach $\approx 30,000$ A/cm^2, although their average overall J_c compares well with a pressed sample, which has a lower superconductor fraction (27 vs. 20%). We have also been able to increase the superconductor fraction, thereby increasing overall J_c. Billets are now being processed with thinner silver sheaths, undergoing a total reduction in overall area of 100 times without any problems in mechanical deformation.

Table 1.   Data for short and long tapes at 77 K

Tape Sample	I_c (A)	J_c (A/cm^2)		SC (%)
		Core	Overall	
Short Sample (Pressed)	51	$\approx$45,000	9,000	20
Short Sample (Rolled)	33	$\approx$21,000	5,000	24
Long Length (34 m)	16	$\approx$11,000	2,500	24
Long Length (70 m)	23	$\approx$15,000	3,500	24
Short Sample	51	$\approx$29,000	7,800	27

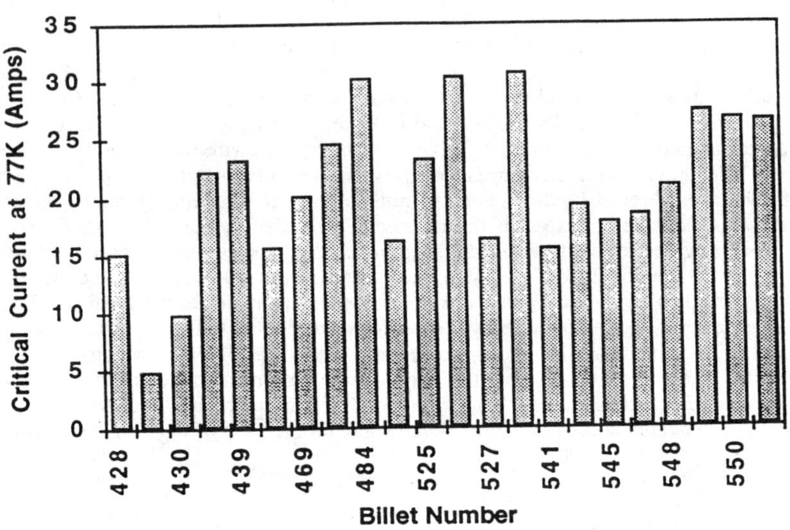

Figure 1. Critical current (I_c) at 77 K for several long lengths (30 to 35 m) of monofilament tape conductor manufactured at IGC.

Continued progress is being made in fabricating long lengths of tape. A 70 m length with a 24% superconductor fraction carried 23 A at 77 K, corresponding to a core J_c of $\approx$15,000 A/cm^2 and an overall conductor J_c of 3,500 A/cm^2. This shows considerable improvement from earlier measurements made on a 34 m length of tape with a similar composite configuration that carried 16 A (see Table 1).

Shown in Figure 1 are critical current data at 77 K for several billets that were processed to ultimately yield 30 to 35 m long HTS monofilament tapes. The billets were all processed similarly. Several short samples were sectioned from various regions of the final conductor lengths and measured. The average values of these samples are plotted to compare the performance of each length of tape. Average value is usually representative of the performance of the entire length. The superconductor fraction for these billets is nominally 24%. The bar chart indicates continued improvement in the critical current data with billet number. Earlier measurements ranged from 5 to 15 A while more recent lengths typically carried 20 to 25 A. Occasionally, a few lengths are fabricated with I_c's exceeding 30 A. Additional work is being performed to fabricate longer lengths of conductor with a reduced fraction of silver sheath. We have successfully processed 100 m continuous lengths of monofilament composite tapes.

Figure 2. Photograph of a recent test magnet assembled with five double-pancake coils.

Figure 2 shows a photograph of a recent test magnet assembled by stacking five double-pancake coils. The dimensions and pertinent parameters of this and other previous magnets and their critical current data are presented in Table 2. At 77 K, a Hall probe was placed in the bore of these magnets to obtain their field constants at the mid-plane. The central field for 4.2 K was calculated from the field constant determined at 77 K. Maximum value for the critical current corresponds to the highest critical current of a pancake coil in the assembly, and the maximum field generated is that determined for the entire assembly the maximum critical current for the best-performing pancake coil. Similar determinations were made for the minimum critical current and the minimum field generated. By improving performance of the HTS conductor and coil-winding techniques, we have improved the performance of our HTS magnets over the past few months (see Table 2). The most recent magnet generated at 77 K a maximum and minimum self-field of 0.36 T and 0.21 T; the values at 4.2 K correspond to 2.6 T and 1.8 T.

Table 2.   Parameters and critical current data of test magnets constructed at IGC

Parameter	Magnet 1 (June 93)	Magnet 2 (Aug. 93)	Magnet 3 (Sep. 93)	Magnet 4 (Oct. 93)
Winding Inner diameter (cm)	2.50	2.50	2.50	2.50
Winding Outer diameter (cm)	7.50	9.65	11.30	11.30
Coil Height (cm)	5.33	9.84	6.35	6.35
Number of Co-Wound Tapes per Pancake	3	5	3	3
Total Length of Tape in Magnet (m)	153	570	480	480
Total Number of Turns in Magnet	330	612	700	700
Overall Winding Cross Sectional (cm^2)	0.0348	0.0532	0.0363	0.0363
Number of Pancake Coils	6	10	10	10
Magnet Constant (G/A)	62.9	64.7	109.7	111.2
77 K I_c Max (A)	19	41	22	32
B_o Max (T)	0.12	0.26	0.24	0.36
4.2 K I_c Max (A)	170	255	167	234
B_o Max (T)	1.01	1.65	1.83	2.60

Summary

The set of results presented here indicates that considerable improvements are being made in the performance of HTS conductors and magnets. These component materials are promising for high-field magnets and electric power devices operating at temperatures below $\approx$30 K; operation above 4.2 K could improve refrigeration efficiency and reliability. Higher-temperature operation (above 30 K) is possible for applications requiring lower operating fields.

Acknowledgments

Work at Argonne National Laboratory and part of the work at Intermagnetics General Corporation is supported by the U.S. Department of Energy (DOE), Energy Efficiency and Renewable Energy, as part of a DOE program to develop electric power technology, under Contract W-31-109-Eng-38.

References

1. K. Sato, T. Mikata, H. Mukai, M. Ueyama, T. Kato, T. Masuda, M. Nagata, K. Iwata and T. Misui, "High-J_c silver sheathed Bi-based superconducting wire," <u>IEEE Trans. Magn.</u> **27** (1991), 1231.

2. Y. Yamada, B. Obst, and R. Flükiger, "Microstructural study of Bi(2223)/Ag tapes with J_c (77 K, 0 T) values of up to 3.3 X 10^4 A/cm^2," <u>Supercond. Sci. Technol.</u> **4** (1991), 165.

3. P. Haldar, J.G. Hoehn, Jr., J.A. Rice, and L.R. Motowidlo, "Enhancement in critical current density of Bi-Pb-Sr-Ca-Cu-O tapes by thermomechanical processing: Cold rolling versus uniaxial pressing," <u>Appl. Phys. Lett.</u> **60** (1992), 495.

4. P. Haldar, J.G. Hoehn, Jr., J.A. Rice, L.R. Motowidlo, U. Balachandran, C.A. Youngdahl, J.E. Tkaczyk, and P.J. Bednarczyk, "Fabrication and properties of high-T_c tapes and coils made from silver-clad Bi-2223 superconductors," <u>IEEE Trans. Appl. Supercond.</u> **3** (1993), 1127.

FABRICATION AND PROPERTIES OF LONG Bi2223

Ag–SHEATHED SUPERCONDUCTING TAPES

S.X. Dou, H.K. Liu, R. Bhasale, Q.Y. Hu,
J. Yau, N. Savvides[*] and C. Andrikidis[*]

Centre for Superconductivity and Electronic Materials,
University of Wollongong, Northfields Av. 2522,
Wollongong, Australia

[*] CSIRO Division of Applied Physics, Lindfield, NSW 2070, Australia

Abstract

Long length Ag–sheathed Bi–based superconducting single and multifilamentary tapes up to 43 meters have been fabricated using powder–in–tube technique. A I_c up to 15 A and J_c of 7,500 A/cm^2 at 77 K over the entire length has been achieved. For short multifilament tapes, a J_c of 17,200 A/cm^2 at 77 K and 140,000 A/cm^2 at 4 K has been achieved in zero field and weak links are improved. Multiprobe measurements for a tape show that the J_c is rather uniform along the length of the tapes with a variation of 10%. There is no indication of accumulative grain boundary resistance effect over the entire length at high temperatures. The critical strains for bend tests show a dependence on the tape thickness, the number of filaments and Ag/oxide volume ratio, attributable to the improved strength and flexibility of the special interfacial layer between Ag and oxides formed during the thermomechanical processes.

The authors gratefully acknowledge to Metal Manufactures Ltd. and Commonwealth Department of Industry, Technology and Commerce for financial support.

Processing of Long Lengths of Superconductors
Edited by U. Balachandran, E.W. Collings and A. Goyal
The Minerals, Metals & Materials Society, 1994

Introduction

High critical current densities (J_c) of the Ag–clad $(Bi,Pb)_2Sr_2Ca_2Cu_3O_{10+x}$ (Ag/Bi–2223) tapes have been achieved using powder–in–tube technique [1–6]. Practical applications such as motors, fault current limiter and superconducting magnets are feasible if long length wires or tapes can be fabricated. Long Ag/Bi–2223 tapes up to several hundreds of meters with a J_c approaching 10^4 A/cm^2 at 77 K and zero field have been fabricated by a number of groups [6–8]. Multifilament tapes are desirable for the robust winding operation in construction of devices. In this paper, we report results of fabrication and properties of the long multifilament tapes.

Experimental Details

A powder–in–tube method was used to prepare the Ag–sheathed Bi–2223 wires. The details of sample preparation were given previously [9]. Powders were prepared by codecomposition of metal nitrate solutions having the cation ratio Bi:Pb:Sr:Ca:Cu = 1.85:0.35:1.90:2.03:3.05. The powders were calcined at 820°C for 10 h, uniaxially pressed into pellets, and sintered at 840°C for different times in order to alter the phase assemblage. It was found that the solution route provided highly reactive, homogeneous, and carbonate–free powders from which the 2223 phase can be formed effectively. It should be, however, pointed out that at this stage the powders contain a small fraction of the 2223 phase and the 2212 is the major phase. This incompletely reacted precursor will facilitate the liquid phase formation in the subsequent processes. The average powder particle size is 2 to 3μm. Large powder particles cause inhomogeneity within the tapes.

The precalcined powders were pressed into round bars of dimensions 50 mm length and 8 mm diameter. The bars were loaded into silver tubes of 13 mm outer and 8 mm inner diameters, and the composites were swaged and drawn to a final diameter of 1.3 to 2.0 mm. Alternatively, a loose powder packing was also used in some cases. A certain number of 1 m long wires were stacked into a silver tube of 10 mm outer and 8 mm inner diameters and drawn to a diameter of 1.0 to 2.0 mm. The same process may be repeated once or twice. The multifilament wires were then rolled into tapes of overall thickness ~0.2 mm and width ~3 to 4 mm.

It has been realised that a liquid phase sintering is necessary to improve the J_c – H behaviour in all high T_c materials. However, in contrast to $YBa_2Cu_3O_{7-x}$ and Bi–2212, melt processing can not be simply used to treat the 2223 phase since it causes 2223 decomposition and phase segregation. Annealing temperatures from 820°C to 860°C have been used in the heat treatment of Ag/BPSCCO wires depending on the starting composition, atmosphere, and annealing time. When the composition is in the vicinity of Bi:Pb:Sr:Ca:Cu = 1.8:0.4:2:2:3 the optimum temperature range is 830°C to 838°C for annealing in air and 820°C to 830°C for annealing in low oxygen partial pressure (Po_2 = 0.05 ~ 0.1 atm). Three or more annealing cycles with duration of 60 to 100 h for each is necessary to achieve better J_c.

For a continuous observation of the wire processing, some tapes were annealed at 832°C to 834°C for up to 280 h with an intermediate pressing at every 60 h. At this temperature range a partial melting is achieved in order to have sufficient grain growth. For this series of experiments, the J_c, the fraction of 2212 phase, the degree of texture and the morphology of the samples were recorded with annealing time.

The transport critical current density for short tapes was determined from the current/voltage curve using a 1 μV/cm criterion, while for long tapes a criterion of $10^{-13}\Omega$m was used to determine the I_c. Microstructural and compositional characterisations were performed with a JEOL JXA–840 scanning electron microscope (SEM) equipped with Link Systems AN10000 energy dispersive spectrometers (EDS).

Results and Discussion

Table 1 lists basic data for four multifilament tapes. It is seen that the total current, critical current density and the bend strain are related each other. A large critical current requires a large proportion of superconductor core, whereas a high ratio of oxide/Ag will reduce the critical bend strain. A desirable configuration of the multifilamnt will have to compromise these three parameters. It appears that the ratio of oxide/Ag should be within a range of 15% to 20%. A J_c of 18,500 A/cm^2 at 77 K and 147,000 A/cm^2 at 4 K has been achieved for a single core, short tape which was taken from the same batch of wires for producing the single core tape. A J_c of 6,000 A/cm$_c$ at 77 K and zero field was obtained for this single core tape over 12 meters long length.

Table 1. The I_c, J_c, bend strain and the ratio of oxide/Ag of Ag–clad superconducting tapes

Filaments	I_c at 77K (A)	J_c at 77K (A/cm^2)	I_c at 4K (A)	J_c at 4K (A/cm^2)	Length (m)	% oxide/Ag	Critical bend strain (%)
1	15.0	18,500	115	147,000	12	25	0.2
19	7.6	13,000			6	9	0.8
21	10	15,000	60	89,000	6	10	0.8
27	19.7	9,900	128	64,000	43	20	0.3
49	9.7	17,200	56	140,000	12	4	1.5

Figure 1 shows the dependence of the J_c on magnetic field at 77 K and 4 K for the 21–filament Ag–clad tape. It is noticed that in the direction of field parallel to the tape surface (normal to c axis) the J_c for this multifilament tape is 15,000 A/cm^2 at 77 K and zero field and 2,300 A/cm^2 in 1 T. In the low field region, the J_c – H behaviour has been improved compared with results reported for the multifilament tapes [10]. A J_c of 89,000 A/cm^2 at 4 K and zero field has been achieved. For the 49–filament tape, a J_c of 17,200 A/cm^2 at 77 K and of 140,000 A/cm^2 at 4 K and zero field has been achieved for a short tape sample. It has been known that the J_c for multifilament tapes are considerably lower than that for the single tapes. J_c greater than 60,000 A/cm^2 at 77 K [11] has been reported for a single tape whereas the best result of J_c for multifilament tapes is 18,500 A/cm^2 [1].

Furthermore, the J_c as a function of increasing field and decreasing field shows no irreversible loop which is usually observed for the Ag–clad tapes. This indicates that weak links in this multifilament has been improved. At 4 K, this 21–filament tape exhibits large irreversible loops in two directions of the applied field, suggesting that grain boundary weak links become relatively pronounced at low temperatures because the flux pinning within the grains is very strong at low temperatures.

Although the individual filaments in the 61–filament tape are rather thin (about 5μm), they retain a good continuity and a smooth interface between superconductor core and Ag sheath in longitudinal direction.

Figure 3 shows V (μv) versus I_c for the 27–multifilament tape over the whole 43 meters long length. A J_c of 7,500 A/cm^2 at 77K and zero field has been achieved over the 43 meters long length. Multi–probe measurements over 10 sections of this tape show that the J_c does not drop when the length of the tape increases to 43 meters with a fluctuation less than 10% (Figure 4). The total critical current is limited by the 'bottle neck' along the length of the tapes. Although the temperature fluctuation along the length during the heat treatment is much greater than that for treating the short tapes, the reproducibility of the long tapes is good because of the high thermal conductivity of the silver sheath.

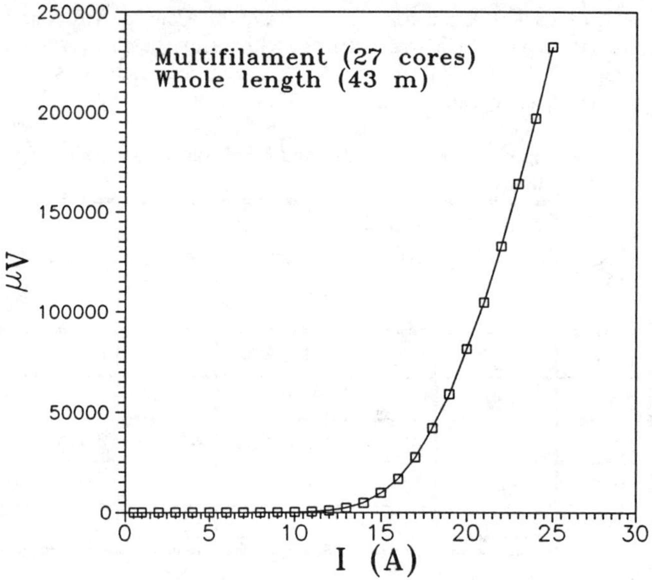

Figure 3. V (μv) versus I_c for the 27–multifilament tape over the whole 43 meters long length

Figure 5 shows J_c versus annealing time during which an intermediate rolling was performed for three times for the 27–multifilament tape. The tape was heat treated at 832°C for 220 h. The proportion of 2212 phase, estimated from the ratio of the intensity for the 008 peak of 2212 phase to that of the 0010 peak of 2223 phase, was also incorporated in the Figure 5. The I_c increases with annealing times and with decreasing 2212 proportion. At an initial period the I_c increased little because the 2223 phase has not formed a matrix at this stage. At the second stage annealing, the 2223 phase increased to over 80% and the I_c showed a significant increase during this period. A further increase in the I_c is mainly attributable to the grain growth.

The critical current density over long length is determined by a number factors in the tedious processes. It has been established that the powder quality and heat treatment conditions are critical to the final J_c. It was found that incompletely reacted powder, a

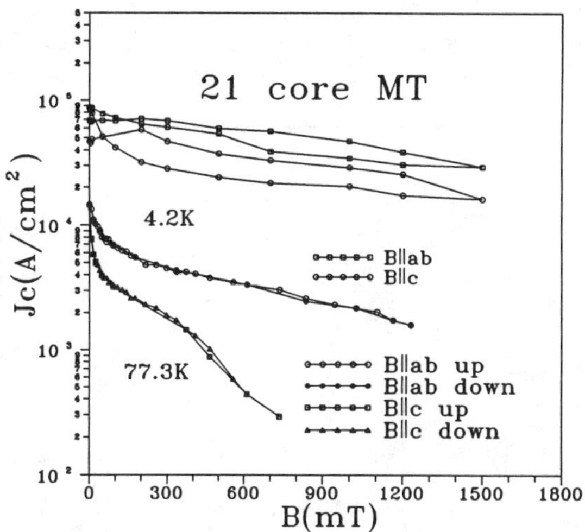

Figure 1. J_c – H curves for the 21 filament tape with field parallel and normal to the tape surface at 77K and 4 K

Figure 2 shows the SEM images of the transverse cross section for a 27–filament tape and the longitudinal cross section for a 61–filament tape. The individual filaments in the 27–filament tape has a thickness of 10 to 20 μm and are uniformly distributed within the tape.

Figure 2. Photomicrographs for the transverse cross section (a) of a 27–filament tape and the longitudinal cross section (b) of a 61–filament tape

small particle size in the starting powder, large grain size and good alignment are important to achieve homogeneity over long tape, and high J_c over the long tape. Although higher fraction of 2223 phase could be obtained through a solid reaction, it does not guarantee a high J_c.

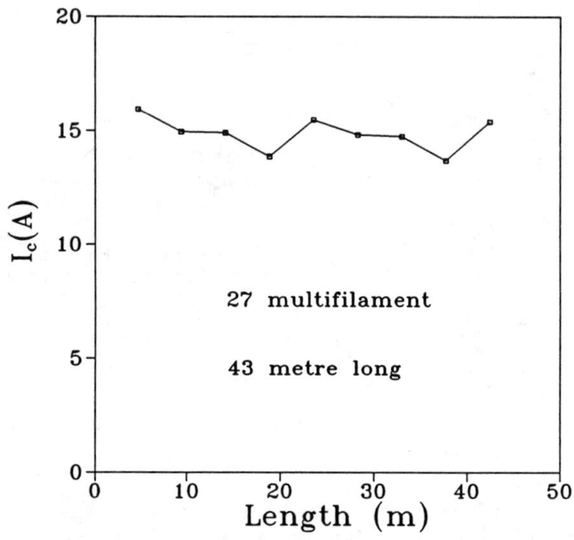

Figure 4. Variation of the I_c along the length of the 27 multifilament at 77 K

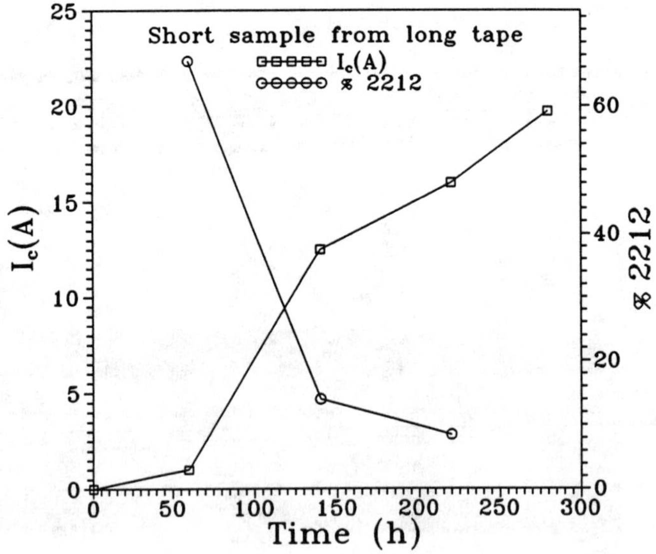

Figure 5. Critical current and fraction of 2212 phase versus annealing times with intermediate rolls for the 27–filament tape treated at 832°C in air for 220 h

Figure 6 shows the normalised J_c versus bend strain for tapes with different number of filaments. For the bend tests, tapes of 3 cm lengths were bent around a series of cylindrical metal tubes with different diameters in one direction. The bend curvature, R, is the same as the diameter of the individual metal tube. The J_c was then determined using the four point contacts on the tape surface with a 1 cm gauge length. As the bending curvature is deceased the change in J_c is recorded after each bend. Continuous bending tests in one direction were performed for the single and multifilamentary tapes.

The J_c is the critical current density at a certain strain and J_{co} is the zero strain critical current density. Here, the bend strain is defined as $\varepsilon = r_t/R$ where r_t the half thickness of the tape (including silver) and R is the radius of the bend curvature. Since the bend property of the silver is far better than the ceramic oxide, the bend strain for the silver/oxide composites will strongly depend on the volume ratio of silver to oxide and the configuration as evidenced in Figure 6.

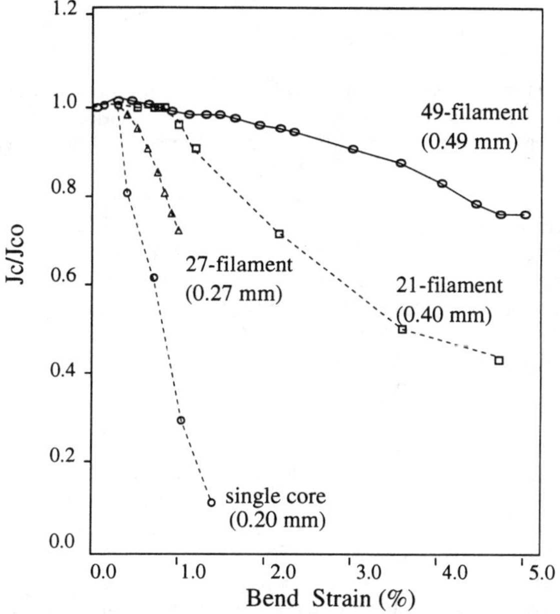

Figure 6. Normalised critical current density versus bend strain Ag–clad $(Bi,Pb)_2Sr_2Ca_2Cu_3O_{10+y}$ single core tape and multifilament tapes.

It is seen that for the single bend tests, the critical strain for the single tape is 0.2% while, the critical bend strain for the 21– and 49–multifilamentary tapes is 0.8% and 1.5% respectively. The critical strain for the 49–multifilamentary tape is much higher than those reported previously(0.7%) by Otto et al [7]. Especially, the normalised J_c drops very slowly with increasing bend strain for the 49–filament tape. Sato et al [4] has reported that the bend strain increased from 0.1% to 0.7% when the numbers of filaments in the tape increased from 1 to 1296. Our bend strain results do not follow this simple pattern since the critical bend strain for the 27–filament tape is only slightly better than that for single tape and is far lower than that for 21–filament tape and 49–filament tape. The bend strain correlates well with the ratio of oxide/Ag. It is evident that the volume ratio of oxide to silver and the configuration are important factors to determine the bend properties.

It is argued that this difference in critical bend strain reported by various group may be attributable to the different volume ratio of the superconducting oxide to the silver. For example, the oxide/silver volume ratio is 10% for the 21- and 4% for the 49-filament tapes, it is 9 to 13% for Otto's tape [7], whereas the 27-filament tape has an oxide to Ag ratio of 20% and the single tape with bend strain of 0.2% has an oxide/Ag volume ratio 25%. This strong dependence of the bend strain on the ratio of oxide/Ag seems to suggest that the interfacial layer has better strength than the bulk oxides since the higher volume ratio of silver versus oxide means higher volume rat'ɔ of the interfacial layer to the bulk oxide. It is proposed that an interfacial composite layer was formed through a strong bonding between the oxide and Ag due to the eutectic reaction during the heat treatment[13]. SEM examination reveals that although cracks usually penetrate the entire superconductor core, the interface between the oxide and silver sheath is still intact when the tape was bent to a strain greater than the critical strain.

CONCLUSION

Long length Ag-sheathed Bi-based superconducting single and multifilamentary tapes up to 43 meters have been fabricated using powder-in-tube technique. A solution route for powder processing, small particle size, large grain size and good alignment are found to be important for homogeneity and hence high J_c over long length tapes. A I_c up to 15 A and J_c of 7,500 A/cm^2 at 77 K over the entire length has been achieved. For short multifilament tapes, a J_c of 17,200 A/cm^2 at 77 K and 140,000 A/cm^2 at 4 K has been achieved in zero field and weak links are improved. Multiprobe measurements for a tape show that the J_c is rather uniform along the length of the tapes with a variation of 10%. The critical strains for bend tests show a dependence on the tape thickness and Ag/oxide volume ratio, attributable to the improved strength and flexibility of the special interfacial layer between Ag and oxides formed during the thermomechanical processes.

REFERENCES

[1]   M. Ueyama, T. Hikata, T. Kato, and K. Sato, Jpn. J. Appl. Phys., 30, L1384(1991)

[2]   S.X. Dou, H.K. Liu and Y.C. Guo, Appl. Phys. Lett., 60, 2929(1992)

[3]   R. Flukiger, B. Hensel, A. Jeremie, M. Decroux, H. Kupfur, W. Jahn, E. Seibt, W. Goldacker, Y. Yamada, and J.Q. Xu, Supercond. Sci. Technol. 5, S61 (1992)

[4]   K. Sato, T. Hikata, H. Mukai, M. Ueyama, N. Shibuta, T. Sato, T. Masuda, M. Nagata, K. Iwata and T. Mitsui, IEEE Transactions on Magnetics, Vol. 27, No. 1231(1991)

[5]   Y.C. Guo, H.K. Liu and S.X. Dou, Appl. Supercond. 1, 25 (1993)

[6]   P. Halder, J.G. Hoehn, J.A. Rice and L.R. Motowidlo, Appl. Phys. Lett., 60, 495 (1992)

[7].   A. Otto, L.J. Masur, J. Gannon, E. Podtburg, D. Daly, G. J. Yurek and A.P. Malozemoff, IEEE Transactions on Applied Superconductivity, 3, 915 (1993)

[8]   T. Kato, M. Ueyama, T. Hikata and K. Sato, Extended abstract of the International Workshop on Superconductivity, June 23-26 1992, Honolulu, Hawaii, USA

[9]   S.X. Dou, Y.C. Guo and H.K. Liu, Physica C, 194, 343-50(1992)

[10]   Q.Y. Hu, H.K. Liu and S.X. Dou, Cryogenics 32, 1039 (1992)

[11]   Y. Yamada, M. Satou, S. Murase, Proc. 5th Int. Symp. on Superconductivity (ISS92), Eds. Y. Bando and H. Yamaauchi, 717, (1993)

[13]   S.X. Dou, K.H. Song, H.K. Liu and C.C. Sorrell, M.H. Apperley, and N. Savvides, Appl. Phys. Lett. 56, 493 (1990)

FABRICATION OF $(Bi,Pb)_{2.2}Sr_2Ca_2Cu_3O_{10}$ Ag-SHEATHED SUPERCONDUCTING

WIRES FROM NON-PHASE-PURE POWDERS*

B.C. Prorok, S.E. Dorris, R.B. Poeppel, S. Sinha**, and M. Hagen

Argonne National Laboratory
Energy Technology Division
Argonne, Illinois 60439

**University of Illinois at Chicago
Department of Civil Engineering, Mechanics, and Metallurgy
Chicago, Illinois 60680

Abstract

A two-powder processing scheme for forming the Pb-doped 2223 phase in the Bi-Sr-Ca-Cu-O system has shown remarkable improvements in phase purity over conventional practice. This process controls the reaction pathway by forming separate intermediate phases ($Bi_{1.8}Pb_{0.4}Sr_{2-x}Ca_{1+x}Cu_2O_8$ and $Sr_xCa_{1-x}CuO_2$) and then combining them to react in situ during the oxide powder-in-tube process. In studying this processing scheme, the effects of two variables were examined. The first variable studied was the Sr/Ca ratio as determined by x in the above formulae (x = 0.0, 0.25, 0.5, 0.75, and 1.0). The second variable was the amount of lead incorporated in the intermediate phase, as determined by the subscript y in the formulae $Bi_{1.8}Pb_ySr_2CaCu_2O_8$. The most favorable processing parameters were: 8% partial pressure of O_2 atmosphere; 815° C firing temperature; mixtures containing the phases 2212, Ca_2CuO_3, and CuO; where all Pb is incorporated into the 2212. Nearly phase pure 2223 (≤ 2 volume, percent second phase) can be obtained in as few as 50 hours firing time. To date, the maximum critical current density achieved was 4.0×10^4 A/cm^2 (I_c=60 A) at 77K and 18.0×10^4 A/cm^2 (I_c=270 A) at 4.2K.

* Work supported by the U. S. Department of Energy, Conservation and Renewable Energy, as part of a program to develop electric power technology, under Contract W-31-109-Eng-38.

Processing of Long Lengths of Superconductors
Edited by U. Balachandran, E.W. Collings and A. Goyal
The Minerals, Metals & Materials Society, 1994

Since the discovery of superconductivity in the Bi-Sr-Ca-Cu-O (BSCCO) system [1], many researchers have concentrated their efforts on synthesizing the various phases in this system. By far, the 2223 phase yields the highest current carrying capacity at 77 K giving it the greatest potential for commercial application [2]. For this reason a considerable amount of effort has been focused on devising a process to stabilize its pure form. However, this goal is not easily accomplished since numerous second phases can form during processing, some of which are very stable and resist subsequent reaction to form 2223.

Because of the promising results that have already been obtained, (Jc as high as 6.9 x 10^4 A/cm^2 [3]) commercial application of the Bi-Sr-Ca-Cu-O 2223 superconductor will most likely be via the powder-in-tube process (Figure 1). This process involves packing superconductor powder into metallic sheaths and then mechanically working it into a composite tape. The overall goal is to obtain a thin superconductor core of uniform cross-section encased in a metallic sheath. Another goal is to have a nearly single phase core or at least have very small, finely dispersed second phases.

In the powder-in-tube process the mechanical and thermal processing play important roles. The mechanical processing compacts the core and aligns the grains. During this processing, the metal sheath work hardens and helps to retain the form of the core. After an intermediate heat treatment, the sheath becomes soft while the core remains relatively hard. At this point incorrect mechanical processing can yield cores that are too thick or are pinched off into segments by the sheath. Mechanical processing may also facilitate formation of 2223 by breaking up the reactant phases and bringing them into contact with one another [4].

The thermal processing is equally important, because it converts the powder into the desired 2223 phase. Thermal treatment needs to be optimized to obtain a liquid phase in the core. This is a critical step in the powder-in-tube process since it enhances the kinetics of 2223 formation, facilitates grain growth and alignment, and heals damage from mechanical processing. It is clear that the liquid should not be formed by melting the 2223 phase since it undergoes peritectic decomposition. This helps to explain why pure 2223 powders do not perform as well

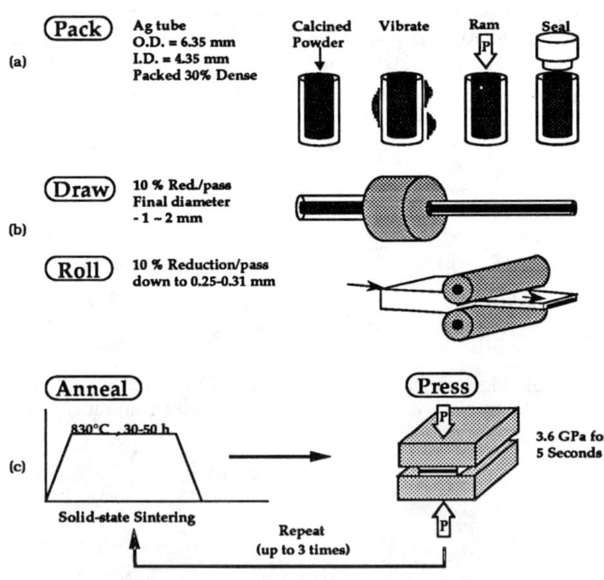

Figure 1. Powder-In-Tube process.

as multi-phase mixtures. The overall goal of the present study is to control the partial melting temperature and amount of liquid phase present by properly manipulating the composition of the powder initially placed in the tube.

Conventional phase formation of 2223 in the oxide powder-in-tube process involves forming the 2223 phase by combining and calcining all constituents in a single step before it is loaded into a silver tube (see Figure 2.(a)). By varying the calcination conditions, it is possible to control the dominant phase in the mixture, however numerous secondary phases usually on the order of 10-20 μm in size remain in the mixture. With such a large number of phases present, it is difficult to bring the appropriate phases into contact with one another in a uniform and reproducible fashion during powder-in-tube processing. Most of these phases are not superconducting and thus, degrade the properties of the entire sample. It then becomes difficult to produce long lengths of wire with reproducible electrical results.

The formation of the second phases results from poor control of the initial phase assemblage. Several researchers claim that 2223 formation is favored by a low temperature, partial melting of calcium rich 2212 [5,6]. Grader et al. [7] showed the melting point of 2212 can be decreased by varying the Sr/Ca ratio. On the basis of these studies, a so-called two-powder process was developed (Figure 2.(b)). In this process, separate intermediate phases of $Bi_{1.8}Pb_{0.4}Sr_{2-x}Ca_{1+x}Cu_2O_8$ and $Sr_xCa_{1-x}CuO_2$ are combined and then reacted in situ during the oxide powder-in-tube process. With this processing scheme, it is possible to control the melting behavior of 2212 by varying the Sr/Ca ratio, as determined by x in the above formulae. In order to stabilize large amounts of the 2223 phase, lead must be substituted for bismuth[8]. Forming intermediate phases in the two-powder process brings about the question of where to place the lead in the mixture. In one process, Arendt et al. [9] adds lead in as PbO. Others speculate that the presence of Ca_2PbO_4 is important for 2223 phase formation [10], but the method for introducing lead is not well understood at this point. This study also evaluates various methods for introducing lead into the two-powder process.

Experimental Methods

Powder Synthesis

A solid state reaction scheme was chosen for the powder synthesis. PbO, Bi_2O_3, $SrCO_3$, $CaCO_3$ and CuO were combined in the cation ratios given in Table I.(a) and I.(b). The powders were placed in polyethylene jars containing ZrO_2 milling media with isopropanol as a liquid medium and ball milled for a period of 24 hours. They were then pan dried, ground with an agate mortar and pestle, and passed through a 50 micron sieve. Finally the powders were calcined in a dynamic vacuum ($\approx$2-3 torr of flowing O_2) to aid in the decomposition of the carbonates and removal of the CO_2 generated in the process. This is important since the presence of carbon decreases the J_c [11,12]

After the vacuum calcination, the powder was placed in another polyethylene jar containing ZrO_2 grinding media with isopropanol and vibratory milled until the average particle size reached 2 microns ($\approx$24 hours). It was then pan dried and passed through a 50 micron sieve. The entire calcination process was repeated to ensure complete precursor conversion.

Subsequent to individual calcinations, the corresponding BSCCO and SCCO compositions (with like values of x) were mixed together to produce an overall stoichiometry of $Bi_{1.8}Pb_{0.4}Sr_2Ca_2Cu_3O_{10}$ (Table I.(c)). Silver is known to lower the melting point of BSCCO [13], so fifteen wt% silver platelet powder (10x10x1 μm) was added to each mixture to give more homogeneous melting behavior throughout the core. After adding the silver each mixture was placed into polyethylene jars containing ZrO_2 grinding media with isopropanol and ball milled for 6 hours, pan dried, and then passed through a 50 μm sieve.

Composite Wire Fabrication and Heat Treatment

The composite wire was formed using the powder-in-tube process shown in Figure 1(a) and 1(b). The wire was cut into 3.8 cm long sections. The width was found to vary between 0.95 and 1.05 cm and the thickness was measured to be 0.02 to 0.03 cm. The oxide core occupied approximately 25 % of the total cross-sectional area of the tapes.

(a) **One-Powder Process**

Mix, mill starting materials
Bi_2O_3, PbO, $SrCO_3$, $CaCO_3$, CuO

⇓

Calcine at reduced pressure
<u>Phases Present:</u>
2201 $(Sr, Ca)CuO_2$
2212 $(Sr, Ca)_2CuO_3$
2223 $(Sr, Ca)_{14}Cu_{24}O_{41}$
Plumbates   Bismuthates
<u>Overall Composition:</u>
$Bi_{1.8}Pb_{0.4}Sr_{2.0}Ca_{2.0}Cu_{3.0}O_{10}$

⇓

Powder–in–tube process

(b) **Two-Powder Process**

<table>
<tr><td align="center">

<u>**Powder 1**</u>
Mix, mill starting materials
Bi_2O_3, PbO, $SrCO_3$, $CaCO_3$, CuO

⇓

Calcine at reduced pressure
<u>Phases Present</u>
2212, 2201
Ca_2PbO_4
<u>Overall Composition</u>
$Bi_{1.8}Pb_{0.4}Sr_{2-x}Ca_{1+x}Cu_{2.0}O_8$

⇓

</td><td align="center">

<u>**Powder 2**</u>
Mix, mill starting materials
$SrCO_3$, $CaCO_3$, CuO

⇓

Calcine at reduced pressure
<u>Phases Present</u>
$(Sr, Ca)CuO_2$
[For x = 0.0: Ca_2CuO_3 + CuO]
<u>Overall Composition</u>
$Sr_xCa_{1-x}CuO_2$

⇓

</td></tr>
</table>

Mix Powder 1 and Powder 2, mill for second time
<u>Overall Composition:</u> $Bi_{1.8}Pb_{0.4}Sr_{2.0}Ca_{2.0}Cu_{3.0}O_{10}$

⇓

Powder–in–tube process

Figure 2.   Processing flow-chart for (a) One-Powder Process and (b) Two-Powder Process.

     The approximate temperature region for processing the tape sections was based on the DTA analysis of the powder mixtures. Atmospheres of 8% oxygen (balance nitrogen) and ambient air (21% O_2/79% N_2) were chosen. Firing temperatures for the atmospheres were 820, 825, and 830°C for 8% O_2 and 835, 840, and 845°C in air. Accuracy of the actual temperature varied by ± 3° C as monitored by type K thermocouples positioned near the samples. All tapes were treated for an initial 50 hours in their respective atmospheres and temperatures before undergoing thermo-mechanical processing. After the initial heat treatment, 3 samples were removed for characterization. The remaining tapes were uniaxially cold pressed for 5 seconds at a pressure of 3.6 GPa. The tapes were subsequently treated for an additional 100 hours at their original conditions. This thermo-mechanical process was repeated

TABLE I. CATION STOICHIOMETRY OF DIFFERENT POWDERS

(a) CATION STOICHIOMETRY FOR BSCCO COMPOSITIONS

	x	Pb	Bi	Sr	Ca	Cu	Nominal Composition
B-1	0.00	0.4	1.8	2.00	1.00	2	2212
B-2	0.25	0.4	1.8	1.75	1.25	2	
B-3	0.50	0.4	1.8	1.50	1.50	2	
B-4	0.75	0.4	1.8	1.25	1.75	2	
B-5	1.00	0.4	1.8	1.00	2.00	2	2122

(b) CATION STOICHIOMETRY FOR SCCO COMPOSITIONS

	x	Sr	Ca	Cu	Nominal Composition
S-1	0.00	0.00	1.00	1	$CaCuO_2$
S-2	0.25	0.25	0.75	1	
S-3	0.50	0.50	0.50	1	
S-4	0.75	0.75	0.25	1	
S-5	1.00	1.00	0.00	1	$SrCuO_2$

(c) CONSTITUENTS OF THE MIXTURE COMPOSITIONS.

	x		BSCCO		SCCO		Relative Phase Mixture
M-1	0.00	=	B-1	+	S-1	→	$2212 + CaCuO_2$
M-2	0.25	=	B-2	+	S-2		
M-3	0.50	=	B-3	+	S-3		
M-4	0.75	=	B-4	+	S-4		
M-5	1.00	=	B-5	+	S-5	→	$2122 + SrCuO_2$

(d) VARIATION OF LEAD-CONTENT IN INTERMEDIATE 2212

	Ca_2PbO_4	Ca_2CuO_3	CuO	PbO	2212	xPb in 2212
P-1	0.0	0.5	0.5	0.4	1.0	0.0
P-2	0.0	0.5	0.5	0.0	1.0	0.4
P-3	0.1	0.4	0.6	0.0	1.0	0.3
P-4	0.2	0.3	0.7	0.0	1.0	0.2
P-5	0.3	0.2	0.8	0.0	1.0	0.1
P-6	0.4	0.1	0.9	0.0	1.0	0.0

until the total time of heat treatment reached 350 hours (see figure 1.(c)).

After each heat treatment, three tapes were set aside for characterization. The J_c of each tape was measured by the four point probe technique with a voltage criterion of 1 μV/cm. The sample with the highest J_c was set aside for SEM characterization. The sample with the second highest J_c was peeled along its length to expose the superconductor core for XRD.

DTA analysis on the powders (with 15 wt% silver) was performed in 8% O_2 with a ramp rate of 5° C per minute. SEM micrographs were taken using a JEOL 5400 microscope in secondary electron mode operating at an acceleration voltage of 30 kV. XRD patterns were obtained by a 2-theta step of 0.005° with a range from 3 - 60°. The voltage and current on the x-ray tube were 30 kV and 15 mA, respectively.

Results

X-Ray diffraction (XRD) of the BSCCO powders showed nearly single-phase 2212 with a trace of Ca_2PbO_4 for x < 0.5. Increasing amounts of Ca_2PbO_4 and 2201 appeared in the range from x = 0.5 to x = 1.0. The SCCO results showed a 2:1 Ca:Cu ratio phase with a balance of CuO for x = 0.0. This was due to the instability of a 1:1 phase in a calcium-rich mixture [14]. For x ≥ 0.5 XRD results indicated essentially pure 1:1 phase. Figure 3 shows the x-ray analysis for x = 0.0.

Differential Thermal Analysis (DTA) performed on the BSCCO and Mixture powders is shown in Figure 4. The large endotherm is presumed to be the partial melting of 2212 which

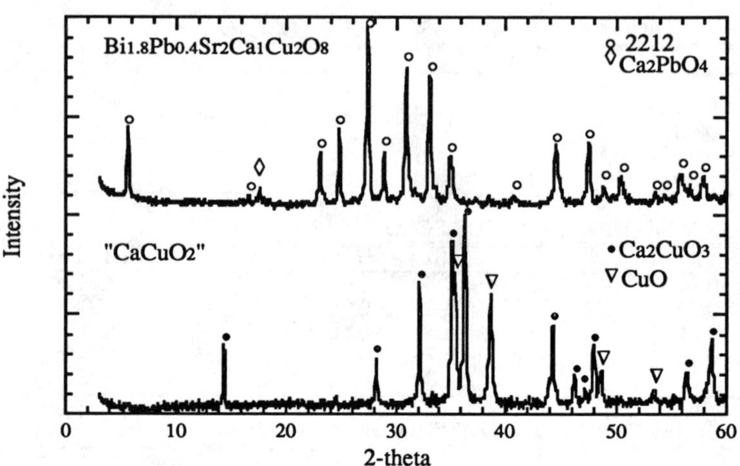

Figure 3. X-ray analysis for x=0.0.

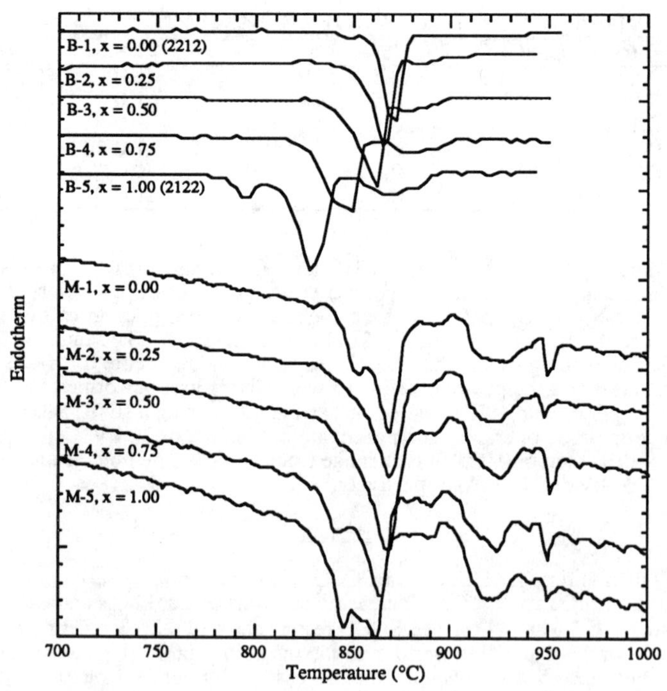

Figure 4. DTA analysis for BSCCO and Mixture powders.

decreases in temperature from ~865°C (x = 0.0) to ~830°C (x = 1.0) as calcium is substituted for strontium. This is consistent with the findings of Grader et al. [8]. DTA analysis of the SCCO powders only revealed phases with melting points well above processing temperatures (~950-1050°C). In the mixture powders the DTA traces vary insignificantly with x. This does not follow the trend of the BSCCO powders. Since the DTA rate was rapid (5°C/min) the reactions occurring in the powders were fast implying that 2223 formation is fast[15].

Figure 5.(a) is a secondary electron image of the x = 0.0 sample after 50 hours of heat treatment in 8% O_2 at 820°C. It shows nearly phase pure 2223 with very little secondary phase (< 10 μm in diameter). The large light gray grains are the 15 wt% silver added to the mixtures. In the remaining samples (x = 0.25 to x = 1.0) the size and number of secondary phase increase rapidly. The J_c for x = 0.0 was measured as 3,500 A/cm^2 and decreased to 500 A/cm^2 for x = 1.0. After thermomechanical processing to a total treatment time of 350 hours, the secondary electron image of the x = 0.0 sample, Figure 5.(b), shows no discernible second phase (J_c = 3.0 x 10^4 A/cm^2). Samples in the remaining mixtures have second phase which also increase in number and size from x = 0.25 to x = 1.0. The quality of 2223 produced in the x = 0.0 sample

(a)

(b)

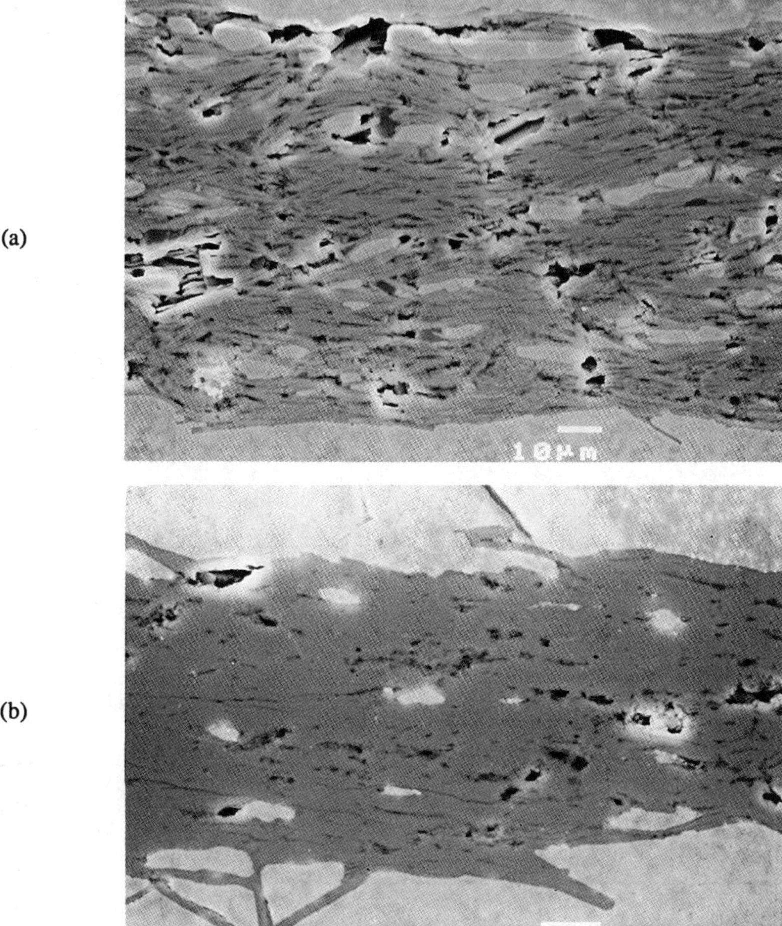

Figure 5. Secondary Electron images of (a) x=0.0 at 50 hrs and (b) x=0.0 at 350 hrs.

can be reflected in AC susceptibility measurements, Figure 6. The curve is smooth, showing only one sharp transition at 109 K. The lack of a second transition indicates no trace of 2212 due to complete transformation to 2223 [16].

Samples were also treated in air at 835-845°C, but yielded electrical properties inferior to those samples treated in 8% O_2. The best results in both atmospheres occurred at the onset of partial melting as determined by DTA. Processing at temperatures either above or below the optimal temperature resulted in degraded electrical and microstructural properties due to decomposition of 2223 above[17] and slow reaction kinetics below.

Figures 7 (a), (b) and (c) are secondary electron images of tapes where all lead was added in as Ca_2PbO_4 (P-6), PbO (P-1), and placed in the 2212 (P-2) phase respectively. They were heat treated in 8% O_2 at 815°C for a total thermo-mechanical processing time of 250 hours. The sample which contained Ca_2PbO_4 as the lead bearing phase shows a large amount of 2212 with a J_c of 1.0 x 10⁴ A/cm². The sample with PbO has a much more aligned microstructure, but also contains large amounts of 2212 as well as rather large grains of secondary phase. The J_c for this sample was 1.0 x 10⁴ A/cm². When all lead was contained in the 2212 phase, resulting microstructures were well aligned and nearly phase pure yielding a J_c of 2.2 x 10⁴ A/cm². This aspect of the two-powder process is discussed in more detail in reference [18]

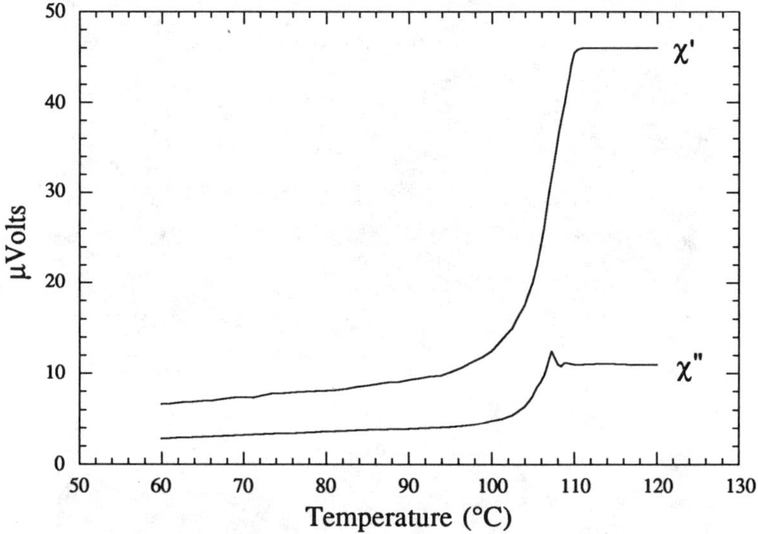

Figure 6. Temperature dependence of AC susceptibility for x=0.0 for 350 hrs.

Conclusions

Control of the reaction pathway has been shown to enhance the phase purity of lead doped 2223 and consequently its electrical properties. A two-powder processing scheme has been shown to orchestrate some measure of control on the initial phase assemblage and yield a highly pure form of 2223. By this method the two powders are formed as separate intermediate phases and then combined to react in situ during oxide powder-in-tube processing. The best powder mixture contained 2212 with a small amount of Ca_2PbO_4, powder 1, and a two phase form of $CaCuO_2$ (Ca_2CuO_3 & CuO), powder 2, showing virtual disappearance of secondary phase after only 50 hours treatment.

Samples were fired in both air and 8% partial pressure of O_2 atmospheres. The 8% O_2 samples had improved microstructure over the air samples as well as the highest J_c values. A comparison of the micrographs for the two indicated the air samples contained higher percentages of secondary phase.

The presence of increasing amounts of Ca_2PbO_4 has been shown to hamper formation of 2223 which resulted in significant amounts of secondary phase and hence, lower J_c values. Addition of lead in the mixture as PbO was found to produce a higher density core, but left significant amounts of 2212 and yielded low critical current densities.

The ability to reproduce a process while sustaining quality properties is a necessity for commercially producing superconducting wire. The two-powder process has proved to meet this requirement (using the M-1 mixture). This scheme has been repeated several times with each one producing the same quality microstructure as well as quality critical current densities. To date the maximum value is 4.0×10^4 A/cm^2 (I_c=60 A) at 77K and 18.0×10^4 A/cm^2 (I_c=270 A) at 4.2K. Increases in J_c values are expected when mechanical processing techniques are optimized. For further details on all subjects contained in this paper please consult reference [19].

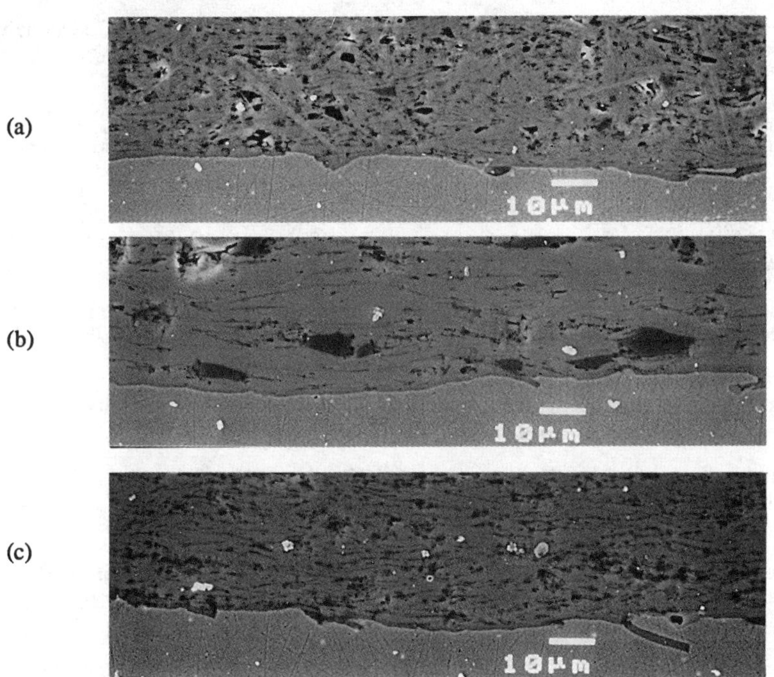

Figure 7.   Secondary Electron images of x=0.0 for 250 hrs where Lead was added in as (a) Ca_2PbO_4, (b) PbO and (c) incorporated into 2212.

Acknowledgments

This work was supported by the U. S. Department of Energy, Conservation and Renewable Energy, as part of a program to develop electric power technology, under Contract W-31-109-Eng-38.

The authors wish to thank E.E. Hellstrom and D.C. Larbalestier (Applied Superconductivity Center at the University of Wisconsin) for the helpful discussions and AC susceptibility measurements.

References

[1] H. Maeda, T. Tanaka, M. Fukotomi and T. Asano, "A New High Tc Oxide Superconductor Without a Rare Earth Element," Jpn. J. Appl. Phys. 27 (1988), L209.

[2] M. Veyama, T. Hikata, T. Kato and K. Sato, "Microstructures of Jc-B Characteristics of Ag-Sheathed Bi-Based Superconductor Wires," Jpn. J. Appl. Phys. 30 (1991), L1384.

[3] Q. Li, K. Broderson, B. Nissen, H.A. Hjuler and T. Freltoft, 1992 Fall Meeting, Materials Research Society, Boston, MA.

[4] D.Y. Kaufman, M.T. Lanagan, S.E. Dorris, J.T. Dawley, I.D. Bloom, M.C. Hash, N. Chen, M.R. Deguire and R.B. Poeppel, "Thermomechanical Processing of Reactively Sintered Ag-Clad (Bi,Pb)2Sr2Ca2Cu3O10 Tapes," App. Supercon, 1 (1992), 34.

[5] K. Aota, H. Hattori, T. Hatano, K. Nakamura and k. Ogawa, "Growth of the 2223 Phase in Leaded Bi-Sr-Ca-Cu Oxide under Reduced Oxygen Partial Pressure," Jap. J. Appl. Phys., 28 (1989), L2196.

[6] J. Tsuchiya, H. Endo, N. Kijima, A. Sumiyama, M. Mizuno, and Y. Oguri, "Reaction Mechanism of High-Tc Phase (Tc=110 K) Formation in the Bi-Sr-Ca-Cu-O Superconductive System," Jap. J. Appl. Phys., 28 (1989), L1918.

[7] G.S. Grader, E.M. Gyorgy, P.K. Gallagher, H.M. O'Bryan, D.W. Johnson, S. Shushine, S.M. Zahurak, S. Jin, and R.C. Sherwood, "Effect of Sr/Ca Stoichiometry of $Bi_2Sr_2CaCu_2O_8$," Phys. Rev. B, 38 (1988), 757.

[8] M. Takano, K. Takada, K. Oda, H. Kitaguchi, Y. Miura, Y. Ikeda, Y. Tomii and H. Mazaki," Effects of Pb Content on the Formation of the High-Tc Phase in the (Bi,Pb)-Sr-Ca-Cu-O System," Jpn. J. Appl. Phys, 27 (1988), L1476.

[9] R.H. Arendt, M.F. Garbauskas and P.J. Bednarczyk, "An Alternate Preperation for (Bi, Pb)2Ca2Sr2Cu3O10," Physica C, 176 (1991), 126.

[10] T. Uzumaki, K. Yamanaka, N. Kamehara and K. Niwa, "The Effect of Ca_2PbO_4 Addition on Superconductivity in a Bi-Sr-Ca-Cu-O System," Jpn. J. Appl. Phys, 28 (1989), L75.

[11] J. Tenbrink, M. Wilhelm, K. Heine and H. Drauth, "Development of High-Tc Superconductor Wires for Magnetic Applications," presented at the Appl. Supercon. Conf., Snowmass Village, CO Sept 24-28, (1990)

[12] R. Fluekiger, B. Hensel, A. Jeremie, M. Decroux, H. Kuepfer, W. Jahn, E. Seibt, W. Goldacker, Y Yamada and J.Q. Xu, "Effect of Carbon on Critical Current Density of High-Temperature Superconductors," Supercond. Sci. Tech., 4 (1991), 124.

[13] K. Osamura, S. Soo Oh and S. Ochila, "Effect of Thermo-Mechanical Treatment on the Critical Current Density of Ag-Sheathed B(Pb)SCCO Tapes," Supercond. Sci. Technol., 3 (1990), 143.

[14] R.S. Roth, C.J. Rawn, J.J. Ritter and B.P. Burton, "Phase Equilibra of the System SrO-CaO-CuO, "J. Am. Ceram. Soc., 72 (1989), 1545.

[15] S.E. Dorris, B.C. Prorok, M.T. Lanagan, S. Sinha and R.B. Poeppel, "Synthesis of Highly Pure Bismuth-2223 By A Two-Powder Process," Physica C, 212 [1-2] (1993), 66.

[16] A. Umezawa, Y. Feng, H.S. Edelman, Y.E. High, D.C. Larbalestier, Y.S. Sung and E.E. Hellstrom, "Electromagnetic Granularity, Critical Current Density, and Low-Tc Phase Formation at the Grain Boundaries in (B1,Pb)2Sr2Ca2Cu3Ox Silver-Sheathed Tapes," <u>Physica C</u>, 198 (1992), 261.

[17] J.S. Luo, N. Merchant, V.A. Maroni, D.M. Gruen, B.S. Tani, W.L. Carter, G.N. Riley and K.H. Sandhage, "Thermostability and Decomposition of the $(Bi,Pb)_2Sr_2Ca_2Cu_3O_{10}$ Phase in Silver-Clad Tapes," <u>J. Appl. Phys.</u>, 72 (1992), 2385.

[18] S.E. Dorris, B.C. Prorok, M.T. Lanagan, N.B. Browning, M.R. Hagen, J.A. Parrell, Y. Feng, A. Umezawa and D.C. Larbalestier, "Methods For Introducing Lead into Bismuth-2223 and Their Effects on Phase Development and Superconducting Properties," Submitted to <u>Physica C</u>. Oct. 1993.

[19] B.C. Prorok, "Formation of the BSCCO (2223) Superconducting Phase By A Two-Powder Process" (M.S. thesis, University of Illinois at Chicago, 1993).

PROCESSING FOR BOTH HIGH-T_c AND HIGH-J_c

(Tl,Pb,Bi)(Sr,Ba)$_2$Ca$_2$Cu$_3$O$_9$ SUPERCONDUCTING TAPES

R. S. Liu , S. F. Wu, C. H. Tai and D. S. Shy

Materials Research Laboratories, Industrial Technology Research Institute, Hsinchu,
Taiwan, R.O.C.

Abstract

We have found an efficient and highly reproducible method for the preparation of the
homogeneous Tl-1223 powders by the partial substitution of Ba^{2+} into the Sr^{2+} sites in the
(Tl$_{0.6}$Pb$_{0.2}$Bi$_{0.2}$)(Sr$_{2-x}$Ba$_x$)Ca$_2$Cu$_3$O$_9$ system. Superconducting tapes (sheathed in silver) based
on the titled system have been fabricated by using the powder-in-tube (PIT) method. Typical
critical temperatures (T_c) of around 120 K and transport critical current densities (J_c) of about
1.05×10^4 A/cm^2 at 77 K in a zero magnetic field have been routinely obtained on short lengths ($\sim$
3 cm) of the sintered Tl-1223 tapes after rolling. Moreover, a prototype superconducting
(Tl,Pb,Bi)(Sr,Ba)$_2$Ca$_2$Cu$_3$O$_9$ magnet (with three pancake coils, each containing four 3 m
lengths of rolled tapes) generated field of 240 G at 77 K was obtained.

Processing of Long Lengths of Superconductors
Edited by U. Balachandran, E.W. Collings and A. Goyal
The Minerals, Metals & Materials Society, 1994

<u>Introduction</u>

Flux pinning in bulk high-T_c cuprate superconductors is essential for achieving the high-J_c values required for a variety technological applications. It has been found that it is possible to increase J_c strongly by the introduction of fine Y_2BaCuO_5 particles into the $YBa_2Cu_3O_7$ (YBCO) matrix through a quench and melt growth (QMG) process [1]. In an important development, Kim et al. [2] proposed that strong interlayer coupling between the CuO_2 planes is a major factor in controlling the intrinsic flux pinning of the cuprate materials. The flux motion associated with the magnetic field induced broadening of the resistive transitions of highly anisotropic high-T_c superconducting cuprates is thus determined by an intrinsic dimensional crossover from three dimensional vortex lines to two dimensional vortex pancakes. Moreover, the authors [2] observed an exponential dependence for the Josephson tunneling between the longest distance of the superconducting CuO_2 planes. Recently, we [3] have found that a high J_c exists in the $(Tl_{0.5}Pb_{0.5})Sr_2Ca_2Cu_3O_9$ (Tl/Pb-1223) material which has only one insulating layer, $(Tl_{0.5}Pb_{0.5})O$, between the superconducting CuO_2 planes. In Fig. 1 we show the crystal structures of $YBa_2Cu_3O_7$ (YBCO), $(Tl_{0.5}Pb_{0.5})Sr_2(Ca_{0.8}Y_{0.2})Cu_2O_7$ (Tl/Pb-1212), $(Tl_{0.5}Pb_{0.5})Sr_2Ca_2Cu_3O_9$ (Tl/Pb-1223) and $Tl_2Ba_2Ca_2Cu_3O_{10}$ (Tl-2223) which encompass the longest distance (d) between the superconducting CuO_2 planes, of 8.32 Å, 8.80 Å, 8.72 Å and 11.28 Å, respectively [4]. It is worth pointing out that the structures of YBCO and Tl/Pb-1212 are somewhat similar except that the CuO chains have a perovskite-type structure in YBCO and the $(Tl_{0.5}Pb_{0.5})O$ layers have a rock salt-type structure in Tl/Pb-1212. The structures of Tl/Pb-1212 and Tl/Pb-1223 differ in that the latter material has one excess superconducting CuO_2 layer (Fig. 1). The significant difference in structure between Tl/Pb-1223 and Tl-2223 is that the former material has only one insulating layer of $(Tl_{0.5}Pb_{0.5})O$ (three-dimensional like) but the latter compound has two insulating layers of Tl_2O_2 which may lead the later material to have a two-dimensional character and a longest d among those four superconductors.

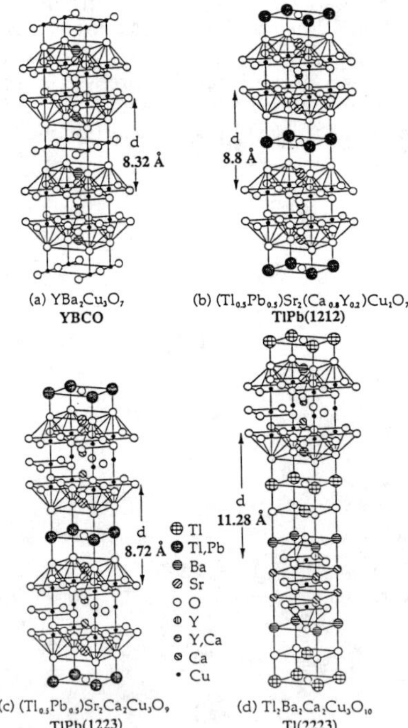

Figure 1- Crystal structures of $YBa_2Cu_3O_7$ (YBCO), $(Tl_{0.5}Pb_{0.5})Sr_2(Ca_{0.8}Y_{0.2})Cu_2O_7$ (Tl/Pb-1212), $(Tl_{0.5}Pb_{0.5})Sr_2Ca_2Cu_3O_9$ (Tl/Pb-1223) and $Tl_2Ba_2Ca_2Cu_3O_{10}$ (Tl-2223) with the longest distance (d) between the superconducting CuO_2 planes of 8.32 Å, 8.80 Å, 8.72 Å and 11.28 Å, respectively.

52

The magnetic field dependence of J_c up to 12 T at 77 K as measured by hysteresis loop for YBCO, Tl/Pb-1223 and Tl-2223 has been reported [3]. The critical current densities at T = 77 K and H = 1 T for the YBCO, Tl/Pb-1223 and Tl-2223 materials are estimated to be 1.04 x 10^5, 1.24 x 10^5 and 0 A/cm^2, respectively. However, these three materials have a comparable J_c value ($\sim$ 1 x 10^7 A/cm^2) at T = 4.2 K and H = 1T. The significant flux pinning ability at high fields and temperatures in the single insulating $(Tl_{0.5}Pb_{0.5})O$ planes of Tl/Pb-1223 was observed. The J_c of Tl/Pb-1223 is far greater than that measured for double insulating Tl_2O_2 layer material Tl-2223 and comparable with single CuO chains of YBCO. It means that an increase in the number (and distance) of insulating layers in the superconducting cuprates weakens the Josephson coupling between conducting CuO_2 planes along the c direction and also results in a decrease in J_c. This decoupling between planes causes the vortex lattice to break into pancakes which are easily thermally activated. Importantly, the low J_c in Tl-2223 is a result of broadening the irreversibility line rather than a lack of pinning centers. Clearly, these sites in Tl-2223 are known to be present as shown by the very strong pining observed at 4.2 K. Therefore, both high T_c and J_c were observed in the Tl/Pb-1223 phase which may be of great promise for technological applications because the material has both a J_c and an irreversibility line comparable to those of YBCO (T_c = 92 K) but the material has a higher T_c value (124 K) [3,5].

Here, we demonstrate an efficient and highly reproducible method for the preparation of the homogeneous Tl-1223 powders by the partial substitution of Ba^{2+} into the Sr^{2+} sites in the $(Tl_{0.6}Pb_{0.2}Bi_{0.2})(Sr_{2-x}Ba_x)Ca_2Cu_3O_9$ system. On the basis of the applications point of view, superconducting tapes of $(Tl,Pb,Bi)(Sr,Ba)_2Ca_2Cu_3O_9$ have been produced by using the powder-in-tube (PIT) method. Moreover, the results on $(Tl,Pb,Bi)(Sr,Ba)_2Ca_2Cu_3O_9$ pancake coils and prototype magnet will also be given.

Experimental

The preparation of the $(Tl_{0.6}Pb_{0.2}Bi_{0.2})(Sr_{2-x}Ba_x)Ca_2Cu_3O_9$ powders

Bulk samples were synthesized by the solid-state reaction. Appropriate amounts of high purity $SrCO_3$, $BaCO_3$, $CaCO_3$ and CuO powders were weighted stoichiometrically and ground in an agate mortar. The well-mixed oxides were calcined at 920 °C in air for 10 h to obtain a mixture Sr-Ba-Ca-Cu-O precursor powders. The Sr-Ba-Ca-Cu-O powders were then mixed with appropriate amount of Tl_2O_3, PbO and Bi_2O_3 to yield mixtures with nominal compositions of $(Tl_{0.6}Pb_{0.2}Bi_{0.2})(Sr_{2-x}Ba_x)Ca_2Cu_3O_9$ (x = 0, 0.1, 0.2, 0.3, 0.4 and 0.5). The mixtures were ground and pressed into a cylindrical pellet, 2 mm in thickness and 10 mm in diameter, under a pressure of about 2 ton/cm^2. The pellets were then wrapped in a Au foil and sintered in flowing O_2 to alleviate possible volatilization of Tl, Pb and Bi at 920 °C for 3 h, then cooled down to room temperature at a rate of 2 °C/min.

Powder X-ray diffraction (XRD) analyses were performed using a Philips X-ray diffractometer with a Ni-filtered CuK_α radiation. Chemical compositions of the specimens were examined by energy dispersive X-ray spectrometry (EDS) from a JEM-2010 electron microscope operating at 200 kV. A standard four-point probe was used to measure the electrical resistance of the samples. Low field magnetization data were obtained using a superconducting quantum interference devices (SQUID) magnetometer (Quantum Design).

The preparation of the $(Tl,Pb,Bi)(Sr,Ba)_2Ca_2Cu_3O_9$ tapes by the PIT method

The $(Tl_{0.6}Pb_{0.2}Bi_{0.2})(Sr_{1.8}Ba_{0.2})Ca_2Cu_3O_9$ sintered powders were loaded into a silver tube with outer diameter of 6 mm and inner diameter of 4 mm. The tube assembly was swaged and drawn to form a wire having an outer diameter of 1.5 mm. The resulting wire was formed into a tape having a thickness of $\sim$ 0.15 mm by rolling, and optionally by pressing at 200 kg/m^2 to a thickness of $\sim$ 0.08 mm. Subsequently, the tapes were heated at the temperatures between 850 °C and 930 °C in oxygen for several hours and then cooled down to room temperature.

The series XRD spectra of the samples $(Tl_{0.6}Pb_{0.2}Bi_{0.2})(Sr_{2-x}Ba_x)Ca_2Cu_3O_9$ with x varying from 0 to 0.5 are shown in Fig. 2. For the stoichiometric nominal composition of the $(Tl_{0.6}Pb_{0.2}Bi_{0.2})Sr_2Ca_2Cu_3O_9$ (x = 0) sample, i.e., without the addition of Ba in the Sr sites, it was found that only the $(Tl_{0.6}Pb_{0.2}Bi_{0.2})Sr_2CaCu_2O_7$ (Tl-1212) phase (as denoted by "o" in Fig. 2) and some impurities were observed in the XRD pattern. This result indicates that the formation of the Tl-1223 phase is rather difficult under the sintering condition of 920 °C for 3 h in oxygen. However, the partial substitution of Ba^{2+} into the Sr^{2+} sites in $(Tl_{0.6}Pb_{0.2}Bi_{0.2})(Sr_{2-x}Ba_x)Ca_2Cu_3O_9$ under the same sintering condition as the x = 0 sample which results in a gradually disappearance of the Tl-1212 phase and forming of the Tl-1223 phase (as denoted by "•" in Fig.1) in the x = 0.1 sample. The nearly single phase of the series $(Tl_{0.6}Pb_{0.2}Bi_{0.2})(Sr_{2-x}Ba_x)Ca_2Cu_3O_9$ was obtained for the x = 0.2 ~ 0.3 samples. Almost all the XRD peaks in Fig. 2 for the x = 0.2 and 0.3 samples can be fitted using a space group of P4/mmm with a tetragonal unit cell of a ~ 3.82 Å and c ~ 15.3 Å. The chemical compositions of twenty small crystallites from the x = 0.2 sample were examined by EDS; the average cation compositions were less than 8 % deviation from the nominal composition. Clearly, the addition of 20 ~ 30 % of Ba into the Sr sites can accelerate the formation of the Tl-1223 phase which is similar to the effect of the partial substitution of Ca into the Sr sites in $(Tl_{0.5}Pb_{0.5})Sr_{1.6}Ca_{2.4}Cu_3O_y$ [5]. Such results lead us to understand that the formation of the Tl-1223 phase is kinetic control but the partial substitution of Ba into the Sr sites will promote the reaction to form the monophasic Tl-1223 phase. For x > 0.3 samples, it is found that an unidentified impurity phase (as denoted by "x" in Fig. 2) was increased with increasing x except the Tl-1223 phase.

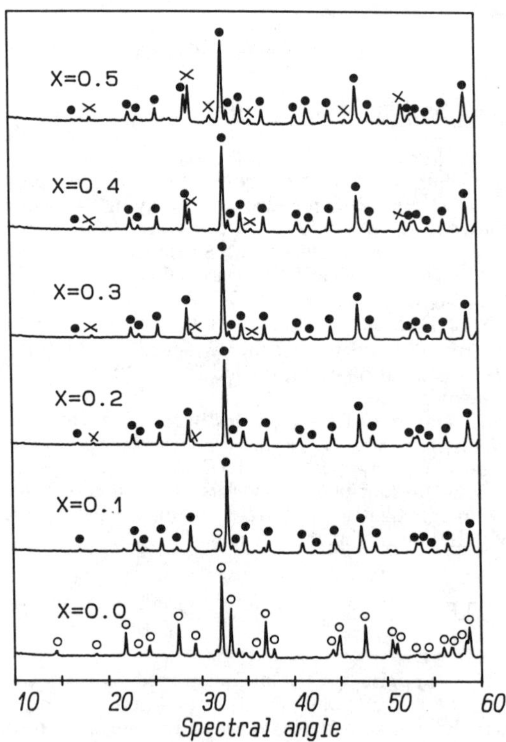

Figure 2 - XRD spectra of the samples $(Tl_{0.6}Pb_{0.2}Bi_{0.2})(Sr_{2-x}Ba_x)Ca_2Cu_3O_9$ with x varying from 0 to 0.5.

In Fig. 3 we show the zero resistance temperature [$T_{c(zero)}$] as measured by electrical resistance as a function of x in the as-synthesized and post-annealed samples of $(Tl_{0.6}Pb_{0.2}Bi_{0.2})(Sr_{2-x}Ba_x)Ca_2Cu_3O_9$. All the Ba substituted samples with x = 0.1 ~ 0.5 have $T_{c(zero)}$'s around 113 K which is identical to the previous founding for the as-synthesized Tl-1223 phase [6,7]. Kaneko et al. [6] have applied a process to elevate the $T_{c(zero)}$ up to 123 K in $(Tl_{0.64}Pb_{0.2}Bi_{0.16})Sr_2Ca_3Cu_4O_z$ when the as-synthesized samples were post-annealed at around 800 °C for 80 h in O_2. Moreover, we have investigated a method to enhance the $T_{c(zero)}$ up to 124 K in $(Tl_{0.5}Pb_{0.5})Sr_{1.6}Ca_{2.4}Cu_3O_y$ by post-annealing the as-synthesized samples in a sealed evacuated quartz tube at 750 °C for 10 days [5]. Therefore, it is worth trying to post-anneal the $(Tl_{0.6}Pb_{0.2}Bi_{0.2})(Sr_{2-x}Ba_x)Ca_2Cu_3O_9$ samples. The as-synthesized samples were annealed at the temperature of 860 °C for 20 h in O_2 and then quench in air. The $T_{c(zero)}$ of the annealed $(Tl_{0.6}Pb_{0.2}Bi_{0.2})(Sr_{2-x}Ba_x)Ca_2Cu_3O_9$ samples can be further increased by 2 ~ 8 K, with a maximum at x = 0.2 [$T_{c(zero)}$ = 122 K] as shown in Fig. 3.

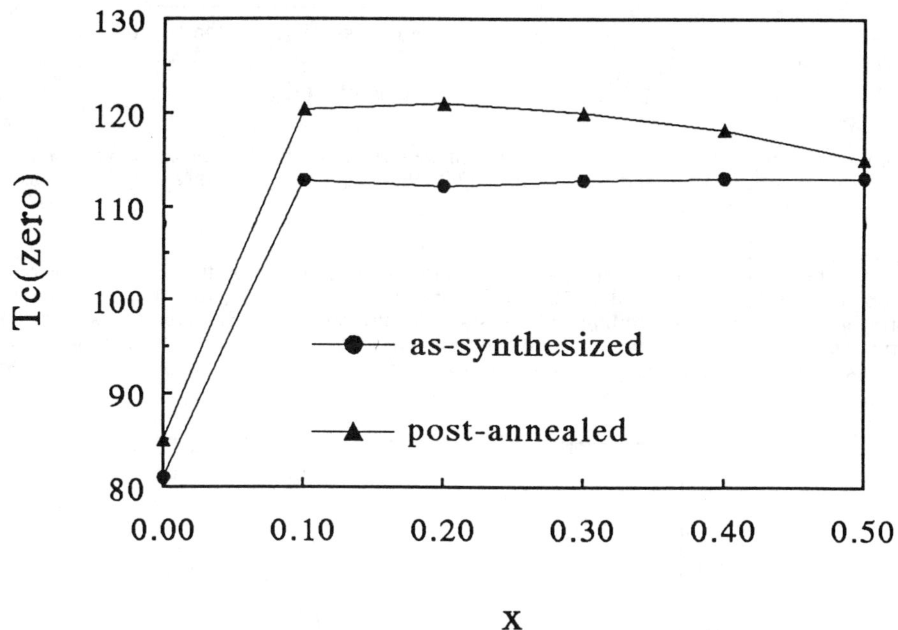

Figure 3 -   Zero resistance temperature [$T_{c(zero)}$] as measured by electrical resistance as a function of x in the as-syntheszed and post-annealed samples of $(Tl_{0.6}Pb_{0.2}Bi_{0.2})(Sr_{2-x}Ba_x)Ca_2Cu_3O_9$.

In Fig. 4 we show the temperature dependence of the low-field magnetization (10 Oe, field cooled) of the as-synthesized and post-annealed $(Tl_{0.6}Pb_{0.2}Bi_{0.2})(Sr_{1.8}Ba_{0.2})Ca_2Cu_3O_9$ (x = 0.2) samples. The onset diamagnetism appears at 115 K and 122 K for the as-synthesized and post-annealed samples, respectively. These results are consistent with the electrical resistance measurements (as shown in Fig. 3). We estimate the superconducting (Meissner) volume fraction of 46 % and 50 % of $-1/4\pi$ at 5 K for the as-synthesized and post-annealed samples, respectively. Not only the T_c but also the superconducting Meissner volume fraction can be enhanced in the post-annealed sample.

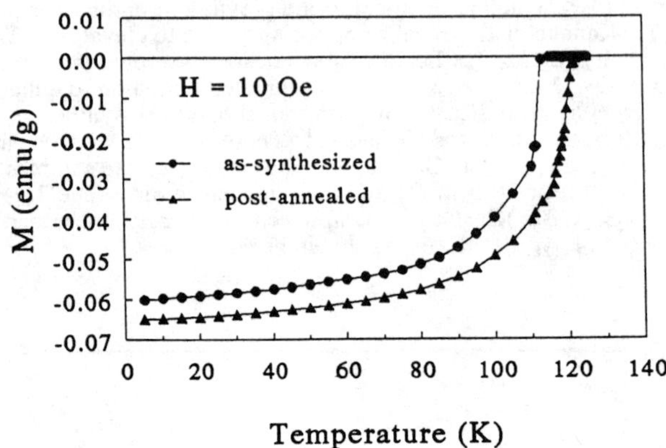

Figure 4 - Temperature dependence of low-field magnetization (10 Oe, field cooled) of the as-synthesized and post-annealed $(Tl_{0.6}Pb_{0.2}Bi_{0.2})(Sr_{1.8}Ba_{0.2})Ca_2Cu_3O_9$ (x = 0.2) samples.

In Fig. 5 we show the voltage-current relationship measured by the standard four-point probe method for the sintered Tl-1223 short tapes (~ 3 cm) after (a) rolling, (b) rolling and pressing one time and (c) rolling and pressing six times at 77 K. Based on an electric field criterion of 1 µV/cm for determining the critical current, the rolled, pressed one time and pressed six times Tl-1223 tapes have J_c values of 10500 A/cm^2, 11200 A/cm^2 and 13300 A/cm^2, respectively.

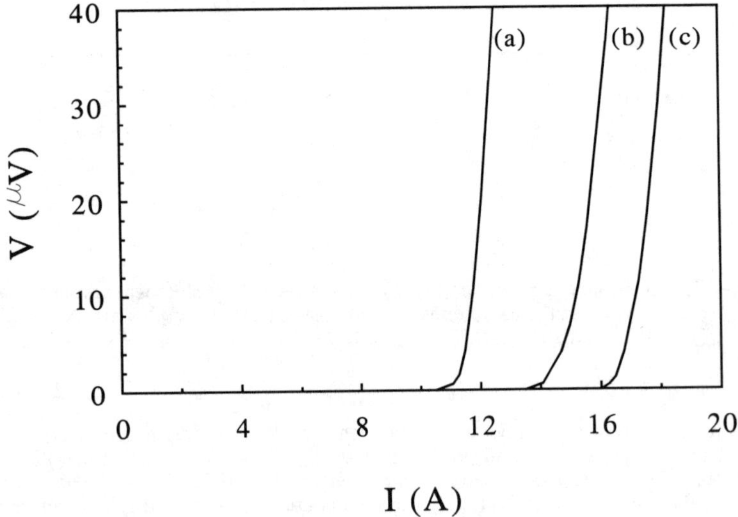

Figure 5 - Voltage-current relationship measured by the standard four-point probe method for the sintered Tl-1223 short tapes (~ 3 cm) after (a) rolling, (b) rolling and pressing one time and (c) rolling and pressing six times at 77 K.

In Fig. 6 we show the dependence of the magnetic moment on the applied magnetic field (B) for the pressed Tl-1223 tape (B // c) with J_c (77 K) = 13300 A/cm^2. This shows clearly two successive transitions at 120 K (superconducting transition temperature, T_c) and 116 K ~ 104 K (decoupling temperature, T_d); the later is strongly dependent on the applied magnetic field as shown in Fig. 7. This phenomena is attributed to a bulk superconducting transition at 120 K and a coupling by weak links across grain boundaries, which gives rise to a multiconnected Josephson network with a transition temperature corresponding to T_d (116 K ~ 104 K).

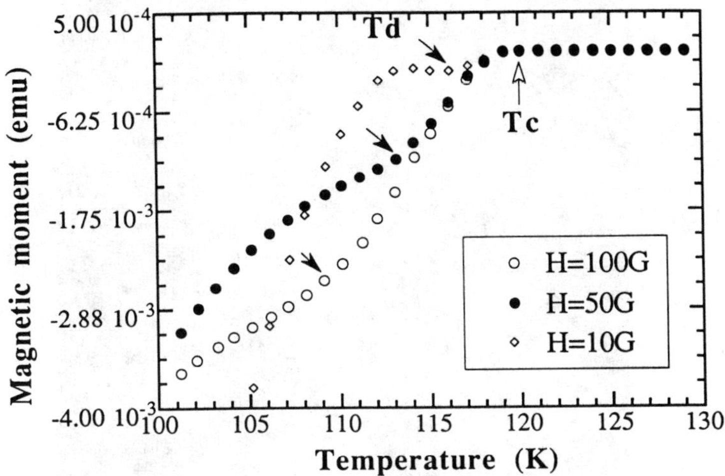

Figure 6 - Dependence of the magnetic moment on the applied magnetic field (B) for the pressed Tl-1223 tape (B// c) with J_c (77 K) = 13300 A/cm^2.

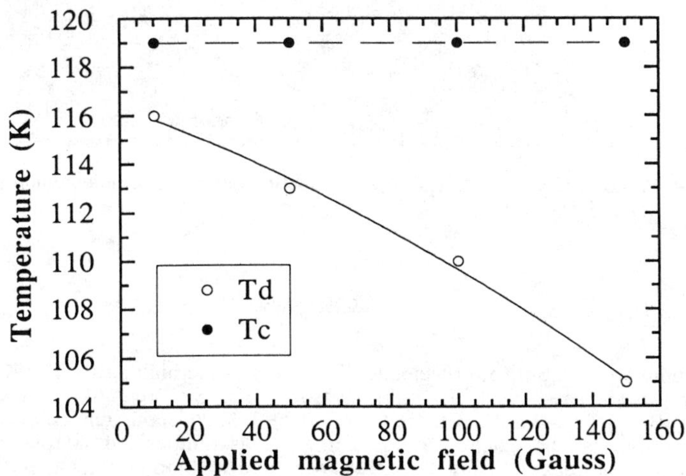

Figure 7 - Dependence of intragrain superconducting transition temperature (T_c) and intergrain decoupling temperature (T_d) on the applied magnetic field for the corresponding pressed Tl-1223 tape derived from Fig. 6.

In Fig. 8 we show a photograph of a "wind-and-react" prototype superconducting magnet assembled by stacking three individual pancake coils, each containing four 3-m lengths of rolled Tl-1223 tapes) which can generate magnetic field up to ~ 240 G at 77 K and zero applied field.

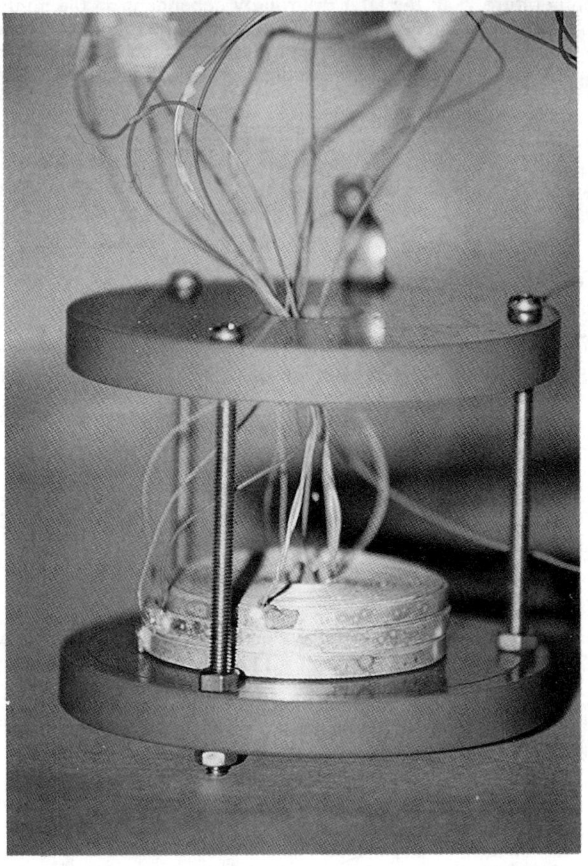

Figure 8 - Photograph of a prototype superconducting magnet assembled with three pancake coils of Tl-1223.

Conclusions

In summary, a significant accelerated formation of the high purity Tl-1223 phase can be achieved via the partial substitution of Sr by Ba in the system $(Tl_{0.6}Pb_{0.2}Bi_{0.2})(Sr_{2-x}Ba_x)Ca_2Cu_3O_9$ for $x = 0.2 \sim 0.3$. A maximum T_c up to 122 K for the $x = 0.2$ sample can be reached by annealing the as-synthesized sample in an oxidizing environment. Moreover, using the powder-in-tube method, we have processed high-J_c ($\sim 10^4$ A/cm^2) short lengths of the corresponding Tl-1223 tape conductors. Long lengths of conductors have also been processed and fabricated into pancake coils. A prototype supercondicting magnet assembled by stacking three individual pancake coils, each containing four 3-m lengths of rolled Tl-1223 tapes) generated magnetic field up to ~ 240 G at 77 K and zero applied field has been obtained.

Acknowledgements

The authors would like to thank the Ministry of Economic Affairs of the Republic of China for the financial support.

References

1. M. Murakami et al., "A new Process of High J_c in Oxide superconductors", Jpn. J. Appl. Phys., 28 (1989), 1189-1194.
2. D. H. Kim et al., "Effect of Cu-O Layer Spacing on the Magnetic Field Induced Resistive Broadening of High-Temperature Superconductors," Physica C 177 (1991), 431-437.
3. R. S. Liu et al., "High Critical-Current Densities in $(Tl_{0.5}Pb_{0.5})Sr_2Ca_2Cu_3O_9$ with T_c up to 115 K", Appl. Phys. Lett. 60 (1992), 1019-1021.
4. D. M. Ginsberg, ed., Physical Properties of High Temperature Superconductors II (Singapore : World Scientific, 1990), 121-198.
5. R. S. Liu et al., "Superconductivity at 124 K in $(Tl,Pb)Sr_2Ca_2Cu_3O_9$", Physica C 198 (1992), 318-322.
6. T. Kaneko et al., "$(Tl,Pb,Bi)Sr_2Ca_2Cu_3O_z$ Superconductors with Zero Resistance at 120 K, Appl. Phys. Lett. 56 (1990), 1281-1283.
7. Y. T. Huang et al., "Accelerated Formation of the Three Cu-O Layered Tl-(Pb,Bi)-Sr-Ca-Cu-O Compound", Appl. Phys. Lett. 57 (1990), 2354-2355.

INFLUENCE OF INITIAL COMPOSITION AND PROCESSING PARAMETERS ON THE

CRITICAL CURRENT DENSITY OF BPSCCO PIT TAPES

A. D. Damodaran[1], K. G. K. Warrier[1], P. S. Mukherjee[1], T. V. Mani[1], M. S. Sarma[1],
S. Augustine[1], G. Swaminathan[2], K. Venugopal[2], and M.V.T. Dhananjayan[2]

[1]Regional Research Laboratory (CSIR), Trivandrum, India
[2]BHEL (R&D) Hyderabad, India

ABSTRACT

Studies on bulk and Ag sheathed tapes of BPSCCO superconducting materials have been conducted. The results show that $Bi_{1.8}Pb_{0.4}Sr_2Ca_{2.2}Cu_3O_y$ prepared through the sol–gel process and vacuum calcined under O_2–rich dynamic flow conditions give most favorable results. Importance of AC susceptibility (x')–temperature (T) measurements became evident in evaluating the quality of the tapes.

Processing of Long Lengths of Superconductors
Edited by U. Balachandran, E.W. Collings and A. Goyal
The Minerals, Metals & Materials Society, 1994

<u>Introduction</u>

Fabrication of high-T_c tapes by powder-in-tube (PIT) process has assumed crucial significance over the past few years due to its potential in commercial applications. Among the high-T_c oxides, the bismuth based superconductor is the most favored one in view of its better stability, easier handling, relatively simpler ambient heat treatment procedures, and larger J_c values even at higher magnetic fields. The J_c values of $(Bi,Pb)_2Sr_2Ca_2Cu_3O_x$ (BPSCCO) tapes range from 15,000 – 60,000 A/cm^2 at T = 77 K (self-field) [1–6]. There appears to be difficulties in reproducing the high J_c values. Based on the well-reported literature and from considerations of the possible applications, a feasible value is set at 25,000 A/cm^2 at 35 K and 2 T with technologically acceptable reproducibility. The present and future efforts are directed to obtain precisely such a target, first in the laboratory and then in the industry. We describe the summary of results obtained during the course of our investigations with the above aims in view. The studies involved both bulk and Ag-sheathed tapes. Optimization of stoichiometry, standardization of powder preparation including choice of appropriate synthesis routes, and thermomechanical procedures for tape fabrication are considered in this work. In all cases the reaction sequences were followed by x–ray diffraction, ρ–T, x'–T, and J_c–B measurements, both for powders and the tapes.

<u>Experimental</u>

Superconducting powders were prepared by solid-state synthesis route and acrylate method. In the solid state route, stoichiometric amounts of Bi_2O_3, PbO, $CaCO_3$, $SrCO_3$ and CuO were carefully weighed and mixed in wet media in an agate mortar and pestle for $\approx$4 h. For quantities larger than 15 g, zirconia balls were used in acetone media in plastic containers of appropriate sizes and mixed for $\approx$30 h. The powder was dried in an oven and again dry mixed for $\approx$1 h. In the acrylate method [7], bismuth nitrate was dissolved in an acrylic acid–water (5:1) mixture to form golden yellow solution. Sr, Ca and Cu nitrates were added one by one to result in a greenish blue solution. Minimum water in the composition was used. The solution is then concentrated by heating at 100°C, where fumes of NO_2 starts evolving. A green gel is formed at the end, which is further decomposed at 300°C resulting in a black precursor powder. Calcination of powders were carried out both in air and under dynamic vacuum conditions. Air calcination was done at 800°C for 12 h, whereas the dynamic calcination was carried out at 750°C for 8 h. The dynamic calcination was carried out in flowing gas of air, nitrogen, and/or oxygen at a pressure of 8-10 mm Hg. Heat treatment of powder was done between 830 - 850°C for periods varying between 120 and 200 h with at least three intermediate grinding and pressing. The tapes were heat treated at temperatures $\approx$12°C lower than that used for bulk samples.

The powder was mechanically filled into silver tubes (OD 10 mm and ID 7.5 mm). Ultrasonic vibration was used to improve the packing density to 3.2 – 3.5 gm/cc. The open side was sealed with a silver plug. It was then groove rolled to 1 mm dia in steps and then rolled to tape form to achieve a final core thickness of $\approx$30 μm with average reduction $\approx$10% per pass. Relevant parameters such as reduction per pass, annealing conditions, etc. were studied as function of the final J_c values.

X-ray powder diffraction of the bulk as well as powders extracted from the tapes were measured at regular intervals for detailed characterization in a Phillipps PW 1710 Automatic Diffractometer using CuK_α radiation with a graphite monochromator. Quantization of 2212 and 2223 phases were carried out using standard method used in the laboratory. 2201 and Ca_2PbO_4 phases were treated as impurities.

AC susceptibility measurements were carried out both for bulk powders and the tapes. A dual phase lock-in-amplifier was used (EG & G AR 5210) for detecting the signals from secondary coil. The built-in oscillator excited the primary cell at a frequency of 28.4 Hz. All measurements were done during warm-up at a very low rate to avoid thermal hysteritic effects. For both resistivity and J_c measurements, a four probe technique was used. A calibrated coil was used for varying magnetic fields up to 200 Oe. All measurements were interphased through a PC with appropriate software. All resistivity and AC susceptibility measurements were recorded at low temperatures using an APD Cryostat during the warm-up cycle. J_c values were measured only at liquid nitrogen temperature.

<u>Results and Discussion</u>

Figure 1 shows the effects of stoichiometric variation of Ca, Sr and Cu on T_c (0) and the high T_c fraction, (HTf). The optimum composition is found at Sr = 1.8 and Ca = 2.2. Critical current J_c as a function of composition is shown in Fig. 2. Highest J_c value is obtained at Ca = 2.2. Figures 3 and 4 give impurity fraction as a function of stoichiometry. The relatively high values of J_c for Ca = 2.2 is attributed to not only the high phase purity with respect to 2223 but also due to the presence of Ca_2CuO_3, though small in quantity, as flux pinner (Fig. 4). Based on the above results, the stoichiometric composition was fixed as $Bi_{1.8} Pb_{0.4} Sr_2 Ca_{2.2} Cu_3 O_x$, generally matching with those reported in literature.

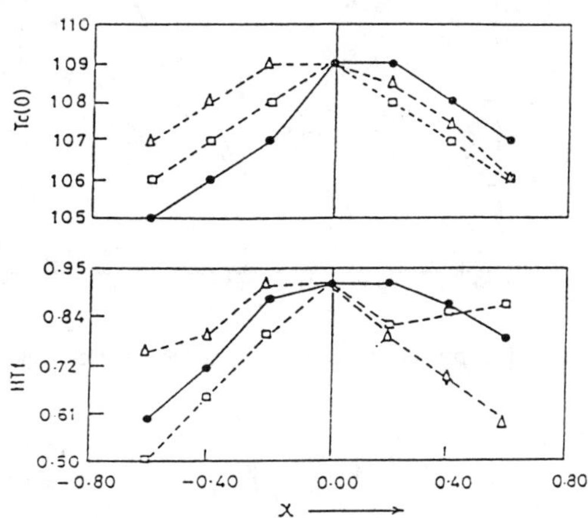

Figure 1 -   Effect of stoichiometric variation on T_c(O) and high-T_c phase fraction (HTf) [•-Ca, Δ-Sr, □-Cu]

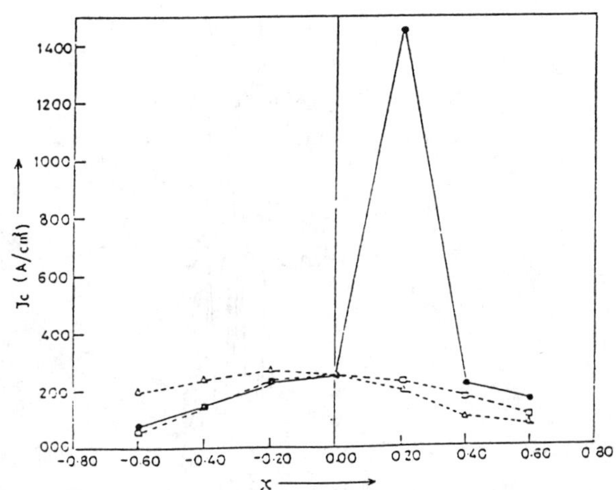

Figure 2 - Variation of J_c as a function of stoichiometry in bulk samples [•-Ca, Δ-Sr, -Cu]

63

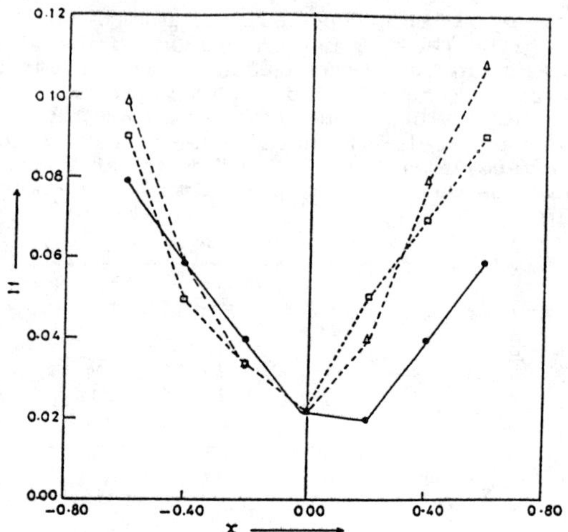

Figure 3 Variation of impurity fraction (If) with composition [•-Ca, Δ-Sr, □-Cu]

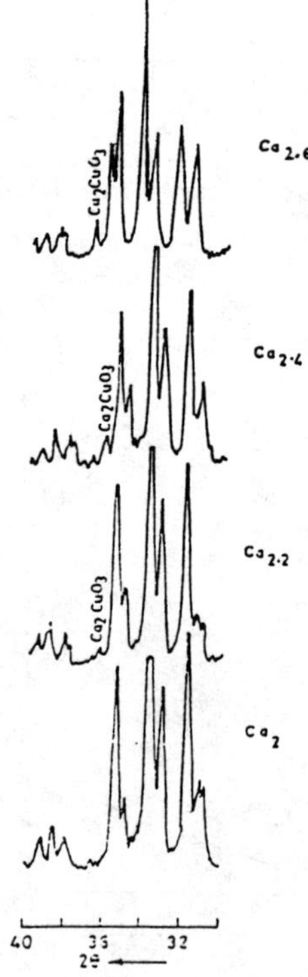

Figure 4 - XRD patterns of samples with varying Ca stoichiometry

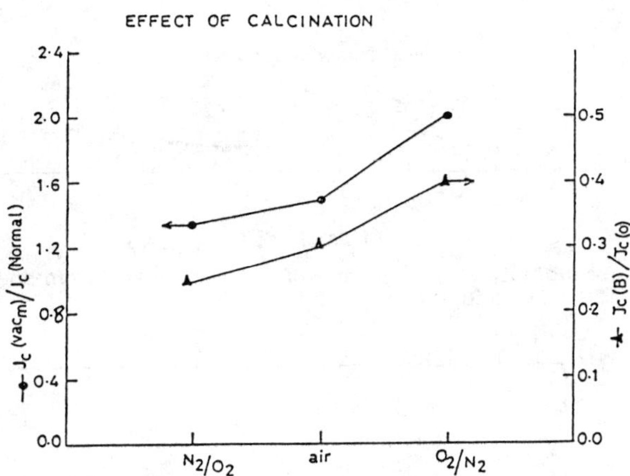

Figure 5 - Effect of gas environment during vacuum calcination on J_c

Table 1 - Effect of preparation and sintering time on transition temperatures

ROUTE	TIME OF SINTERING	HIGH T_c FRACTION	T	T'
CHEMICAL	120 h	96.2	109.3	105.5
SOLID STATE	200 h	89.8	109.5	104.5

Figure 5 describes the effect of dynamic vacuum calcination at varying O_2/N_2 partial pressure ratios on J_c. Results clearly indicate that dynamic vacuum calcination at relatively higher O_2 partial pressures gives the best results, possibly due to better elimination of residual carbon. Optimum method of preparation of powders is thus standardized as sol–gel synthesis to ensure compositional homogeneity followed by oxygen rich dynamic vacuum calcination to eliminate residual carbon in the best possible way.

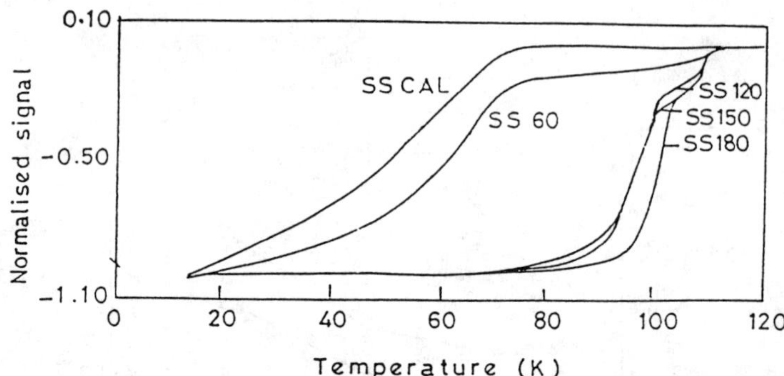

Figure 6 - AC susceptibility vs temperature of samples derived from solid state route and sintered for 60, 120, 150, and 180 h

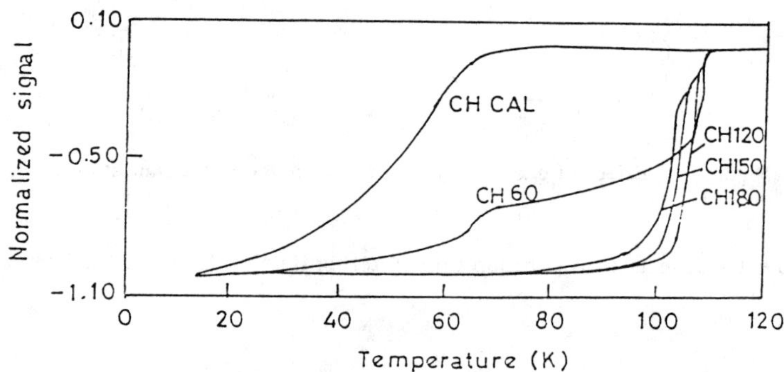

Figure 7 - AC susceptibility vs temperature of samples derived from wet chemical route and sintered for 60, 120, 150, and 180 h

Having fixed the stoichiometry and the method of calcination, powders prepared by the solid state and the sol-gel methods were evaluated. Table 1 shows the results obtained from bulk studies on the above powders in which T and T' represent the first and second transition temperatures in the x' - T curves with T' falling intermediate between the transition temperatures of 2212 and 2223 phases. When T' reaches the value of T, the material is pure 2223 (Figs. 6 and 7). The effect of T' on $J_c(B)$ is also clear from Table I. The chemical route gives the best quality powder. The effect of the precursor phase composition on J_c was studied systematically by X-ray measurements which showed that 2212 rich powders gave the best J_c value of the tapes (Fig. 8).

Results of the optimization experiments are shown in Figs. 9-11. Typically, J_c ≈12,000 A/cm^2 have been obtained at 77 K in self-field in 30 µm thick tapes. The optimum steps of thermomechanical processing are given below:

Reduction per pass	10 – 15 µm
Annealing time for Chemical Powder	170h
Final thickness of the core	20 – 40 µm
Number of intermediate heat treatment and rolling	3–4

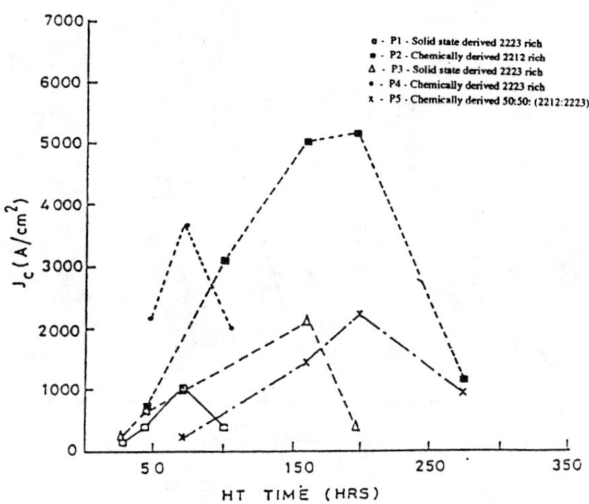

Figure 8 -   Effect of various synthesis routes and percursor phase compositions on J_c of tapes

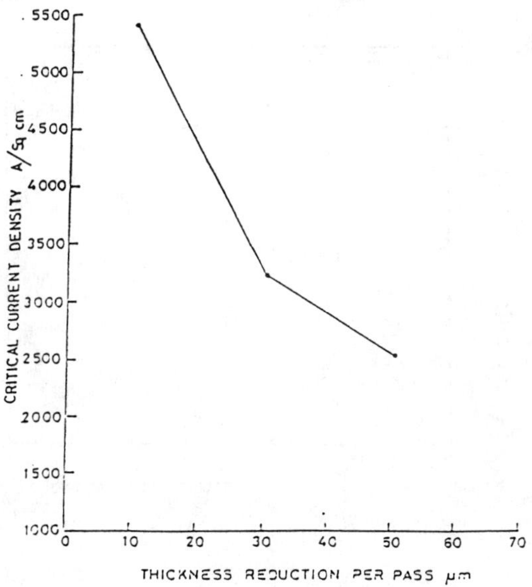

Figure 9 -   Effect of reduction per pass on Jc of tapes

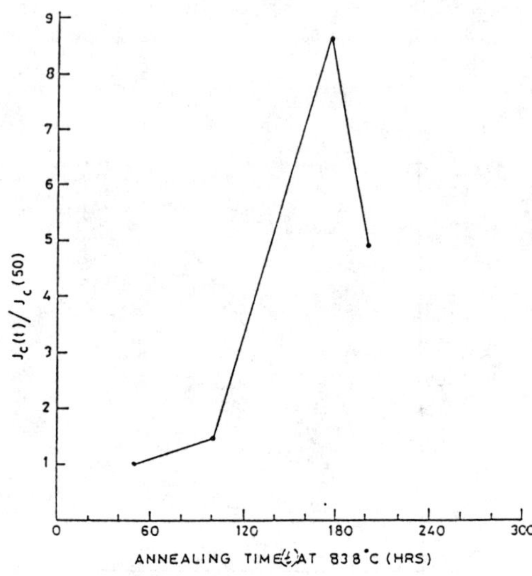

Figure 10 -   Effect of annealing time on J_c of tapes

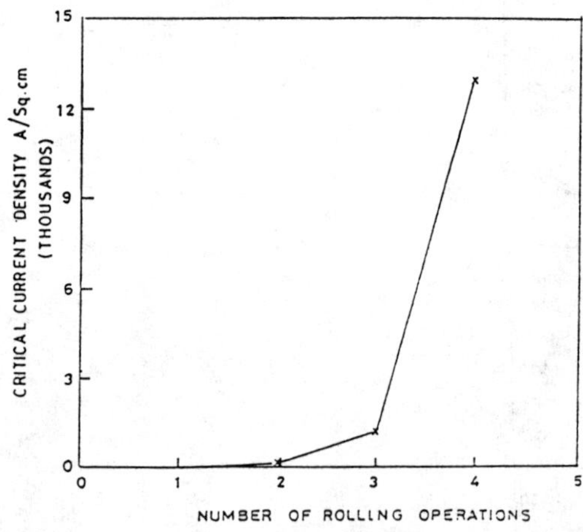

Figure 11 -   Effect of intermediate heat treatments on J_c of tapes

Summary

Powders having the chemical composition $Bi_{1.8}Pb_{0.4}Sr_2Ca_{2.2}Cu_3O_x$ prepared through the sol–gel process and vacuum calcined under O_2 rich dynamic flow conditions gave the most favourable results for conversion to Ag sheathed tapes by the PIT method.
In line with what was reported earlier by Umezawa et al. [8], AC susceptibility measurements as functions of T' have given valuable information with respect to the evolution of 110 K phase. x' – T measurements of the tape/polycrystalline pellet are able to bring out clearly the deleterious influence of trace quantities of the 2212 phase. Since the behavior of J_c vs magnetic fields are different for 2212 and 2223 phases, presence of any 2212 phase would certainly influence the J_c values. This factor may perhaps be the real determinant of final J_c value.

References

1. T. Hikata, T. Nishikawa, H. Mukai, K. Sato and H. Hitotsuyanagi, <u>Jpn. J. Appl. Phys.,</u> **28**, 1024 (1989).

2. M. Ueyama, T. Kikata, K. Sato and Y. Iwasa, <u>Jpn. J. Appl. Phys.,</u> **30**, 1271 (1991).

3. M. Wilhem, H. W. Neumüller and G. Ries, <u>Physica C,</u> **185–189**, 2339 (1991).

4. K. Heine, J. Tenbrink and M. Thoner, <u>Appl. Phys. Lett.,</u> **55**, 2339 (1989).

5. K. Sato, N. Shibuta, H. Mukai, T. Hihata, M. Ueyama and T. Kato, <u>J. Appl. Phys.,</u> **70**, 67484 (1992).

6. S. X. Dou, H. K. Kiu, Y. C. Guo and C. C. Sorrell, <u>Supercond. Sci. Technol.,</u> **5**, S471 (1991).

7. T. V. Mani, H. K. Varma, K. G. K. Warrier, and A. D. Damodaran, <u>Brit. Ceram. Trans.,</u> **91**, 120 (1992).

8. A. Umezawa, Y. Feng, H. S. Edelman, Y. E. High, D. C. Larbalestier, Y. S. Sung, E. E. Hellstrom and S. Fleshler, <u>Physica C,</u> **198**, 261 (1992).

DEVELOPMENT AND PROPERTIES OF Bi-2212/Ag SUPERCONDUCTING

WIRES AND TAPES

R. Wesche, A. M. Fuchs, B. Jakob, and G. Pasztor

Paul Scherrer Institute (PSI), CH-5232 Villigen, Switzerland

Abstract

Superconducting Bi-2212/Ag wires and tapes have been fabricated by the "powder in tube" method. Using three-stage annealing, the scatter in j_c data has been reduced to typically 10 % for the wires. Average critical current densities up to 79000 A/cm^2 (B = 0) and 28000 A/cm^2 (B = 12 T) have been reached at 4.2 K. For textured Bi-2212/Ag tapes j_c values of about 40000 A/cm^2 at 4.2 K and B = 12 T have been achieved. X-ray diffraction indicates imperfect grain alignment.

One of the authors (R. W.) wishes to acknowledge the financial support of the Swiss National Foundation (NFP-30) and of the Swiss Priority Program "Materials Research and Engineering" (PPM).

Processing of Long Lengths of Superconductors
Edited by U. Balachandran, E.W. Collings and A. Goyal
The Minerals, Metals & Materials Society, 1994

<u>Introduction</u>

The realization of magnet and energy applications using high T_c superconductors requires a continuous processing method leading to reproducible high critical current densities for long lengths of wires and tapes. The most promising results were so far achieved for Bi based high T_c superconductors. Critical current densities well above 10000 A/cm^2 at 77 K and B = 0 were reported for Bi-2223/Ag tapes [1-6]. The j_c values of Bi-2212/Ag and Bi-2223/Ag high T_c superconductors at 4.2 K in high magnetic fields are well above the corresponding j_c values of Nb$_3$Sn superconductors. Recently several groups have reported successful fabrication of multifilamentary high T_c superconductors [4-7].

In our laboratory the development of Bi-2212/Ag wires has been continued and an attempt has been made to reduce the scatter of the j_c data. The previously reported beneficial effect of a three stage annealing [8] could be confirmed for further wire lengths and the results will be discussed. To enhance the connectivity of the grains, textured tapes were fabricated. The factors which influence the critical current density of the tapes were examined and first results are reported. Finally, the performances of Bi-2212/Ag tapes and wires prepared from the same batch of powder will be compared.

<u>Experimental Details</u>

The Bi$_2$(Sr,Ca)$_3$Cu$_2$O$_8$ powder was prepared by the oxide carbonate route. Appropriate amounts of high purity oxides and SrCO$_3$ were thoroughly mixed and calcined at temperatures between 840 and 880°C in air. Using x-ray powder diffraction it was confirmed that four calcination steps with intermediate grinding are sufficient to produce nearly single phase Bi$_2$(Sr,Ca)$_3$Cu$_2$O$_8$ powder. The freshly prepared Bi-2212 powder was filled under vacuum into silver tubes of 6 mm inner and 8 mm outer diameter. The Bi-2212/Ag composites were drawn to wires with final diameters of 1.5 mm yielding lengths of about 5 m. Some sections of these wires were rolled to tapes with a final thickness of 0.125 mm.

Two types of heat treatment schedules were used for the Bi-2212/Ag wires, i. e. two- and three-stage annealing. In the case of the two-stage annealing the samples were heat treated at 900°C, leading to partial melting followed by long term annealing for renewed formation of the 2212 phase. In the case of the three-stage annealing the samples were additionally heat treated at a temperature slightly below 880°C before the partial melting. For the Bi-2212/Ag tapes the influence of the cooling rate after partial melting on j_c and grain alignment was studied. All heat treatments were performed in air.

The critical temperature of the wires and tapes was measured resistively. Critical current measurements were performed by a standard four probe method. The j_c values were determined using a 1 μV/cm criterion. In the case of the tapes the silver sheath was removed mechanically to confirm the grain alignment by x-ray diffraction. Polished cross-sections were prepared to examine the microstructure using optical and scanning electron microscopy.

Bi-2212/Ag Wires

Using an additional annealing step before the partial melting of the Bi-2212 phase, the scatter in the critical current data could be reduced. The distributions of the j_c values for two- and three-stage annealings are compared in Figure 1. Groups of 4 - 6 specimens were heat treated together. For each of these groups the j_c values are normalized to the average critical current density within the group. The total number of specimens is 42 and 84 for the two- and the three-stage annealings, respectively. The two-stage annealing leads to a broad distribution of the j_c values which are not grouped around a single maximum. On the other hand, a narrow distribution of the j_c values has been found for the three-stage annealings. The average critical current density is enhanced and only a few specimens show strongly reduced critical current densities.

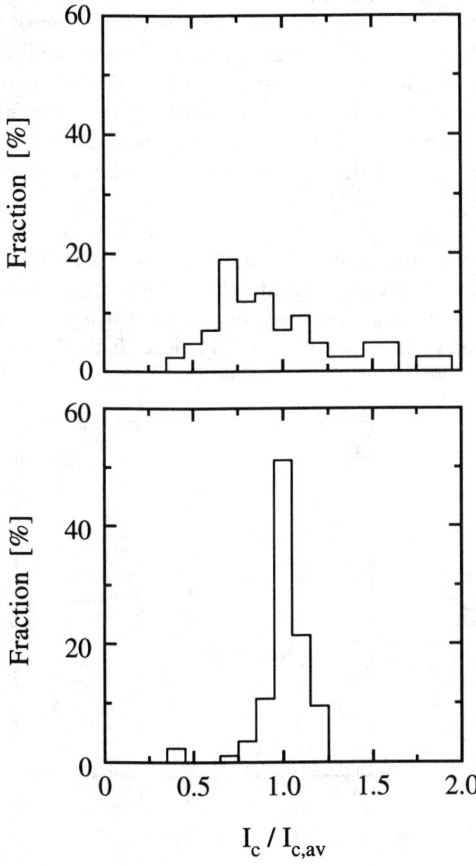

Figure 1 - Distribution of the j_c values at 4.2 K for two-stage (top) and three stage-annealings (bottom).

The effects of oxygen diffusion are probably responsible for this behaviour. For the three-stage annealing the released oxygen has enough time to diffuse through the silver sheath, avoiding the formation of very large voids. In addition the oxygen concentration in the superconducting core may be optimized. This explanation is supported by thermogravimetry data in reference [9] showing release of oxygen from Bi-2212/Ag tapes heat treated in air above 870°C.

Recently the scatter of the j_c data could be further reduced. Two lengths of wire each about 5 m long were fabricated from one batch of Bi-2212 powder. The x-ray powder diffraction revealed nearly single phase composition of the powder. Within groups of specimens cut from these two wires and heat treated together, the scatter of the j_c values at 4.2 K was as low as 10 %. Furthermore, the critical current densities could be enhanced. Maximum j_c values of 82000 A/cm^2 at B = 0 and 30000 A/cm^2 at B = 12 T have been achieved at 4.2 K. At 77 K and zero applied field, j_c values up to 2500 A/cm^2 have been reached. The nominal cross-section of the superconducting core is typically 0.77 mm^2 for the wires; this value was used for the calculation of j_c. The critical temperature of the Bi-2212/Ag wires is typically 88 K. Very probably the reduced scatter of the j_c data is a consequence of a slightly prolonged heat treatment at 900°C and a reduced amount of minor phases in the Bi-2212 powder after four calcination steps. Figure 2 shows the dependence of the critical current density at 4.2 K on the applied magnetic field for the best and the worst specimens within a group of 5 specimens heat treated together.

For NMR magnets and other potential applications of the high T_c superconductors the shape of the resistive critical current transition is an important parameter. The index of resistive transition n was determined by fitting a power law dependence $E \sim I^n$ to the measured data in the electric field range 0.25 μV/cm $\leq$ E $\leq$ 2.5 μV/cm.

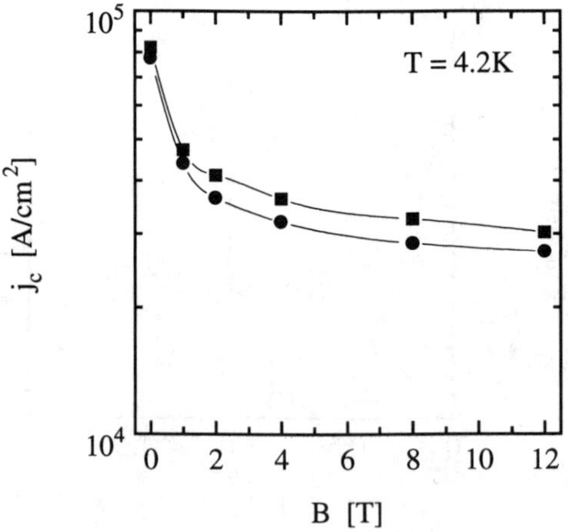

Figure 2 - Maximum and minimum j_c values at 4.2 K in a group of 5 specimens heat treated together.

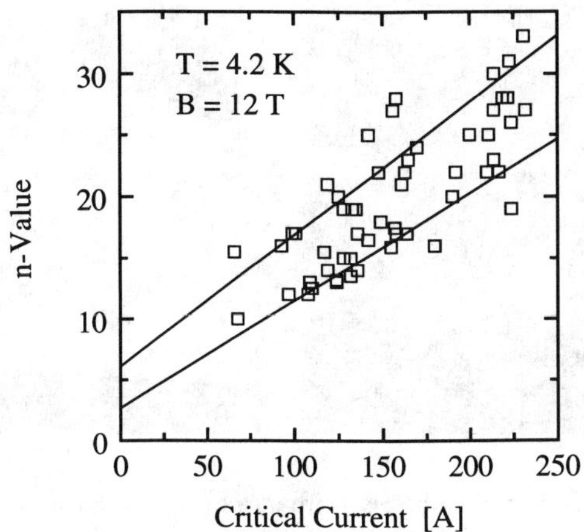

Figure 3 - Relation between n and I_c values at 4.2 K and B = 12 T.

Figure 3 shows the relation between the n values and the critical currents at 4.2 K and 12 T for 55 specimens prepared from two batches of powder. There is a correlation between I_c and n values. Fitting a straight line to the data one obtains for the center of the distribution.

$$n(I_c) = 4.4(1.7) + 9.8(1.0) \cdot 10^{-2} I_c \tag{1}$$

Taking into account the errors for the slope and the intercept with the n coordinate the following straight lines are obtained

$$n_l(I_c) = 2.7 + 8.8 \cdot 10^{-2} I_c \tag{2}$$

$$n_u(I_c) = 6.1 + 10.8 \cdot 10^{-2} I_c \tag{3}$$

which contain 36/55 (65 %) of the data points.

It seems that the n values are correlated with the amount and distribution of voids and non-superconducting precipitates. Figure 4 shows SEM micrographs for two specimens with very different critical currents and n values at 4.2 K and B = 0. Small dark precipitates of alkaline earth cuprates are characteristic of the specimen with I_c = 485 A and n = 33 (left side in Fig. 4), whereas very large gray and dark precipitates are visible for the specimen having I_c = 16 A and n = 3.3. In both specimens only a few voids are present. However, for another specimen with low n value the large non-superconducting precipitates are not present. In this case many large voids were observed.

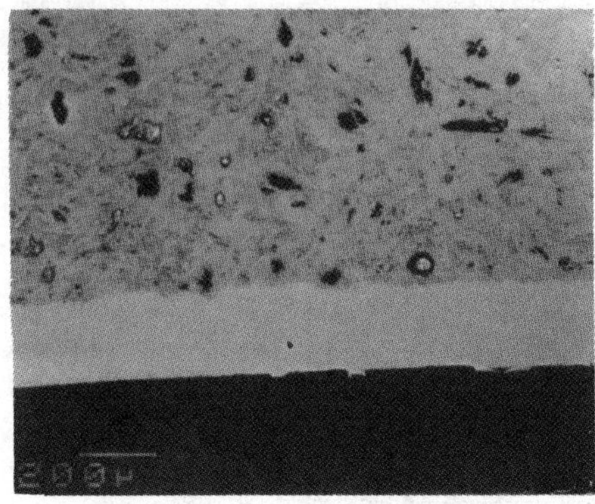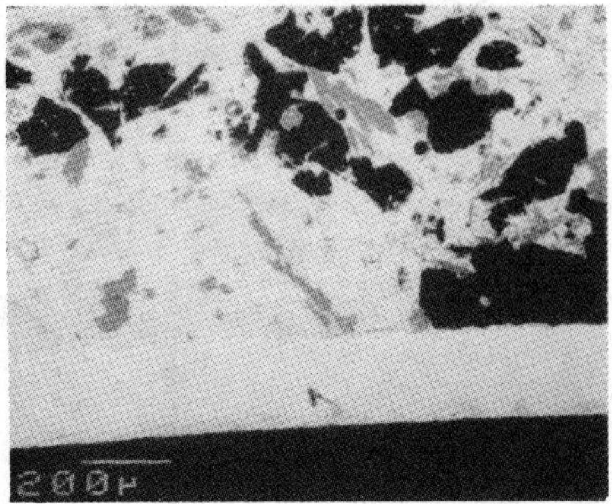

Figure 4 - Comparison of SEM images for specimens with high (left) and low
n values (right).

Bi-2212/Ag Tapes

It is well known that partial melting leads to grain alignment at the interface of silver
and Bi-2212 [9-11]. Sections of the last wires fabricated in our laboratory were rolled
to form tapes with a final thickness of 0.125 mm and a width of 2.6 mm. After partial
melting the specimens were cooled at different rates between 10 and 400°C/h to a temperature of 875°C. The influence of the cooling rate on grain alignment and on critical
current density has been studied. For cooling rates of 10 and 50°C/h the CuKα x-ray
diffraction patterns are shown in Figure 5. The 00l Reflexes are dominant for both
cooling rates. Besides two peaks of silver only the strongest hkl reflexes are visible.
The x-ray diffraction patterns indicate imperfect alignment of the a-b planes parallel to
the tape surface. For the higher cooling rate the intensity of the hkl reflexes increases
corresponding to an increased fraction of misaligned grains. It should be pointed out
that the silver sheath was removed mechanically and as a consequence the region of
highest degree of texture formation at the interface to the silver was not examined.

The maximum critical current densities have been achieved for cooling rates of 50°C/h,
whereas the highest degree of grain alignment corresponds to the lowest cooling rate of
10°C/h. The reason for this unexpected behaviour may be an enhanced homogeneity in
the case of higher cooling rates. The n values show a similar dependence on the cooling
rate as the critical current densities also indicating a reduced sample homogeneity for
small cooling rates. For $Bi_2Sr_2CaCu_2Ag_{0.8}O_y$ silver sheathed tapes large precipitates
of alkaline earth cuprates, which seem to be responsible for very low critical current
densities, were reported for heat treatment in air [10].

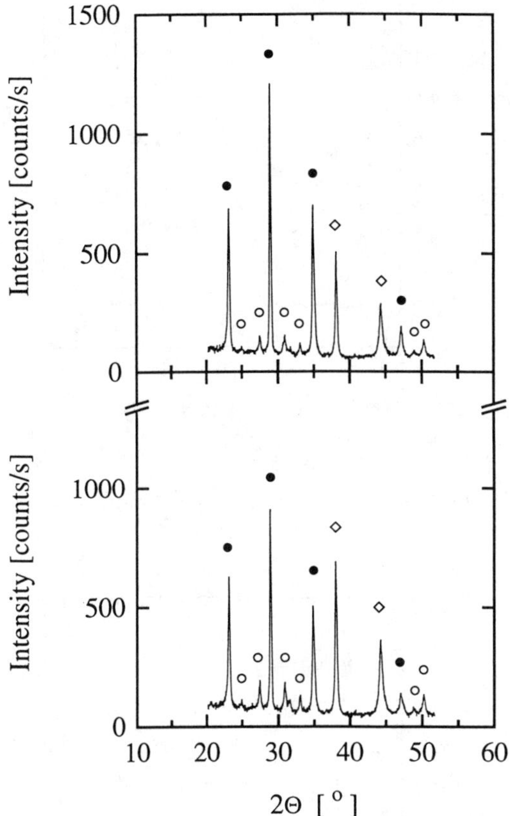

Figure 5 - X-ray diffraction patterns of Bi-2212/Ag tapes for cooling rates of 10 (top) and 50°C/h (bottom) (● 00l reflexes, o hkl reflexes, ◇ Ag).

Polished cross-sections of the Bi-2212/Ag tapes show some variation in the cross-section area of the superconducting core. The minimum value of 0.12 mm² was used to calculate the critical current densities of the tapes. The critical current density depends on the direction of the applied magnetic field with respect to the tape surface. In Figure 6 the maximum critical current densities are shown for the applied magnetic field parallel and perpendicular to the tape surface. The difference in the j_c values for the both field directions is about 25 % at 4.2 K. The observed anisotropy is consistent with an imperfect alignment of the a-b planes parallel to the tape surface. For highly textured Bi-2212/Ag tapes produced by the doctor blade method the corresponding difference in the j_c values is about 50 % [11]. Figure 7 shows the dependence of the critical current density on the angle φ between the surface normal of the tape and the magnetic field direction, for several values of the applied field.

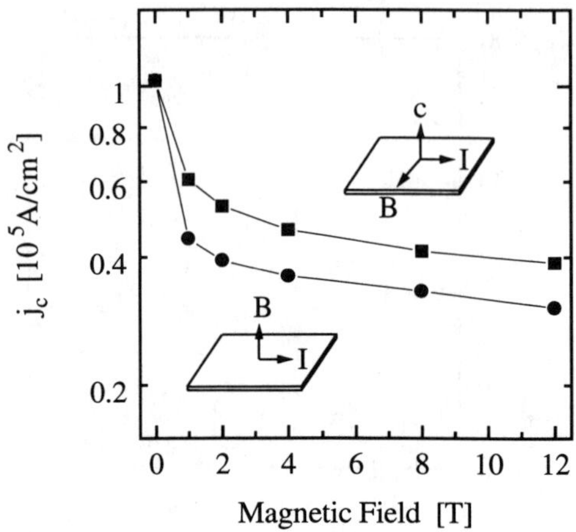

Figure 6 - Maximum j_c values at 4.2 K for the applied magnetic field parallel and perpendicular to the tape surface.

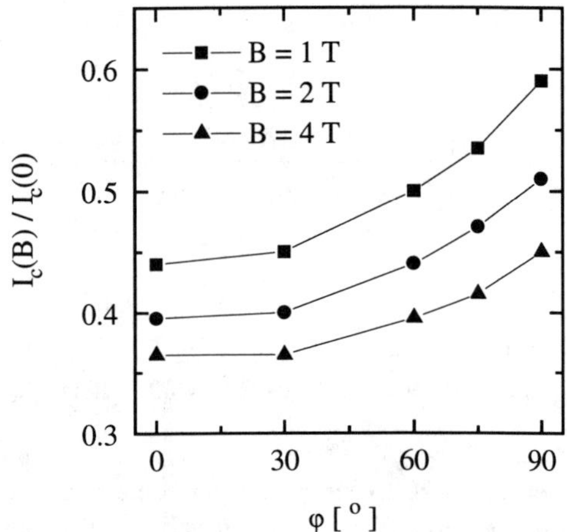

Figure 7 - Anisotropy of the j_c values at 4.2 K.

Comparison of Wires and Tapes

Figure 8 shows a comparison of the overall critical current densities for Bi-2212/Ag wires and tapes prepared from the same batch of powder. For both types of conductors the overall critical current density is well above 10000 A/cm^2 at 4.2 K and 12 T. For the tape only the j_c values for the field direction parallel to the tape surface are shown.

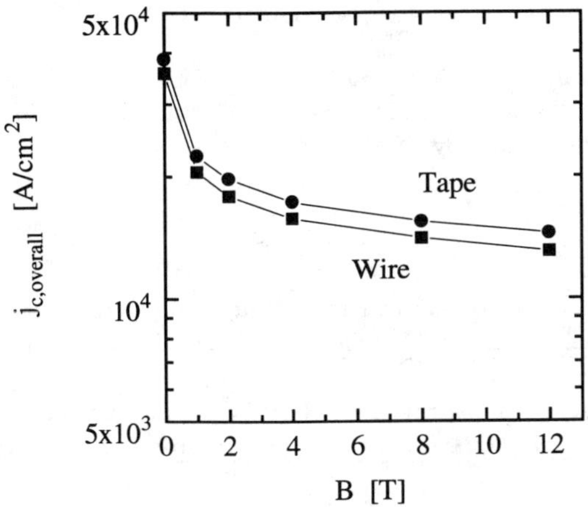

Figure 8 - Comparison of the overall critical current densities of Bi-2212/Ag wires and tapes (B ∥ tape surface).

At 77 K maximum overall critical current densities of 1100 A/cm^2 have been achieved for the wires. The transition temperature is about 88 K for the wires whereas for the tapes it drops to 78 K. The low T_c of the tapes leads to very small critical current densities at 77 K. The difference in the T_c values is due to the thinner silver sheath of the tape allowing a larger oxygen absorbtion during furnace cooling. To enhance the T_c of the tapes an additional heat treatment in a reducing atmosphere would be necessary.

Reasons for the moderate j_c values in the tapes may be that the thickness of the superconducting core exceeds 60 μm. This results in a low ratio of textured to non-textured volume within the core. Additionally, voids and non-superconducting phases are present disturbing the grain alignment. Investigation of the influence of heat treatment conditions and atmosphere on the j_c of the Bi-2212/Ag tapes is the subject of future work.

Recently Bi-2212/Ag wires with diameters well below 1 mm were fabricated. Overall critical current densities of nearly 19000 A/cm^2 at 4.2 K and 12 T have been achieved. Details will be published elsewhere. These preliminary results suggest that very high critical current densities can be reached in Bi-2212/Ag multifilamentary round wires.

<u>Conclusion</u>

Using a three-stage annealing process the scatter in j_c data can be reduced to less than 10 % within groups of Bi-2212/Ag wires heat treated together. Promising results for wires with diameters less than 1 mm indicate that multifilamentary round wires may achieve very high j_c values. A first attempt to produce textured Bi-2212/Ag tapes leads to j_c values slightly above those of round wires. Both types of conductors are potential candidates for high field applications.

<u>Acknowledgment</u>

The authors are grateful to L. Cavegn, W. Hürlimann, F. Roth, J.-C. Sauter, and F. Schraner for technical assistance. Thanks are due to R. Brütsch for the examination of Bi-2212/Ag superconductors by SEM.

<u>References</u>

1. J. Tenbrink et. al., "Development of High-T_c Superconductor Wires for Magnet Applications", <u>IEEE Transactions on Magnetics</u> 27 (1991), 1239-1246

2. B. Hensel et. al., "Microstructure, Resistivity and Critical Currents of Ag-Bi/Pb(2223) Tapes", <u>IEEE Transactions on Applied Superconductivity</u> 3 (1993), 1139-1142

3. P. Haldar et. al., "Transport Critical Current Densities of Silver Clad Bi-Pb-Sr-Ca-Cu-O Tapes at Liquid Helium and Hydrogen Temperatures", <u>Applied Physics Letters</u> 61 (1992), 604-606

4. K. Sato et. al., "Bismuth Superconducting Wires and their Applications", <u>Cryogenics</u> 33 (1993), 243-246

5. Q. Y. Hu, H. K. Liu, and S. X. Dou, "Characterization of Ag-Sheathed $(Bi,Pb)_2Sr_2Ca_2Cu_3O_{10-x}$ Multifilamentary Tapes", <u>Cryogenics</u> 32 (1992), 1038-1041

6. A. Otto et. al., "Multifilamentary Bi-2223 Composite Tapes Made by a Metallic Precursor Route", <u>IEEE Transactions on Applied Superconductivity</u> 3 (1993), 915-922

7. J. Tenbrink et. al., "Development of Technical High-T_c Superconductor Wires and Tapes", <u>IEEE Transactions on Applied Superconductivity</u> 3 (1993), 1123-1126

8. R. Wesche, B. Jakob, and G. Pasztor, "Development and Performance Characteristics of Bi-2212/Ag Superconducting Wires", <u>IEEE Transactions on Applied Superconductivity</u> 3 (1993), 927-930

9. K. Nomura et. al., "Fabrication Conditions and Superconducting Properties of Ag-Sheathed Bi-Sr-Ca-Cu-O Tapes Prepared by Partial Melting and Slow Cooling Process", <u>Applied Physics Letters</u> 62 (1993), 2131-2133

10. A. Endo and S. Nishikida, "Effects of Heating Temperature and Atmosphere on Critical Current Density of $Bi_2Sr_2Ca_1Cu_2Ag_{0.8}O_y$ Ag-Sheathed Tapes", <u>IEEE Transactions on Applied Superconductivity</u> 3 (1993), 931-934

11. H. Kumakura et. al., "Anisotropy of Critical Current Density in Textured $Bi_2Sr_2Ca_1Cu_2O_x$ Tapes", <u>Applied Physics Letters</u> 58 (1991), 2830-2832

OPERATION OF OXIDE SUPERCONDUCTOR MAGNET AT 20K

Seiji Hayashi, Takashi Hase, Kazuyuki Shibutani, Toshio Egi,
Yoshio Masuda, Rikuo Ogawa and Yoshio Kawate

Superconducting and Cryogenic Technology Center, Kobe Steel, Ltd.
Takatsukadai 1-5-5, Nishi-ku, Kobe 651-22, Japan

Abstract

Cryostat which has large room temperature bore of 45mm was fabricated for 20K-operation of oxide-superconductor magnet. Oxide superconductor magnet made of Bi2212 silver sheathed tapes was installed in the cryostat. The magnet succeeded in generating magnetic field of 0.68T at 20K. In order to prepare such practical scale of magnet, reproducibility of the high-Jc in long length silver sheathed tapes was the most important condition. In addition to the previously reported conditions, it was clarified that prolonged calcination and rapid cooling from 885C to 875C during melt process played important role to improve the reproducibility.

Processing of Long Lengths of Superconductors
Edited by U. Balachandran, E.W. Collings and A. Goyal
The Minerals, Metals & Materials Society, 1994

Introduction

A superconducting magnet operated at around 20K has been proposed as one of the practical applications of oxide superconductor (1). This type of application will be very hopeful because it does not need liquid helium coolant for operation. Several groups have reported the operation of the oxide superconductor magnet at 20K, such as generation of 0.53T for Bi2223-magnet by Mukai et al (2) and 0.62T for Bi2212-magnet by present authors (3). However, all of these magnets had small inner diameters of less than 40mm and it was impossible to prepare room temperature bore. In order to fabricate a practical scale magnet, realization of high-Jc over length of larger than 1000m will be inevitable condition. A lot of factors which influence Jc of Bi2212 silver sheathed tape have been extensively investigated. For example, it has been well known that Jc was very sensitive to maximum temperature and cooling rate in the partial melt and slow cooling process (4). Kumakura et al reported that melt-solidification with $Bi_2Al_4O_9$ suppressed the bismuth vaporization and led to improved reproducibility in Jc (5). Hasegawa et al (6) and present authors (3) reported that Bi-rich composition and silver addition improved Jc and relaxed the heat-treatment conditions. However, these factors have not been sufficient to reproduce high Jc value. In this paper, we show that optimization of calcination process improves reproducibility of Jc and rapid cooling from maximum temperature to just above the Bi2212 crystallization temperature in partial melt process is effective to obtain highly aligned structure. By applying all of these optimized conditions, we have prepared the magnet with large bore size of 72mm in diameter. Performance of the magnet at 20K is also described.

Experimental Procedures

Cryostat Fabrication

We designed and fabricated the cryostat for magnet-operation at 20K. Figure 1 shows schematic diagram of the cryostat. This cryostat has large room temperature bore of 45mm in diameter. Inner most vessel made of copper is designed to be evacuated. After evacuation, inside of the vessel is filled with helium-gas in order to maintain good heat transfer between the vessel and the magnet. Using the heater attached to this vessel, temperature inside of this vessel can be controlled. The base plate of the vessel is attached to the second stage of GM-cryocooler (Iwatani D-840), and

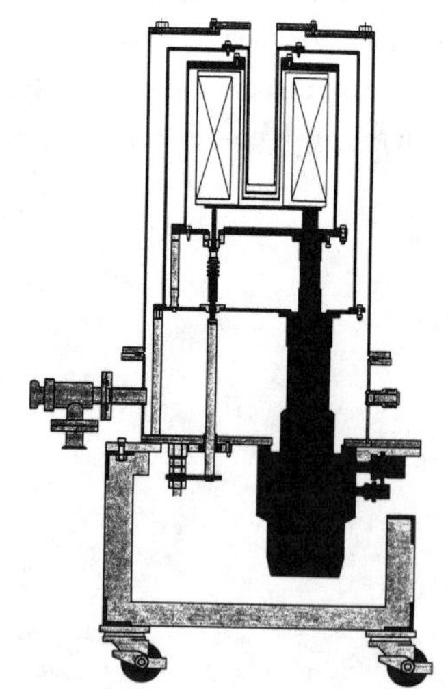

Figure 1 - Schematic diagram of the cryostat equipped with GM-cryocooler

radiation shield made of copper is attached to the first stage of the cryocooler. Temperature distribution inside of the vessel is monitored by four sets of thermo-couple.

Preparation of the Oxide Powder

Oxide powder with nominal composition of $Bi2.1Sr2.0Ca1.0Cu1.9Ag0.1Ox$ was prepared by solid state reaction. In order to investigate how the crystal phases and carbon content do affect to Jc, we prepared the powders heat treated by different calcination conditions as shown in Table I. For these powders, X-ray diffraction analysis and carbon content determination and Differential Scannning Calorimeter (DSC) measurement have been performed.

Table I   Calcination conditions of the powders

Sample	Heat treatment before packing		
	750C	835C	865C
I	12h	10h	1h
J	12h	20h	1h
K	12h	30h	1h
L	12h	60h	1h
M	12h	100h	1h

Heat Treatment of Short Samples

The oxide powders calcined by the same condition as above-mentioned sample-L was packed into silver pipe, and rolled into 10mm-wide tape. Short samples with 4cm length have been heat treated according to the schedule shown in Fig.2. Unique features of the present heat-treatment compared to the conventional process are additions of holding at 835C and rapid cooling from 885C to Tq. Tq in Fig.2 is changed as 885C, 875C, 870C and 860C. Heat-treatment of

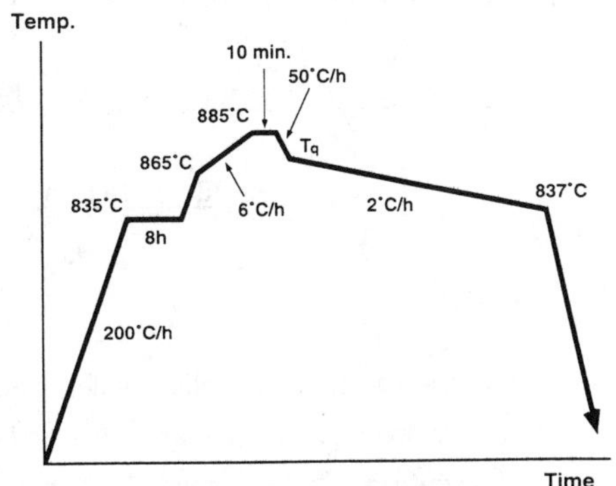

Figure 2 - Heat treatment schedule for the silver sheathed tapes

Tq=885C corresponds to the conventional partial-melt and slow cooling process except for holding at 835C. Critical current measurement at 4.2K and SEM-observation of cross-sectional view for these samples have been performed.

Magnet Fabrication

Three sheets of silver sheathed tapes were stacked and wound around a ceramic tube (76.8/81.0mm in inner/outer diameter and 25mm long). Al2O3 paper was employed for the insulator between every turn of the tapes. The workable ceramics were employed for the insulation between pancake coils. This coil was heat-treated according to the schedule which corresponds to Tq=875C in Fig.2. After the heat-treatment, this magnet was impregnated by the melting wax. Table II and Fig.3 show the specifications and the photo image of the assembled Bi2212 superconducting coil. The numbers in the parenthesis in Table II show the size of coil-form. This coil was installed in the cryostat and tested at around 20K.

Figure 3 - Photo image of Bi2212 superconducting coil

Table II The Specifications of the Bi2212 Superconducting Coil

Tape		Coil	
Thickness of tape	100 μm	Inner diameter	81mm (72 mm)
Width of tape	10 mm	Outer diameter	185 mm (200 mm)
Total tape length	1556 m	Length	218 mm (225 mm)
		Number of turns	1232 turns
		Insulation	Al2O3 paper

Results and Discussion

Cryostat Operation

It took about 35 hours to cool down the magnet (weight about 20kg) from room temperature to 14K. The lowest temperature of the magnet reached 13.5K, and the temperature distribution in the magnet was confirmed to be less than 0.5K. Cooling power of the GM-cryocooler was 8W at 20K, but 1/2 of this power was consumed to compensate heat leak mainly via copper power leads even in zero-operating current. Available cooling power decreases with increasing operating current because of the heat generation in copper power leads. In case of the blank operation without installation of the magnet, maximum applicable current below 20K was 195A.

<u>Evaluation of the Oxide Powder</u>

Irrespective of the melt process, it has been well known that the reproducibility was very poor in superconducting tapes produced by partial melt and slow cooling process. Reproducibility is very important for realization of practical scale magnet, because operation current is limited by the lowest part of Jc in the tape and the longer the tape the higher the probability of some kind of faults which lead to decrease in Jc. As for the reproducibility in Bi2212 tapes, various conditions have been already discussed by many authors as described in the preceding introduction (3,4,5,6). In addition to these conditions, we have found that calcination process also played very important role to improve reproducibility. With increasing the calcination time at 835C, the Jc and its reproducibility were very much improved. According to the XRD-analysis, Bi2212-phase increases with increasing the heat treatment at 835C. Figure 4 shows the comparison of DSC-curves for powders I, J and K. As shown in the figure, melting tempeature increases and temperature region in which liquid phase exists becomes narrower with increasing calcination time. Figure 5 shows the carbon content as a function of calcination time. Carbon content decreases with increasing the calcination time, and saturates for longer calcination time than 20h. Irrespective of the carbon content saturation, the Jc was improved for longer calcination time, therefore, it is considered that the carbon content is not dominating factor of Jc. As one of the reasons for improvement in Jc and reproducibility of Jc, suppression of the segregation during heating-up is plausible, because in case of shorter calcination, liquid appears at lower temperatures and this may lead to harmful segregation.

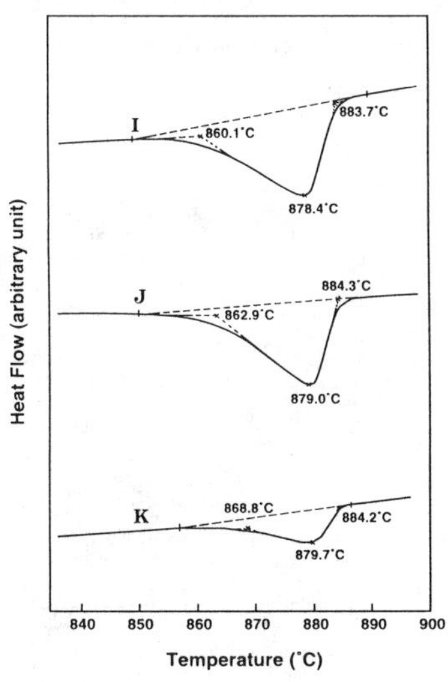

Figure 4 - DSC-curves for powders I, J and K

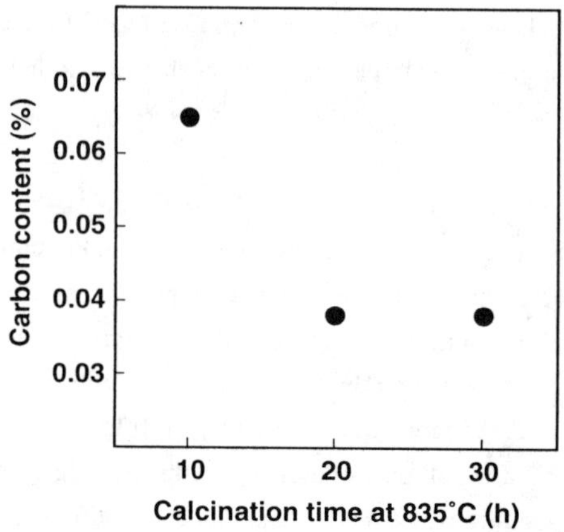

Figure 5 - Carbon content as function of calcination time

<u>Evaluation of Short Samples</u>

The critical current, Ic, which is the average of four samples is shown in Fig. 6 as a function of rapidly cooled temperature Tq. As shown in the figure, the maximum Ic was obtained for Tq=875C and it rapidly decreased for Tq < 870C. Extensive studies on crystal phases of Bi2212-system have been performed by Hasegawa et al (6). and Hellstrom et al (7). Hasegawa et al. studied the solidification phenomenon by using the high temperature XRD technique. Hellstrom et al. investigated the crystal phases of rapidly quenched silver sheathed tapes by EPMA. According to their results, at higher temperature of more than 900C, CaO is the only phase which exists as a solid phase. When the temperature is decreased from this partial-molten state, $(Ca,Sr)2CuO3$ and $(Ca,Sr)CuO2$ starts to crystallize from about 895C. Solidification of the Bi2212 phase takes place in the temperture region from 870C to 835C. The Bi2201 phase appears in the temperature range below 835C.

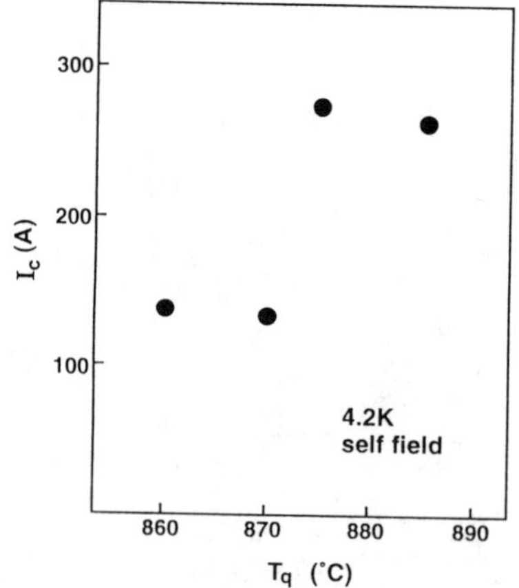

Figure 6 - Critical current as a function of Tq

Figures 7(a,b,c) show SEM-photographs of the cross-sectional view of the silver sheathed samples. The (a), (b) and (c) correspond to the samples of Tq=885C, Tq=875C and Tq=860C respectively. Although the critical currents for Tq=885C and Tq=875C are almost the same value, the structure is very much different from each other. In case of the sample of Tq=885C, grain growth of Bi2212 is not so large compared to the sample of Tq=875C. Very large and highly aligned Bi2212 crystal is obtained for Tq=875C. In case of the sample of Tq=860C, grain alignment is very poor and its size is very small. This is considered to be related to the fact that Bi2212 starts to crystallize below about 870C. In order to obtain large size of grains and highly oriented structure, rapid cooling from partially molten state to the temperature just above the Bi2212-solidification temperature is effective. Solid phases such as (Sr,Ca)2CuO3 and (Sr,Ca)CuO2 produced during cooling process from 885C to Bi2212 solidification temperature restricts the grain growth of Bi2212. We can say that slow cooling process between 885C and 875C in the conventional method is not necessary for large grain growth and aligned structure. However, only one fault of the process employed for the sample Tq=875C, is the existence of large volume fraction of (Sr,Ca)CuO2. Because of the   stability of this phase throughout whole temperature range of the slow cooling process, it is impossible to reduce this phase by

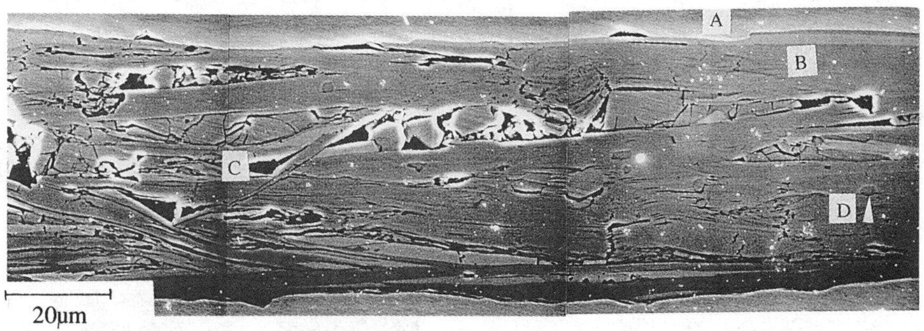

Figure 7(a) - SEM-photograph of the cross-sectional view of the sample of Tq=885C.
A, B, C and D correspond to silver sheath, Bi2212, (Sr,Ca)CuO2 and Bi(Sr,Ca)2Ox.

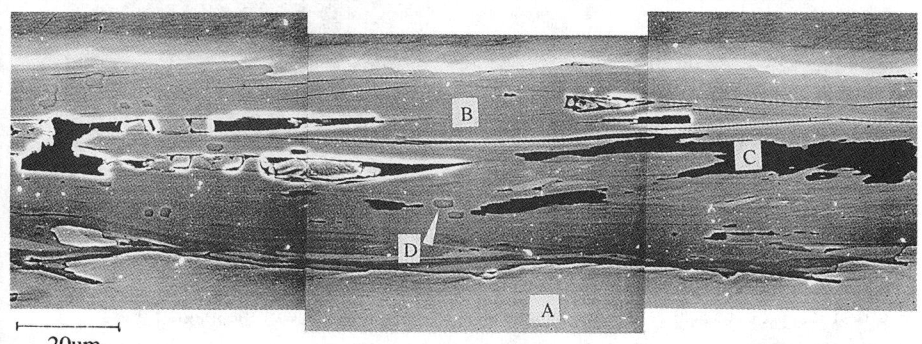

Figure 7(b) - SEM-photograph of the cross-sectional view of the sample of Tq=875C.

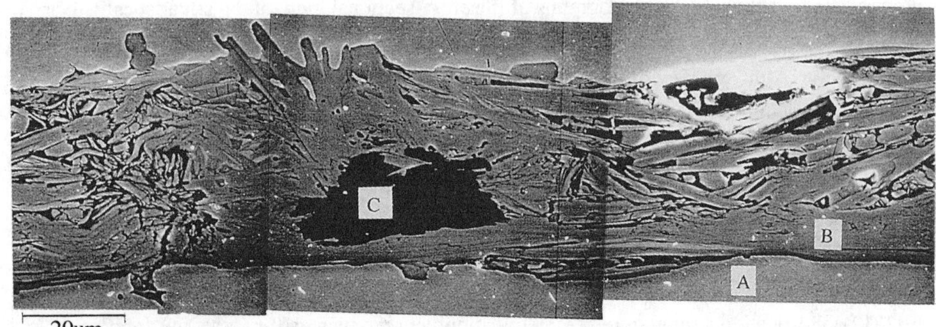

20μm

Figure 7(c) - SEM-photograph of the cross-sectional view of the sample of Tq=860C.

the present method. The causes which lead to reduction of Jc are; small grain size of Bi2212 and thus existence of a number of grain boundaries for Tq=885C, large volume fraction of Bi-free phase for Tq=875C and poor crystal alignment for Tq=860C. Therefore, in order to improve the Jc further, the noble process which can reduce the Bi-free phase for Tq=875C will be anticipated.

<u>Performance of the Magnet at 20K</u>

Figure 8 is the photo of the magnet installed to the cryostat. The Ic was determined using the electric field criterion of $1\mu V/cm$. This value corresponds to the voltage 52mV over the whole length of 519m (1556/3 = 519) tapes. Since each pan-cake coil was connected by normal conducting metal (silver or copper), the point which showed a deviation from a straight line with constant slope was employed for the critical current. Critical current reached 165.8A ($Jc = 1.7 \times 10^4$ A/cm^2) and magnetic field of 0.68T was generated at 20K. The IcL product, which is a measure

Figure 8 - Photo image of the magnet installed to the cryostat

of reproducibility of Jc in long wire, reached 8.6×10^4Am. To our knowledge, this is the highest record of IcL product ever reported. Because of such practical scale of the magnet, field homogenity (B/B0) was confirmed to be less than 1% within the cyrindrical area of 45mm in diameter and 30mm length.

88

Conclusion

The reproducibility of the critical current density in Bi2212 silver sheathed tape was improved by prolonged calcination time and by holding at 835C during heat treatment. Rapid cooling from maximum temperature in the partial melt process to the temperature just above the solidification temperature of Bi2212 was effective to obtain large size of grain and textured structure. Applying these optimized process conditions, superconducting magnet made of Bi2212 silver sheathed tapes with practical bore size of 72mm was fabricated. This magnet was installed in the cryostat equipped with GM-cryocooler, and succeeded in generating magnetic field of 0.68T at 20K.

References

1. Y. Iwasa, "HTS Magnets", Advances in Superconductivity, 1992, no.5: 1205-1210.

2. H. Mukai, K. Ohkura, N. Shibuta, T. Hikata, M. Ueyama, T. Kato, J. Fujikami, K. Munakata and K. Sato, "Bi-Based Silver-Sheathed High-Tc Superconducting Wire and Application", Advances in Superconductivity, 1992, no.5: 679-684.

3. S. Hayashi, T. Egi, T. Hase, K. Shibutani, R. Ogawa and Y. Kawate, "Oxide Superconductor Magnet Operated near 20K", (Paper presented at the ICMC'93, Albuquerque, Newmexico, 15 July 1993), CY-5.

4. J. Kase, K. Togano, H. Kumakura, D.R. Dietderich, N. Irisawa, T. Morimoto and H. Maeda, "Partial Melt Growth Process of $Bi_2Sr_2CaCu_2O_x$ Textured Tapes on Silver", Jpn. J. Appl. Phys., 29 (1990), L1096-1099

5. H. Kumakura, H. Kitaguchi, K. Togano, H. Maeda, J. Shimoyama and K. Nomura, "Properties of Bi-2212 Tapes and Coils Fabricated Applying Dip-Coating Process", (Paper presented at the 5th US-Japan Workshop on High Tc Superconductors, Tsukuba, November 1992).

6. T. Hasegawa, H. Kobayashi, H. Kumakura, H. Kitaguchi and K. Togano, "Phase Studies and Superconducting Properties of Ag-Added $Bi_2Sr_2CaCu_2O_y$/Ag Tapes", Advances in Superconductivity, 1992, no.5: 737-740.

7. E.E. Hellstrom and R.D. Ray II, " Phase Studies in Ag-Clad 2:2:1:2 BSCCO Wires", (Paper presented at the International Workshop on Superconductivity Co-Sponsored by ISTEC and MRS, Honolulu, Hawaii, 23-26 June, 1992), 201-204.

Fabrication of Pancake Coil Magnets from One to Ten Meter Length Powder-in-Tube $Bi_2Sr_2CaCu_2O_{8+x}$/Ag Tapes

T. Haugan, M. Pitsakis, S. S. Li, S. Patel and D. T. Shaw

New York State Institute on Superconductivity
and Department of Electrical Engineering
State University of New York at Buffalo
Amherst, NY 14260 U.S.A.

Abstract

Results of processing one to ten meter length $Bi_2Sr_2Ca_1Cu_2O_{8+x}$/Ag cold-rolled powder-in-tube tapes with uniform properties and high critical current densities are presented.   Tape sections were processed by partial-melting and slow cooling and were free of bubble defects. Pancake coil magnets (ID = 1.25 cm) were fabricated with one to ten meter length tape sections by a wind-react-wind (WRW) method using different configurations, including single layer, multiple layer, single and double pancakes and different initial and final winding dimensions. The central magnetic fields of coil magnets were measured with a Hall sensor at 4.2 K and varied from 0.07 - 0.45 T, in excellent agreement with theoretical values calculated for coil dimensions such as OD, ID and I_c. Critical current densities (J_c's) of one to ten meter length tape sections measured resistively after winding to magnet form were in the range of 32,000 A/cm^2 - 54,000 A/cm^2 at 4.2 K. The bending strains exerted on coil segments from winding were calculated as ~ 0.10 - 0.25% for the range of initial and final coiling dimensions tested. By measuring the J_c vs bending strain properties of short length tapes, the drop in J_c expected from winding strains was estimated as ~ 65 - 75%.   Combined with a measured ~ 3 - 14% drop in J_c's from magnetic self field effects, transport J_c's of one to ten meter tape sections prior to winding were estimated as ~ 55,000 - 80,000 A/cm^2 at 4.2 K, ~ 60 - 85% of J_c's of short length tapes processed under similar conditions. The possible operation of Bi-2212 coils at 20 K are discussed and the temperature dependence of J_c for several processing conditions are presented.

Introduction

Critical current densities (J_c's) of short lengths of $Bi_2Sr_2Ca_1Cu_2O_{8+x}$ (Bi-2212) tapes processed on silver substrates by partial-melt slow-cooling (PMSC) have reached sufficiently high values (~ 5.9 x 10^5 A/cm^2 at 4.2 K and 0 T applied field)[1] for practical applications. What is required further is to demonstrate the ability to process long lengths of Bi-2212 tapes with similar properties. Recently, Bi-2212/Ag/Bi-2212 double sided uncovered (DSU) tape sections 1.5 meter in length were prepared by dip-coating and PMSC with long length transport critical current densities (J_{ct}'s) reduced to 25% of short length values.[2] Using this method, the superconductor layers were exposed to the atmosphere during processing and outgassing problems that occur during or below partial-melting conditions were avoided. A disadvantage of the DSU tape configuration is that the oxide layers break significantly more under bending strain than other tape configurations, such as powder-in-tube or (Ag/Bi-2212/Ag) single layer or multilayer tapes.[3] Alternatively, powder-in-tube methodology has been used to process tape sections up to 300 m in length for the $(Bi,Pb)_2Sr_2Ca_2Cu_3O_{10+x}$/Ag system,[4] similar to the Bi-2212/Ag system. Efforts at applying powder-in-tube methodology to the Bi-2212/Ag system were limited thus far because of outgassing and bubble defect formations that occur during or before the partial-melting process.[5-8] Transport J_c's measured across short lengths of bubble defects are typically reduced by 50% or more. Also because it is difficult to predict the size and effect of bubble defects on superconducting properties, it would be best to completely eliminate bubble formations. Recently, techniques for controlling and eliminating outgassing and bubble formations in long lengths of single and multifilamentary Ag-sheathed Bi-2212 PIT tapes processed with PMSC were developed.[9] As a consequence of that discovery, it became possible to process long lengths of Bi-2212 tapes with uniform properties and high critical current densities. Herein, the results of that work and properties of Bi-2212 tapes and coil magnets made from uniform property one to ten meter length tape sections are described.

Experimental

Powder-in-tube $Bi_2Sr_2Ca_1Cu_2O_{8+x}$ tapes were made by cold-rolling powder filled Ag tubes with dimensions OD ~ 6.35 mm and ID ~ 4.35 mm. Precursor powder with composition (Bi:Sr:Ca:Cu = 2:2:1:2) was packed into Ag tubes by mechanical methods. Powder-filled Ag tubes were pre-annealed by heating: 22°C ---> 820°C at 120 to 300°C/hr, 20 minutes at 820°C, and 820°C ---> 22°C at furnace cooling. Pre-annealing was necessary for eliminating outgassing from organic contaminants in the powder introduced during fabrication and packing processes. Pre-annealing would also increase the 2212 volume phase percentage and affect the oxygen content, depending on the cooling procedure and powder preparation method and particle size distribution.

Following pre-annealing, powder packed Ag tubes were cold-rolled to dimensions of ~ 60 μm thickness and ~ 0.9 cm width using ~ 200 rolling steps with an average thickness decrease of ~ 25 μm at each rolling step. Using a smaller number of rolling steps (i.e. 50 - 100) the superconductor cross-section was not as uniform and breaks in long lengths of tapes were observed. One size of twin rollers (OD ~ 61 mm) were used for each rolling step and rolling speeds were varied from ~ 0.5 - 4 m/min. After this rolling process, the tape cross-sectional area was reduced by a factor of ~ 60 and the superconductor film layer was ~ 15 - 18 μm thick by ~ 0.75 cm wide. The powder packing density after rolling was ~ 70%, nearly the highest packing density that can be achieved using spherical shaped particles (74%).

The mechanical uniformity of tape sections prior to annealing was determined by measuring statistical thickness variations with a micrometer. The mean thickness of one ten meter tape section measured 135 times at uniform intervals was 62.0 μm with a standard deviation of thickness of 0.7 μm.

After rolling, tape sections were cleaned with alcohol and stored in dessicator boxes for periods of up to several weeks to one month before final processing. Tape sections up to ten meters in length were inserted into teflon spacer coil molds made with different insulation thickness spacings between turns (t_{insu}) of 1430 μm and 794 μm. Teflon mold - coil composites were annealed at 150°C for two hours to release the stress in the rolled PIT tape, leaving the Bi-2212/Ag coils free standing after removing the Teflon spacer molds. Final PMSC heat treatments for coils were 22°C ---> 830°C at 120°C/hr, 830°C (20 minutes), 830°C ---> 889°C at 60°C/hr, 889°C (5 minutes), 889°C ---> 874°C at 60°C/hr, 874°C (75 minutes), 874°C ----> 871°C at 60°C/hr, 871°C (75 minutes), 871°C ---> 22°C with furnace cooling. Melting temperatures were ~ 5 - 10°C higher than temperatures used for processing "uncovered" single layer tapes.[10,11]

To measure long length transport properties, tapes were wound into coils (ID ~ 1.25 cm) around Al piece cores using mylar tape (25 μm thick x 1 cm wide) for electrical insulation. Magnetic fields and transport current measurements of coils were measured at 4.2 K in liquid helium. A Hall-field sensor was used to measure magnetic field properties. Transport J_c's were measured across the entire tape lengths by placing voltage taps ~ 2 - 3 cm inside of the current leads placed at the innermost and outermost tape ends. A 1 μv/cm criteria and the superconductor cross-sectional area were used to determine the critical current density.

Results and Discussions

Physical, magnetic and electrical properties of coil magnets made from Bi-2212 PIT tape sections are shown in Table. I. The winding insulation thicknesses (t_{insu}) listed in Table I were calculated from coil dimensions of OD, ID and tape thicknesses after winding. In Table I and Fig. 1, theoretical values for coil core magnetic fields were calculated from coil dimensions and I_c's as a function of tape length and compared to measured values. The agreement between measured and theoretical values shown in Fig. 1 was excellent.

As shown in Table I, the uniformity of processing and consistency of I_c results was excellent. For example, to fabricate a 0.45 T magnet, sixteen 1.5 meter tape sections were chosen arbitrarily from five different rolling batches. After annealing the sixteen tapes four at a time in four days, the tapes were arbitrarily combined by inserting eight coils in parallel. The measured I_c's of the two pancake coils fabricated by this method were quite close (~ 368 A and ~ 380 A), as shown in Table I.

As a consequence of the self-generated magnetic fields, J_c's of tape sections after winding are reduced from their pre-wound values. The maximum magnetic field (B_{max}) value affecting the coils can be determined from graphical analysis for coil parameters $\beta = 2*L/ID$, $\alpha = OD/ID$ and B_{core} values, with L = magnet length.[12] The drop in J_c as a function of applied magnetic fields is shown in Fig. 2 for a typical range of tape configurations and processing methods.[13,14] A J_c drop of ~ 3 - 9% is expected for B_{max} magnetic fields from ~ 0.1 - 0.3 T at 4.2 K.

In addition to magnetic field effects, J_c's of coils are reduced from winding strains. The surface bending strain of a tape bent with radius of curvature r(cm) is estimated as $\varepsilon \cong t_t * k / 2$, where $k = 1 / r(cm)$ is the curvature and t_t is the tape thickness.[15] The strain of a coil segment wound from dimensions r_{init}(cm) to r_{final}(cm) was estimated by computing the difference in strain of a tape bent to different curvatures, as $\varepsilon = \varepsilon(r_{final}) - \varepsilon(r_{init})$.[15]

In Figs. 3(a) and (b), the strain of coiling is calculated for different tape segments as a function of distance from the inner core current tap for different initial winding dimensions (t_{insu} = 1430 μm and t_{insu} = 790 μm) and different final winding dimensions (t_{insu} = 25 -

Table I. Physical, Electrical, and Magnetic Properties of Bi-2212 Coils.

Sample #	TH365	TH366	TH367	TH380	TH381	TH410A	TH410B	TH429
Tape Length (m)	1.0	1.0	1.0	4.5	4.5	1.5 (x 8)	1.5 (x 8)	10.2
ID Before Coiling (cm)	1.7	1.7	1.7	1.4	1.5	1.4	1.4	1.3
OD Before Coiling (cm)	3.8	3.8	3.8	9.2	9.2	5.3	5.3	13.7
ID After Coiling (cm)	1.25	1.25	1.25	1.25	1.25	1.25	1.25	1.25
OD After Coiling (cm)	2.3	2.35	1.77	2.85	2.75	4.2	4.1	4.3
t_{insu} Before Winding (μm)	790	790	790	1430	1430	1430	1430	1430
t_{insu} After Winding (μm)	230	250	65	55	45	305	295	70
Tape (μm) Thickness	62	62	62	58	58	{ 59 -- 64 }		60
Film (μm) Thickness	16.3	16.3	16.3	15.3	15.3	17.8	17.8	16.5
Film X-Sect. Area ($\times 10^{-4}$ cm^2)	1.20	1.20	1.20	1.15	1.15	1.41	1.41	1.35
Tape Width (cm)	0.874	0.874	0.874	0.886	0.886	{ 0.91 -- 0.95 }		0.942
T_c (K)	74.5							
I_{ct} (A) at 20 K	39.6							
J_{ct} (A/cm^2) at 20 K	32300							
Bmeas at 20 K (T)	0.045							
B Theor at 20 K (T)	0.05							
I_{ct} (A) at 4.2 K	63.5	64.6	62.6	41.5	37.8	368	380	42.5
J_{ct} (A/cm^2) at 4.2 K	52800	53700	52000	36000	32800	32600	33700	31500
Bmeas at 4.2 K (T)	0.06	0.044	0.072	0.170	0.150	0.450		0.238
B Theor at 4.2 K	0.076	0.075	0.098	0.173	0.165	0.490		0.237

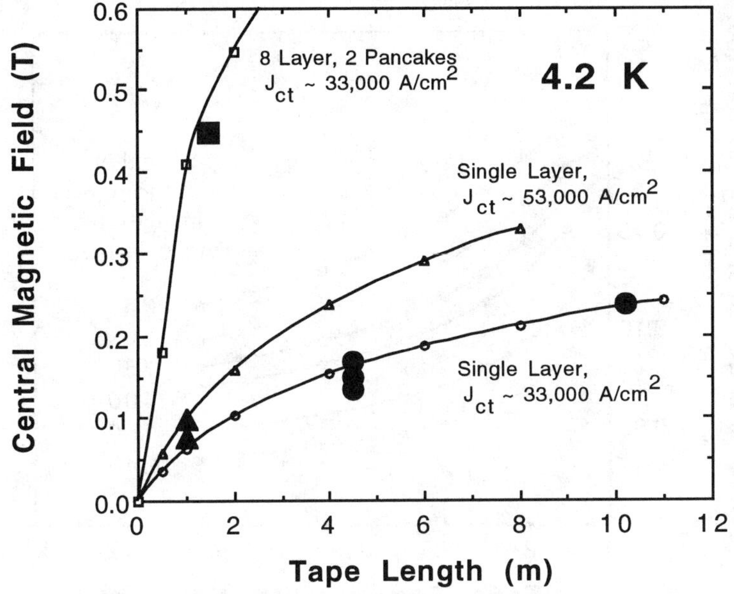

Fig. 1. Measured (●▲■) and calculated (○ ▲ □) values of coil magnetic fields generated for coil dimensions: tape thickness = 60 μm; tape width 0.75 cm, insulation thickness = 50 μm (○ ▲) and tape width = 1 cm, insulation thickness = 310 μm (□).

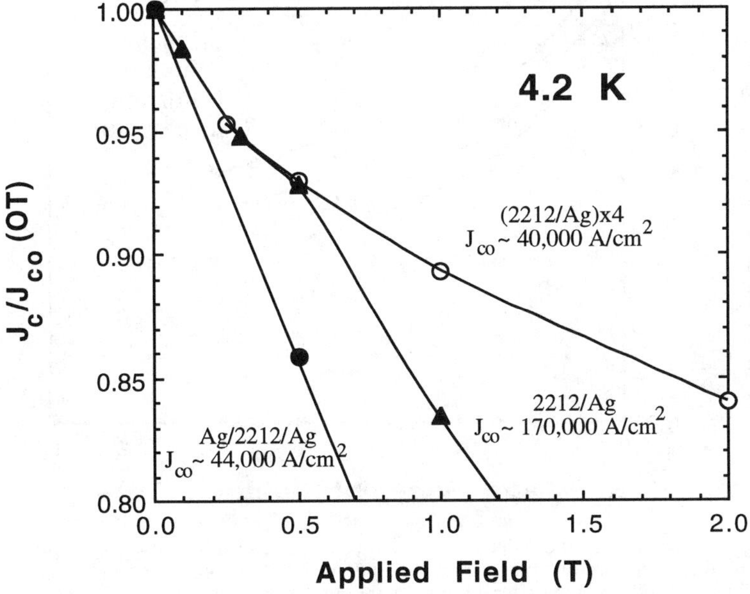

Fig. 2. The reduction of transport J_c's at 4.2 K as a function of applied magnetic field for different processing methods and tape configurations.[13,14]

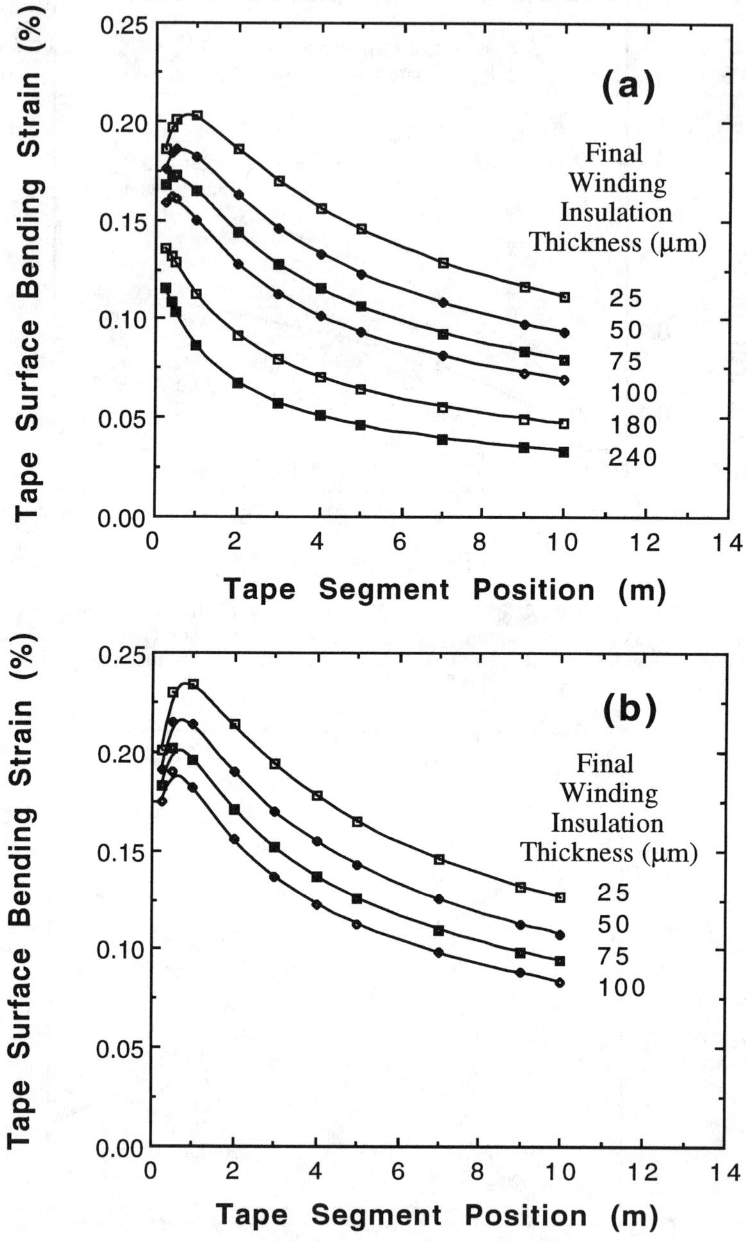

Fig. 3. Strain exerted on tape segments after winding as a function of position from the inner current tap and final winding insulation thickness for a tape thickness = 60 μm and different initial winding dimensions:
(a) t_{insu} = 790 μm, ID = 1.7 cm and (b) t_{insu} = 1430 μm, ID = 1.25 cm.

240 μm) listed in Table I. As expected, if the coil is wound tighter initially (t_{insu} = 790 μm) before processing, the coil experiences less strain from the final winding process than a coil wound from much larger dimensions initially (t_{insu} = 1430 μm) to similar tightness. In general, the innermost tape segments (< 5 m) experience higher strains than the outermost tape segments (> 5 m), regardless of the t_{insu} spacing used. For coiling dimensions of pancake coil magnets listed in Table I, the winding strains were in the range of ~ 0.10 - 0.25%.

To determine the decrease in J_c expected for different winding strains, J_c's of short length Bi-2212/Ag PIT tapes were measured for different bending curvatures. The results are shown in Fig. 4 and compared to results obtained previously for 2212/Ag "uncovered" single layer (USL) tapes.[16]   An irreversibility limit of $\varepsilon \cong 0.06\%$ was observed for PIT tapes compared to $\varepsilon \cong 0.0$ % for USL tapes. For strains ~ 0.12 - 0.30%, a ~ 65 - 70% drop in J_c was observed.

In Table II, the effect of winding strains and self-induced magnetic fields on J_c's are summarized for single layer pancake coils listed in Table I. Values for strain were determined from Figs. 3 (a) and (b) and Fig. 4 using initial and final winding dimensions given in Table I. Magnetic field effects were determined for B_{max} values and J_c - magnetic field properties shown in Fig. 2.

In Fig. 5, J_c's measured after winding are compared to estimated values of J_c before winding computed from Table II.   Estimated values of J_c's of one to ten meter tape sections before winding were ~ 60 - 85% of J_c's of short length tapes processed under similar conditions.

To estimate the performance of Bi-2212 coil magnets at 20 K, the temperature dependence of J_c was measured and shown in Fig. 6 for several tapes processed under different conditions. The decrease in J_c from 4.2 K to 20 K was ~ 60% for tapes processed under these conditions. As shown in Fig. 6, the T_c and J_c could be raised slightly by cooling in an Ar/air mixture and quenching at a higher temperature.

Conclusions

Results of processing one to ten meter lengths of $Bi_2Sr_2CaCu_2O_{8+x}/Ag$ powder-in-tube tapes with uniform properties and high critical current densities were presented. Pancake coil magnets made from tape sections by a wind-react-wind method generated central magnetic fields from 0.07 - 0.45 T measured at 4.2 K. Transport J_c's of one to ten meter sections measured after winding to tight dimensions were in the range of 32,000 - 54,000 A/cm^2. The drop in J_c expected from winding strains and magnetic field effects were calculated as ~ 60 - 70% for the initial and final winding dimensions tested. As a result, J_c's of one to ten meter length sections were estimated as ~ 55,000 - 80,000 A/cm^2 at 4.2 K before winding, ~ 60 - 85% of short length J_c's obtained under similar processing conditions. With further improvements in rolling methods, the use of a more stress tolerant tape configuration, optimization of film and Ag layer thicknesses, improvements in PMSC methods and other factors, J_c's of long length Bi-2212 PIT tapes would be expected to improve further.

Acknowledgements

This work was supported in part by the New York State Institute on Superconductivity.

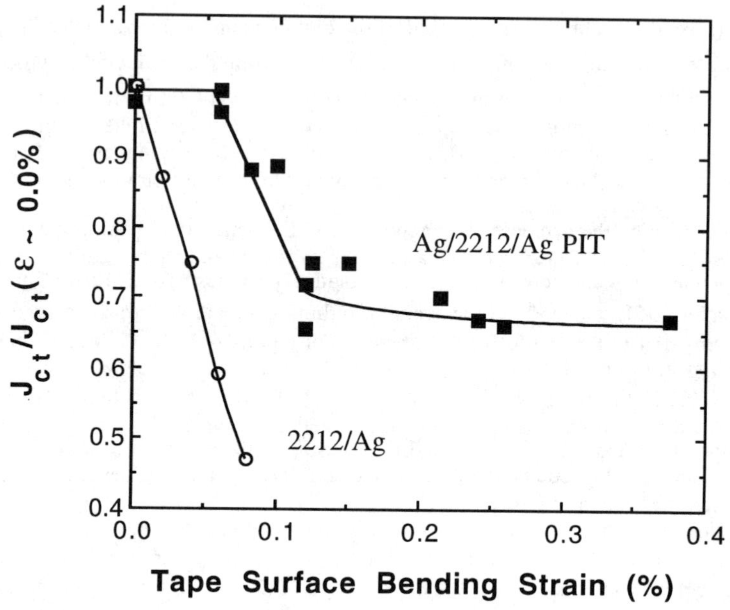

Fig. 4. The reduction of transport critical current densities as a function of strain for Bi-2212/Ag powder-in-tube tapes with thickness 60 μm and 75 μm (■) and Bi-2212/Ag "uncovered" single layer tapes (O).

Table II. Estimated ranges of J_c percentage decreases from winding strains and self-induced magnetic field effects.

Sample #	TH365	TH366	TH367	TH380	TH381	TH429
Coil Tape Length (m)	1.0	1.0	1.0	4.5	4.5	10.2
Coil Winding Strain (%)	0.08 - 0.14	0.07 - 0.12	0.15 - 0.17	0.15 - 0.22	0.16 - 0.24	0.10 - 0.21
J_c % Decrease From Winding Strains $\{J_c/J_c(\epsilon \sim 0.0\%)\}$	0.85 - 0.70	0.85 - 0.70	0.70 - 0.68	0.70 - 0.65	0.70 - 0.65	0.75 - 0.65
Self-Generated Magnetic Fields B_{max} (T)	0.09	0.09	0.12	0.19	0.19	0.27
J_c % Decrease From Magnetic Field Effects $\{J_c/J_c(0\ T)\}$	0.97	0.97	0.97	0.96 - 0.94	0.96 - 0.94	0.95 - 0.92

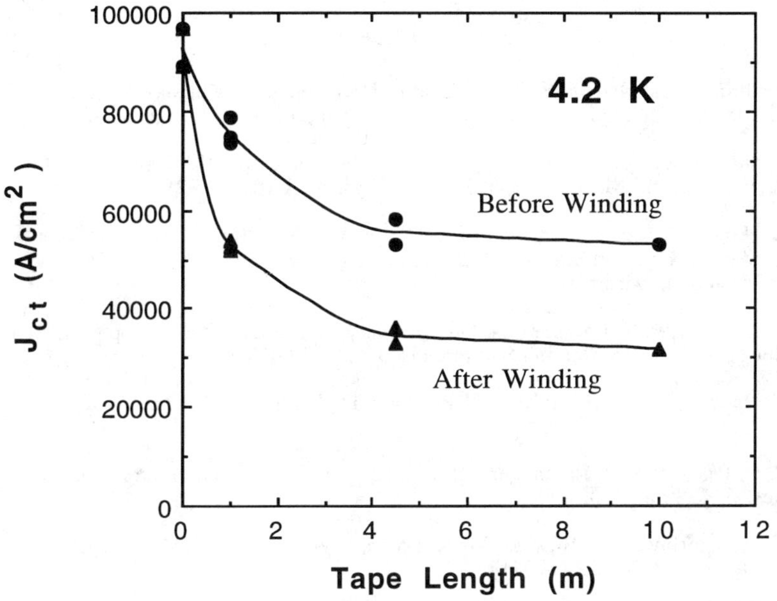

Fig. 5. Transport J_c's of Bi-2212/Ag powder-in-tube long length tapes measured after winding to dimensions given in Table I and estimated values before winding.

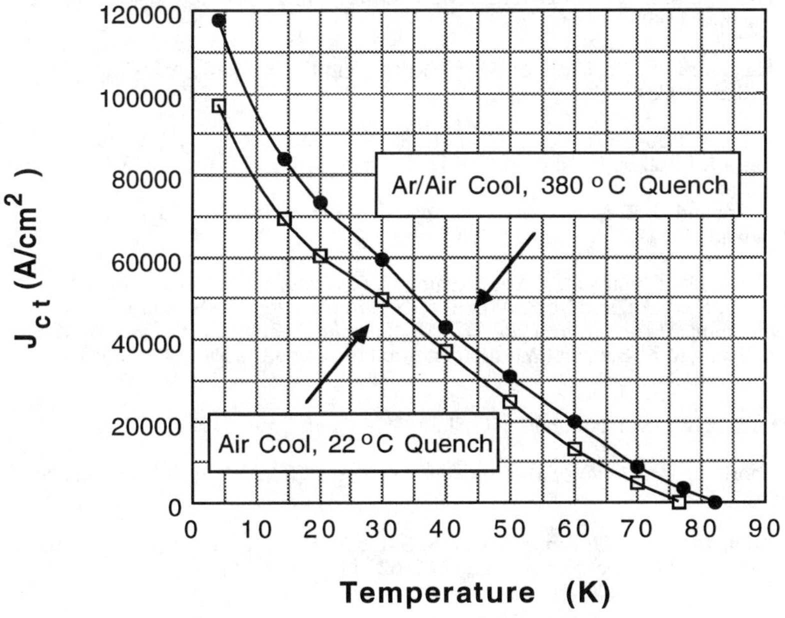

Fig. 6. The effect of temperature on the transport critical current density (J_{ct}) of Bi-2212/Ag powder-in-tube tapes processed with partial-melting and slow-cooling for different cooling conditions.

References

1. J. Shimoyama, N. Tomita, T. Morimoto, H. Kitaguchi, H. Kumakura, K. Togano, H. Maeda, K. Nomura and M. Seido, Jpn. J. Appl. Phys 31, L1328 (1992).

2. J. Shimoyama, T. Morimoto, H. Kitaguchi, H. Kumakura, K. Togano, H. Maeda, K. Nomura and M. Seido, Jpn. J. Appl. Phys. 31, L163 (1992).

3. T. Haugan, J. Ye, S. Chen, S. S. Li, S. Patel and D. T. Shaw (Paper presented at the 6th Annual Conf. on Superconductivity and Its Applications, Buffalo, NY 1992), **P.25**, and shown herein.

4. L. J. Masur, E. R. Podtburg, A. Otto, D. Daly, and C. Craven (Paper presented at the Spring 1993 MRS Meeting,San Francisco, CA 1993), **T6.6**.

5. T. Haugan, J. Ye, M. Pitsakis, S. Patel and D. T. Shaw, AIP Conf. Proc. **251**, 408 (1991).

6. P. N. Peszkin, R. J. Raymakers, R. S. Fiegelson, L. V. Moulton and Z. Lu, J. Mat. Res. **6**, 2280 (1991).

7. G. A. Whitlow, J. C. Bowker, W. R. Lovic, F. A. List and D. M. Kroeger, AIP Conf. Proc. **253**, 559 (1993).

8. S. Patel, T. Haugan, S. Chen, F. Wong, E. Narumi and D. T. Shaw, accepted for publication, Cryogenics, 1993.

9. T. Haugan, S. Patel and D. T. Shaw, manuscript in preparation, to be submitted to Cryogenics (1993).

10. T. Haugan, J. Ye, S. Chen, S. S. Li, S. Patel and D. T . Shaw, AIP Conf. Proc. **253**, 609 (1993).

11. S. Patel, T. Haugan, J. Ye, S. Chen, S. S. Li, A. Shah, C. Li, A. Ardounis, F. Wong, M. Pitsakis, E. Narumi, and D. T. Shaw, AIP Conf. Proc. **253**, 567 (1993).

12. Y. Iwasa and D. B. Montgomery, in *Applied Superconductivity*, ed. by V. Newhouse (Academic Press, New York, 1975).

13. D. T. Shaw, T. Haugan, J. Ye, S. Patel, C. C. Tseui, C. C. Chi, T. Frey, D.B. Mitzi, T. Kazyaka, and M. K. Wu, in *HTS Materials, Bulk Processing & Bulk Applications, Proc. of the 1992 TCSUH Workshop*, ed. by C. W. Chu, W. K. Wu, P.-H. Hor and K. Salama (World Scientific Publishing Co. Pte. Ltd., Singapore, 1992), p. 252.

14. J. Ye, S. Hwa, S. Patel and D. T. Shaw, AIP Conf. Proc. **219**, 524 (1990).

15. F. Chorlton, *Mechanics of Materials for Engineers* (John Wiley & Sons, Inc. New York, 1962).

16. S. Patel, J. Ye, T. Haugan, S. Chen, S. S. Li, F. Wong, D. T. Shaw and C. C. Tseui, Mat. Res. Soc. Symp. Proc. **275**, 621 (1992).

Thermomechanical Processing of Ag–Clad

$Bi_2Sr_2CaCu_2O_x$ Superconductors*

C.–T. Wu, K. C. Goretta, M. T. Lanagan, A. C. Biondo,
and R. B. Poeppel

Argonne National Laboratory, Argonne, IL 60439–4838 USA

Abstract

$Bi_2Sr_2CaCu_2O_x$ (2212) powders were synthesized by solid–state reaction, loaded into Ag tubes, and processed into tapes by various combinations of drawing, rolling, and heat treatment. Critical current densities $> 10^5$ A/cm^2 at 4.2 K were achieved from microstructures consisting of large, highly textured 2212 grains. Optimal microstructures were produced by specific mechanical–working conditions and heat–treatment schedules that incorporated solid–state and partial–melt sintering. The relationships between processing, microstructure, and critical current density will be discussed.

Processing of Long Lengths of Superconductors
Edited by U. Balachandran, E.W. Collings and A. Goyal
The Minerals, Metals & Materials Society, 1994

<u>Introduction</u>

It has been shown that Ag-sheathed $Bi_2Sr_2CaCu_2O_x$ (2212) tapes fabricated by the powder-in-tube method have high current-carrying capability at 4.2–25 K(1,2). Tenbrink et al. (1) used a partial-melting process for 2212 wire fabrication and attained a critical current density (J_c) at 4.2 K that exceeded 10^4 A/cm^2 to fields of 25 T. Kase et al. (2) determined that large, highly c-axis-aligned grains were developed by the use of partial-melt processing, which consisted of melting at $\approx$890°C, followed by prolonged annealing at $\approx$870°C. However, because of characteristics of the peritectic reaction, nonsuperconducting phases, such as alkaline-earth cuprates, alkaline-earth bismuthates, CaO, and CuO were also formed. The alkaline-earth cuprates could grow quite large during cooling. These secondary phases disturb local 2212 alignment and obstruct the current path, and consequently decrease J_c.

The optimum maximum temperature of the partial-melt process is reported to be 880–920°C, depending on the stoichiometry of the 2212 and the atmosphere (1,2). However, the subsequent optimal annealing temperature and time are not so clearly defined. The 2212 phase will crystallize quickly from Bi-Sr-Ca-Cu-O glasses over a board range of temperature and time (3). This implies that the annealing conditions for 2212 formation may not be critical. The purpose of this work was to clarify the relation between heat treatment and the microstructure of 2212 tapes. The growth kinetics of the secondary phases and the effects of rolling conditions on J_c were also studied.

<u>Experimental Details</u>

The nominal cation molar content of the Ag-clad tapes was 2.0:1.7:1.0:2.0 (Bi:Sr:Ca:Cu). The starting powder was made by solid-state reaction of oxides and carbonate (4). The phases after calcination were 2212 and trace amounts of $Bi_2Sr_2CuO_x$ (2201) and CaO. The reacted powder was packed into an Ag tube and drawn and rolled to a tape 0.1–0.2 mm thick and $\approx$3 mm wide.

A series of quench experiments was conducted to monitor the kinetics of nucleation and growth of the superconducting and secondary phases. During heating, specimens were quenched from 850, 860, 870, 880, and 890°C. Two additional specimens were quenched after holds of 0.1 and 0.5 h at 890°C. A 10°C/h cooling rate was used to cool the tapes from 890°C to 880, 870, 865, 860, and 855°C. Tapes were quenched after 1 and 3 h holds at 870, 865, 860 and 855°C. Rolled tapes (5 cm long, 0.4 cm wide, 0.2 mm thick) were heated to selected temperatures and then quenched in ice water. The distance between the quench medium and the outlet of the furnace was short to minimize the possibility of air quenching. Complete partial-melt heat treatments were also performed (Table 1). After partial-melt treatment, some tapes were rolled to $\approx$0.15 mm with various reductions per pass, and exposed to a final annealing treatment at 840°C in air for 100 h.

Onsets of significant melting were determined by differential thermal analysis (DTA) in air. Phase identification and grain alignment of the tapes were determined by X-ray diffraction (XRD) and scanning electron microscopy (SEM) with energy dispersive analysis of X-rays (EDS). Degree of c-axis alignment was defined as the ratio of the intensity of the (00l) peak to the intensity of the (117) peak. Current-voltage characteristics of Ag-clad tapes were measured by the conventional d. c. four-probe method at 4.2 K, and J_c was defined by a 1 μV/cm criterion.

Table 1. Partial-Melt Treatment of 2212 Tapes

Schedules	Partial-melt Temp. (°C)	Cooling Rate (°C/h)	Holding Temp. (°C)	Holding Time. (h)
1	890	10	870	3
2	890	10	865	3
3	890	10	860	3
4	890	10	855	3

<u>Results</u>

Figure 1 shows the DTA curve of a typical Ag-sheathed 2212 tape. The endotherm at $\approx$884°C was due to incongruent melting of the 2212 phase. During cooling, an exotherm was observed between 850 and 865°C. This peak, as will be discussed later, was due to solidification of 2212 phase. The delay in solidification relative to melting occurred because nucleation required undercooling to overcome the nucleation energy barrier to form critical nuclei.

Figure 2a shows XRD data for tapes that had been quenched from five different temperatures during cooling. No 2212 or 2201 phases formed from 880 to 865°C. Small peaks of $(Sr,Ca)_2CuO_3$ (2/1) and $(Sr,Ca)_{14}Cu_{24}O_x$ (14/24) were detected in this temperature range (5). At 860°C, the 2212 phase began to form; its intensity increased substantially at 855°C. No significant texturing of 2212 grains was observed. The 2/1 and 14/24 phases were also detected at 860 and 855°C. XRD data from tapes quenched after 1 h at various temperatures are shown in Fig. 2b. After 1 h at 870°C, no 2212 or 2201 was observed; 2212 formed at 865°C, with a small amount of favorable texture. The texture was strongly developed after 1 h at 860°C, but was less strong after holding at 855°C. Figure 2c shows XRD data from tapes quenched after 3 h at temperature. The results were similar to those shown in Fig. 2b. Again, the strongest c-axis texture was developed at $\approx$ 860°C (Table 2).

Figure 3 shows SEM micrographs for various quenched tapes. Small amounts of needle-like alkaline-earth cuprate phases, 14/24 by EDS, started to form at 860°C (Fig. 3a); average size was $\approx$100 μm long and $\approx$10-20 μm wide. As temperature increased to 880°C, the secondary

103

phases grew rapidly to ≈300–400 μm in length, and the volume fraction increased (Fig. 3b). Above 880°C, the 14/24 phase decomposed and the 2/1 phase formed. After 0.5 h at 890°C, the temperature was decreased at 10°C/h to various temperatures. Figure 3c shows the microstructure of a tape quenched from 880°C during cooling. The size of the secondary phase remained about the same as that at 890°C; however, the volume fraction increased.

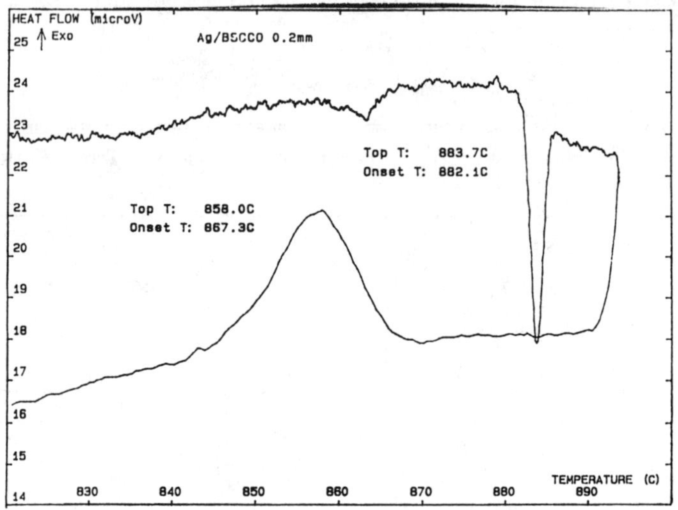

Figure 1. DTA Curve for Ag-Sheathed 2212 Tape

Table 2. XRD Peak Intensity Ratio of (0010) to (117) of Tapes Quenched at Various Holding Temperatures and Times

Holding Temp.(°C)	Holding Time		
	0 h	1 h	3 h
870	0	0	0
865	0	1.28	1.03
860	1.36	1.97	2.03
855	1.17	1.45	1.57

Substantial coarsening of the secondary phases took place during cooling from 880°C to 870°C. In tapes quenched at 870°C, secondary phases 800–1000 μm long and 50–100 μm wide were observed (Fig. 3d). EDS analysis indicated that these phases were 14/24. Smaller 2/1 phases (100–200 μm long, 10–20 μm wide) were also observed. Substantial growth of the platelike phases was observed after holding 3 h at 870°C (Fig. 3e). Large 14/24 phases were also observed in the tapes quenched from 865 and 855°C (Figure 3f-h). In particular, remarkable coarsening occurred after 3 h at 865°C. However, no coarsening occurred after 3 h at 855°C. Little coarsening of the 2/1 occurred on any cooling schedule.

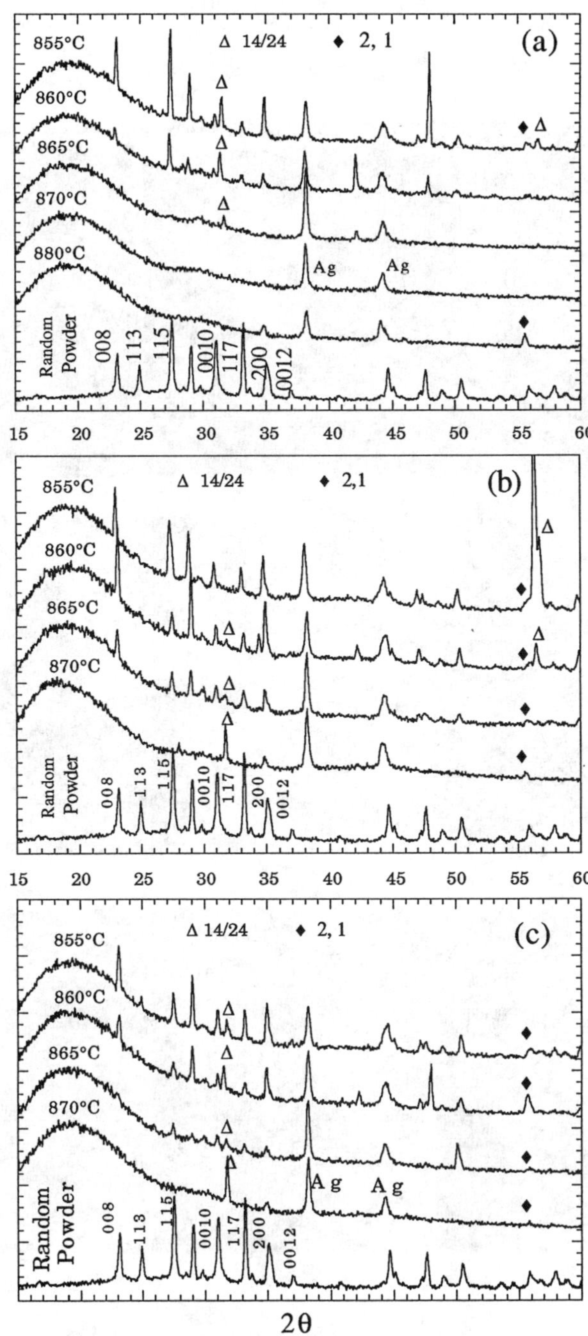

Figure 2. XRD Data for Tapes Quenched After (a) 0 h, (b) 1 h, and (c) 3 h Holds at Various Temperatures, with a Cooling Rate of 10°C/h from 890°C

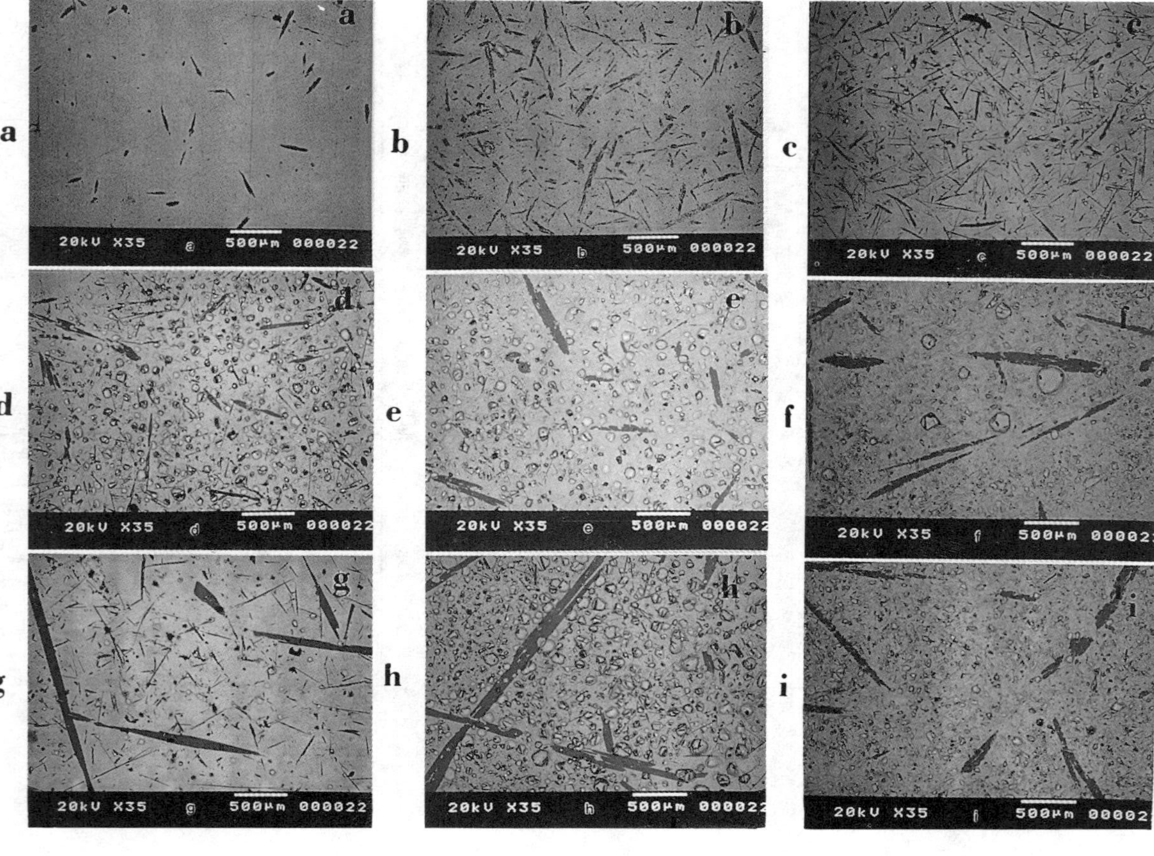

Figure 3. Microstructure of Tapes Quenched from Various Holding Temperatures and Times on Heating and Cooling

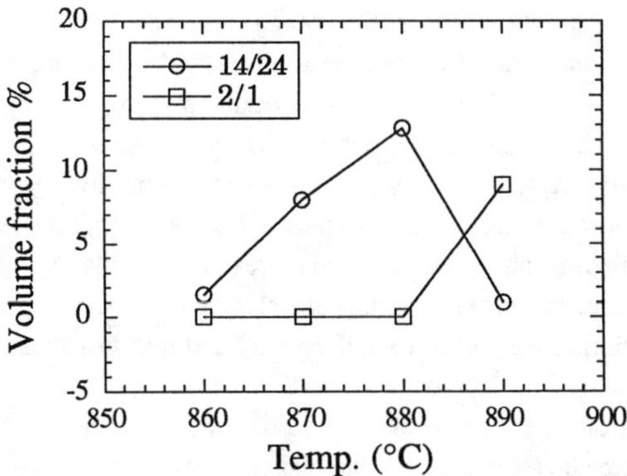

Figure 4. Volume Fraction of Secondary Phases as a Function of Temperature upon Heating

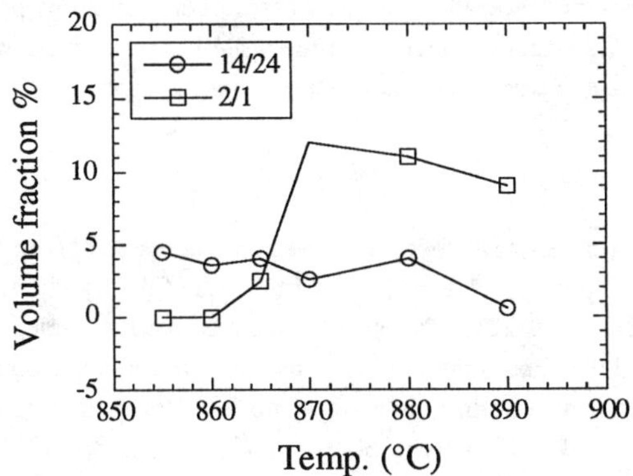

Figure 5. Volume Fraction of Secondary Phases as a Function of Temperature upon Cooling

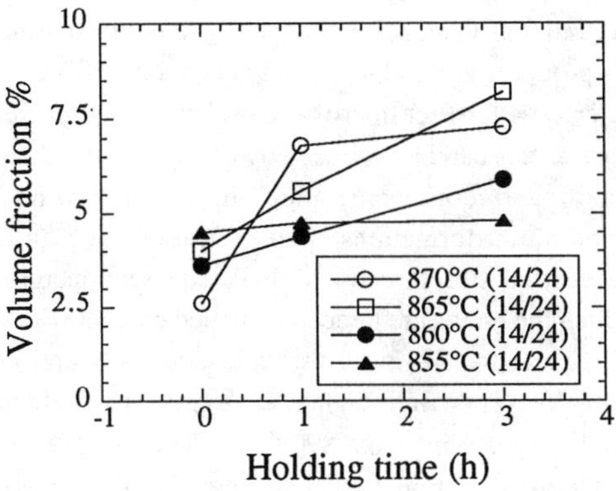

Figure 6. Volume Fraction of 14/24 as a Function of Annealing Time at Temperature

However, some 2212 grains adjacent to secondary phases grew along the interface, and thus the growth direction was not parallel to the rolling direction. In addition, some 2212 grains stopped growing when their growth fronts were obstructed by secondary phase. Presence of large secondary phases reduced the extent of favorable texture. The microstructure of tapes annealed at 860°C is shown in Fig. 7b. Well-aligned grains were present throughout the entire oxide core. Some 2201 (lighter gray) grains were embedded between the 2212 (darker gray) grains. The microstructure of tapes annealed 3 h at 855°C is shown in Fig. 7c. The desired well-aligned grain structure was only present near the Ag interface. Randomly oriented grains were present throughout the central zone of the oxide core. In addition, some 2212 grains penetrated into Ag and disrupted the well-textured structure at the Ag interface.

The J_c values for the tapes after partial-melting treatment, but before final rolling and heat treatment at 840°C, are shown in Fig. 8. The highest J_c values were produced by annealing at 860–865°C. A tape that was held 3 h at 860°C in the partial-melt step and finally annealed 100 h at 840°C yielded the highest J_c, 1.5×10^5 A/cm^2, at 4.2 K. The highest J_c value was produced by rolling with 10% reduction per pass; larger reductions per pass decreased J_c values $\approx$ 50% (4). The effect of annealing time at 840°C on J_c is shown in Fig. 9. In this processing, at least $\approx$100 h was needed to produce a high-J_c tape.

Discussion

The broad exotherm on the DTA curve, Fig. 1 ranged between 850 and 865°C. Maximum heat was released from this system at $\approx$857°C. If the heat of this exotherm is due to latent heat released by solidification of 2212, this curve indicates that the nucleation process of 2212 started at $\approx$865°C and the maximum number of nuclei were formed at $\approx$857°C. XRD analysis of tapes quenched during cooling shows that no 2212 phase forms until 860°C during continuous cooling. 2212 peaks were observed after 1 h at 865°C. No 2212 formed at 870°C. These results confirm that the DTA exotherm was due to solidification of 2212. XRD data for the tapes quenched after 1 h at various temperatures during cooling indicate that a better textured 2212 was obtained at 860°C than at 855°C. In addition, the grains of the tapes annealed 3 h at 860°C and 865°C exhibited better alignment than grains in tapes annealed at 855°C. A high J_c was produced by heat treatment at 865 and 860°C. The XRD, SEM, and J_c results indicate that a tape with better microstructure, primarily maximum c-axis alignment, and better electrical properties can be produced by cooling from 890 to 860–865°C, holding there for $\approx$ 3 h, and cooling to room temperature. In general, low nucleation rates and high growth rates during phase transformations or solidification are required for a large-grained microstructure. The desired structure for high-T_c superconductors with good electrical properties is one in which the grains are large and aligned parallel with the direction of current flow. Kase et al. (2) found that 2212 grains nucleated easily and grew rapidly along Ag interfaces. A 20-μm-thick layer of well-aligned 2212 grains parallel with an Ag interface was formed at the interface during cooling from 890°C to 870°C. It is believed that this thin, well-aligned grain layer is the major region of current flow (2). The thicker this layer, the higher the J_c. Therefore, the assignment is to find a proper partial-melt schedule that will increase the thickness of this well-aligned grain structure near the Ag interface.

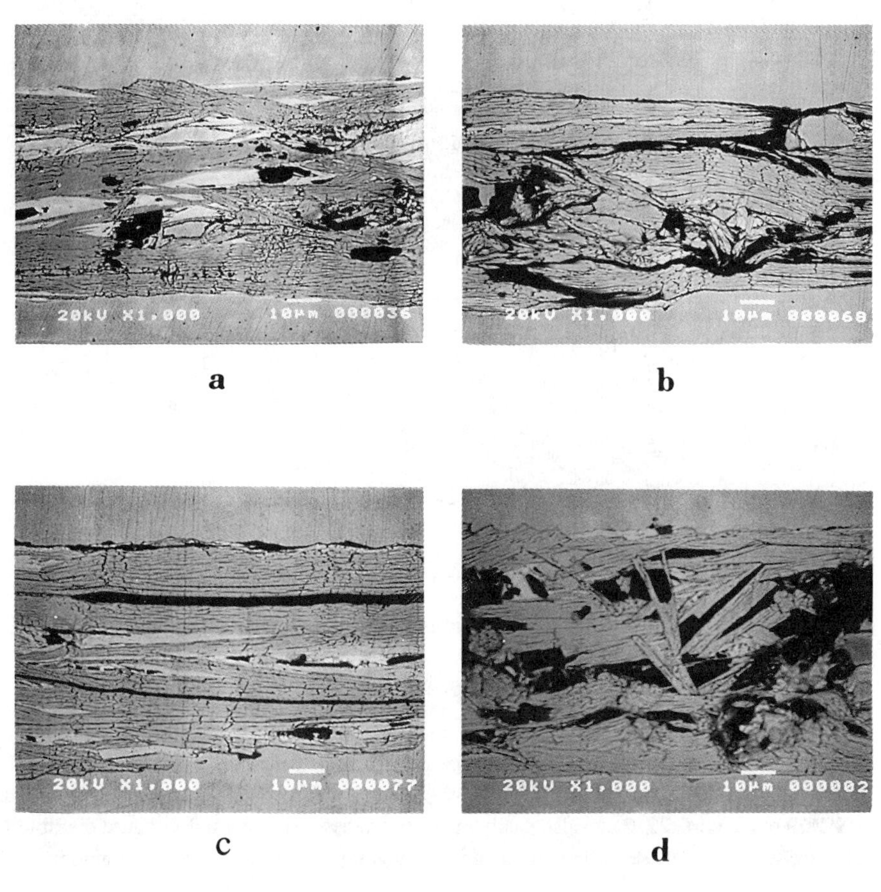

Figure 7. SEM Microstructure of Tapes Partially Melted according to (a) Schedule 1, (b) Schedule 2, (c) Schedule 3, and (d) Schedule 4, as Listed in Table 1

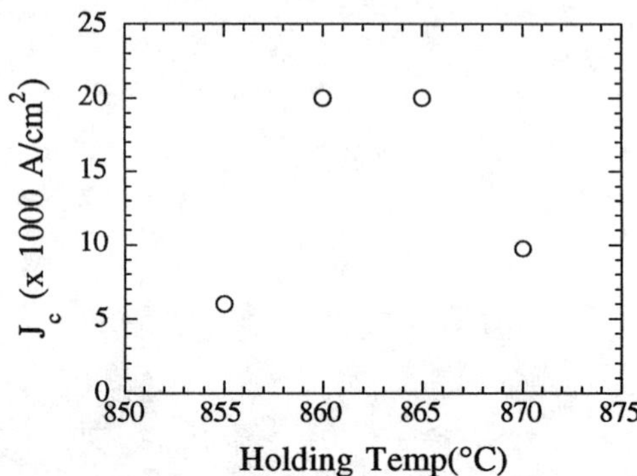

Figure 8. J_c at 4.2 K for Tapes Heat Treated according to Schedules Listed in Table 1

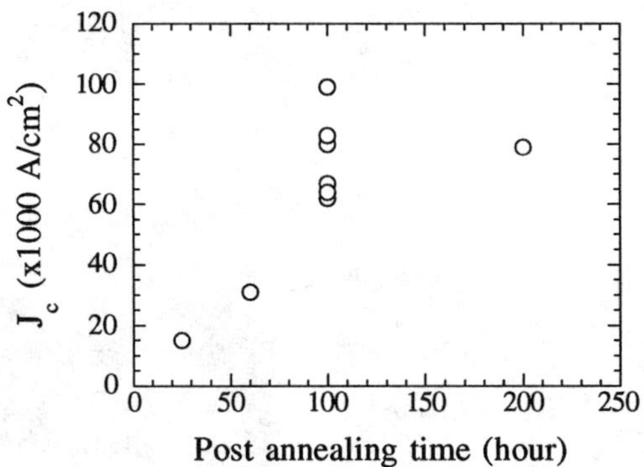

Figure 9. Relationship Between J_c and Sintering Time at 840°C in Air

The Ag, essentially acting as a mold wall in the solidification process, is a favorable site for heterogeneous nucleation of 2212. However, because the secondary phase is present in the BSCCO oxide core, 2212 can also easily nucleate on these particles and, because distribution of these secondary phase is random, the growth directions of the 2212 are also random.

The grain structure produced during partial-melt processing is determined by competition between the growth of 2212 in two regions: first, in the region near the Ag interface, where the 2212 grains are well-aligned, and second in the region near the center of the oxide core where the 2212 grains may grow along interfaces with secondary phases. It has been found that addition of Ag to 2212 can lower the melting point by $\approx$15°C (4). Fang et al. (6) showed that diffusivity of Ag is high at high temperatures. It is possible that there exists a layer of the core near the Ag interface, in which the melting point is lowered because Ag is presenct. As a result, more liquid may be present in this Ag-rich layer than in areas with less Ag. This liquid-rich zone near the Ag interface may promote growth of 2212 after it has nucleated on the Ag. In the partial-melt process used here, 860–865°C is good for the desired grain structure, because the nucleation rate is lowand the growth rate is high duo to the presence of liquid. However, coarsening of secondary phases takes place at ≈865°C and at higher temperatures. These coarse secondary phases have a deleterious effect on the overall alignment of 2212. Therefore, the holding temperature following partial melting should be below that favored for coarsening of secondary phase. At 855°C, more nuclei are formed in the oxide and the thickness of the liquid-rich layer may be thinner than that at higher temperature. As a result, thin layers of well-aligned grain structure near the Ag interface and random structures at the center are produced.

The highest J_c value after partial-melting processing was ≈2 x 10^4 A/cm^2, which is below the value needed for most practical applications. The low J_c may be attributed to several factors. First, because the retrograde densification associated with anisotropic grain growth, the density of the oxide core decreased after partial melting (7). Second, pores may have been generated during heating because of decomposition of residual carbonates or hydroxides (8), or sudden evolution of oxygen upon decomposition of the 2212 (9). Further rolling redensified the core and enhanced c-axis alignment by mechanical action. Post annealing was employed to connect the grains and heal the small cracks that may have been generated during the second rolling. However, severe mechanical working may result in irreparable damage, such as large cracks and sausaging. SEM revealed that sausaging was most pronounced in tapes with high concentrations of secondary phases. It appears that gradual and limited mechanical working is favorable for high current capability.

Assessment

The partial-melting process substantially affects the final microstructure and J_c of Ag-clad 2212 tapes. Parameters to be controlled in this process are atmosphere, maximum temperature, holding time at temperature, and heating and cooling schedules. The structures of 2212 and the secondary phase vary with these parameters, and consequently, electrical properties vary as well. In the second rolling step, severe reduction may cause damage; moderate reduction can densify the core structure without introducing serious damage to the core. However, because of the size of the secondary phase, the thickness after the second rolling can be limited. A post annealing process is essential to heal the cracks that are generated during rolling and to join the superconducting grains.

Acknowledgment

We thank C. A. Youngdahl and J. S. Kallend for advice and experimental assistance. This work was supported by the U.S. Department of Energy (DOE), Energy Efficiency and Renewable Energy, as part of a DOE program to develop electric power technology, and Basic Energy Sciences–Materials Sciences, under Contract W-31-109-Eng-38.

References

1. J. Tenbrink et al., "Development of High-T_c Superconductor Wires for Magnet Applications," presented at the Applied Superconductivity Conf., Snowmass Village, CO (September 24-28, 1990).

2. J. Kase, "Preparation of the Textured Bi-Based Oxide Tapes by Partial Melting Process," presented at the Applied Superconductivity Conf., Snowmass Village, CO (September 24-28, 1990).

3. T. G. Holesinger, D. J. Miller, and L. S. Chumbley, J. Mater. Res. 7, 1658-1671 (1992).

4. C.-T. Wu, K. C. Goretta, and R. B. Poeppel, "Effects of Processing Parameters on Critical Current Density of Ag-Clad $Bi_2Sr_2CaCu_2O_x$ Tapes," Applied Superconductivity 1, 263 -267 (1992).

5. R. D. Ray II and E. E. Hellstrom, "Phase Development in Ag-Clad BSCCO Wires," High Temperature Superconductors: Fundamental Properties and Novel Materials Processing, (Pittsburgh, PA: MRS, 1990), pp. 1291-1294.

6. Y. Fang et al., "Trace Diffusion of ^{110}Ag in $Bi_2Sr_2CaCu_2O_x$," Appl. Phys. Lett. 60 (18), 4 May 1992 pp. 2291-2293

7. D. W. Johnson, and W. W. Rhodes, "Retrograde Densitication in $Bi_2Sr_2CaCu_2O_8$ Superconductors," J. Am. Ceram. Soc., 72, (12) (1989), pp. 2346-2350.

8. Yan, M. F., "Conventional Ceramics Processing of High-T_c Superconductors," Better Ceramics Through Better Chemistry III, Brinker, C. J., Clark, D. E. and Ulrich, D. R., eds (Pittsburgh, PA: MRS, 1988), pp. 385-400.

9. Almond, D. P., Chapman, B,. and Saunders, G. A., "The Properties of Quenched Bi-Ca-Sr-Cu Oxide Superconductors," Supercond. Sci. Technol., 1 (1988), p. 123

MANUFACTURE OF LONG LENGTHS OF

HTSC WIRES BY EXTRUSION

Toru Yamashita, Atit Bhargava, Steve Golden, Rupeng Zhao,
David Page, John Barry and Ian D. R. Mackinnon

Centre for Microscopy and Microanalysis
The University of Queensland
St. Lucia, Queensland 4072
Australia

Abstract

Powders of $YBa_2Cu_3O_7$ were prepared by co-precipitation using acqueous solution of nitrates and oxalic acid. Precipitated powders were then filtered and dried. The dried powders were calcined at 500°C or 800°C and mixed with 2wt% of hydroxypropylmethylcellulose and small amounts of water. Long lengths of superconducting $YBa_2Cu_3O_7$ wires were fabricated by means of extrusion followed by sintering. The maximum critical current density, J_c, 1100 A/cm^2 (in transport) was obtained for samples prepared from 500°C calcined powder and sintered at 920°C. To improve J_c, Y211 phase was also introduced to the calcined powder mix and extruded with binder. The maximum J_c obtained by these techniques was 200 A/cm^2. A long length of wire was also extruded from the mixture of calcined Y123 powder, HPMC and a small amount of water. J_c measurements show sintered density and absence of macro-pores are critical to maintain J_c values. In long lengths of wire for which this condition holds the reduction of J_c value with wire distance is minimal.

Processing of Long Lengths of Superconductors
Edited by U. Balachandran, E.W. Collings and A. Goyal
The Minerals, Metals & Materials Society, 1994

Introduction

One of the most important applications of high T_c superconductors is in the form of superconducting wires. Wires may be used for power transmission, electrical magnets for MRI, scientific equipment, particle accelerators and mineral separators. In these cases, conventional designs indicate that metre-lengths of wires are required for viable applications. Thus, good performance of long lengths of wire i.e., high Jc at various magnetic fields, is required. A wide range of J_c studies have been made for $YBa_2Cu_3O_7$ and $Bi_2Sr_2Ca_2Cu_3O_{10}$ in both wire and bulk forms. Although high J_c's have been obtained for $Bi_2Sr_2Ca_2Cu_3O_{10}$ (BSCCO) by a powder-in-tube method, there is no clear evidence for effective pinning centers in the BSCCO. On the other hand, it has been pointed out that Y_2BaCuO_5 can provide pinning centers [1] and/or can be introduced by specific processing methods. Moreover, studies on the temperature dependence of irreversible fields for various high T_c superconductors have revealed that $YBa_2Cu_3O_7$ has intrinsically superior characteristics for pinning centers than Bi- and Tl-based high T_c superconductors [2]. Thus, in this present paper, emphasis is placed on processing $YBa_2Cu_3O_7$-based wires.

Experimental

$YBa_2Cu_3O_7$ (Y123) and Y_2BaCuO_5 (Y211) powders were prepared by a co-precipitation process in which the appropriate ratios of yttrium oxide, barium nitrate and copper nitrate were weighed out and dissolved in water. Dilute nitric acid is required to reduce the pH in order to obtain a clear solution which is then reacted with oxalic acid solution. The co-precipitated powders were filtered and dried in a vacuum oven at 80°C for 15 hours. Oxalate powders were calcined at 500°C or 800°C for Y123 and 500°C for (Y211) for 10 hours in flowing oxygen. The calcined Y211 powder was sintered at 980°C for 15 hours in air. XRD analysis of the 500°C calcined powder reveals the presence of CuO, $BaCuO_2$, Y_2O_3, etc. while the 800°C calcined powder consists of single phase Y123. Starting materials for the wires were prepared by the following two different methods; (1) the 500 and 800 °C calcined Y123 powders were mixed with 2 wt% hydroxypropylmethylcellulose (HPMC) and a small amount of water and (2) the 500 and 800 °C calcined Y123 powders were mixed with 25 and 50 wt % Y211 powder and 2 wt% HPMC. Using a CO_2 gas meter, it was revealed that HPMC evaporates around 300°C [3]. The extruded wires were sintered at various temperatures under oxygen flow. Oxygenation of all samples was carried out at 475°C for 5 hours in flowing oxygen. In general, the diameter of these sintered wires ranged from 0.6 mm - 0.8 mm. Critical current density (J_c) of the wires were measured at liquid nitrogen temperature using four-point probe with contacts set at least a few mm apart. The wires were characterized by XRD, optical microscopy, SEM, TEM and a.c. susceptometer.

<u>Results and Discussion</u>

<u>500°C Calcined Powders (Y123) with HPMC</u>

Both the 500°C and 850°C calcined Y123 powders were extruded separately and sintered at incremental temperatures between 890 and 990°C for 10 hours followed by an oxygenation step. At liquid nitrogen temperature, the J_c (transport) values of these wires around 1.5 cm length were measured. Figure 1 shows the J_c vs sintering temperature for Y123 which was prepared using the 500°C calcined powder. The highest J_c of about 1100 A/cm^2 was obtained for these samples sintered at 920°C. The J_c values show a broad peak between 900 and 920°C and the temperatures are considered to be at an optimum for high J_c.

114

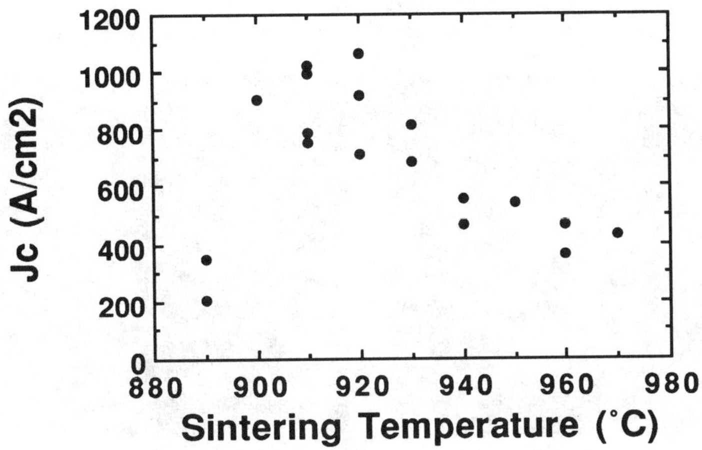

Figure 1 - Temperature dependence of Jc for the wires
prepared from 500°C calcined powder.

Microstructures of the wires sintered at 890, 910, 940 and 960°C are shown in Figure 2. Plate like grain growth is observed for samples sintered at 940 and 960°C, while uniaxial grain growth is observed for samples sintered at 890 and 910°C. Though the microstructures of both samples sintered at 890 and 910°C show a similar morphology, the sample sintered at 910°C appears to be better sintered, while the sample sintered at 890°C is still porous.   High resolution TEM images as shown in Fig. 3 (a), show that the microstructure of most of grain boundaries is clean for the wire sintered at 910°C On the other hand, samples sintered at 940 and 960°C may have liquid formation during sintering and thus, large grain growth was observed.

From this TEM study, both samples sintered at 910 and 960°C show mostly clean grain boundaries and only a small amount of secondary phases along the boundaries. The most prominent difference in microstructure between the two is that the sample sintered at 960°C contains a lot of cracks within a grain (Fig. 3(b)), while the sample sintered at 910°C does not have such features. The magnetic susceptibility of wires sintered at 910 and 960°C were measured to liquid nitrogen temperature as shown in Figure 4. The Meisner signals of both samples were nearly identical at 80K. These similar signals indicate only very small amounts of secondary phases exist in these samples, if at all. These measurements are also consistent with the results obtained from the TEM study i.e., occasional intergrowths of secondary phases can be seen. Figure 4 also shows a sharper transition for the wires sintered at 960°C. This transition may be attributed to the grain size of the wires. For the wire sintered at 910°C, the grain size is 1-3μm. If it is assumed that the penetration depths of a magnetic field for Y123 (λ) are λ_{ab}=100nm and λ_c=700nm [4], λ_{ab} and λ_c are 0.3 and 2.4 μm, respectively, at 90K. These values for λ, given the grain size of these samples, may account for the poor transition of magnetic susceptibility in the wire sintered at 910°C.

<u>800°C Calcined Powders (Y123) with HPMC</u>

The J_c vs sintering temperature for Y123 wires prepared from the 800°C calcined powder is shown in Fig.5. This plot shows that optimum sintering temperature for a high J_c is around 940°C and the highest J_c value is 920 A/cm². This shift for the peak temperature towards

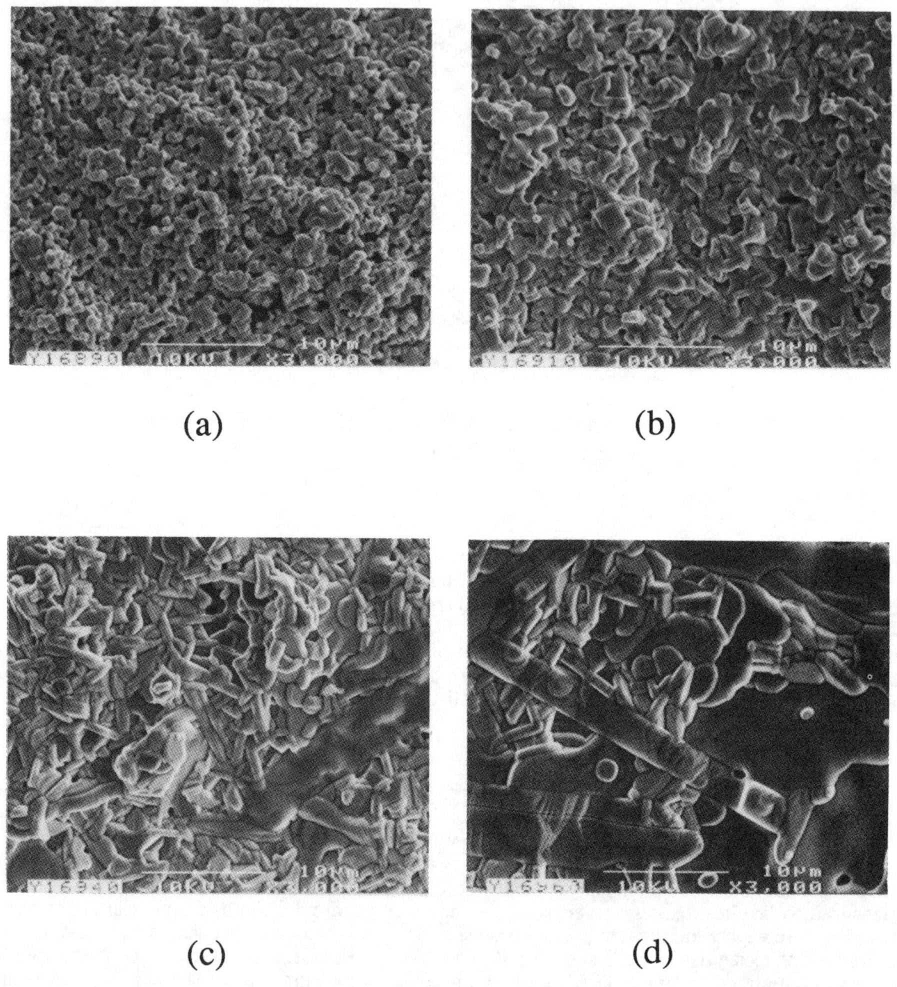

Figure 2   SEM microstructures of wires sintered at (a) 890°C, (b) 910°C, (c) 940°C and (d) 960°C. The wires were prepared from 500°C calcined powder.

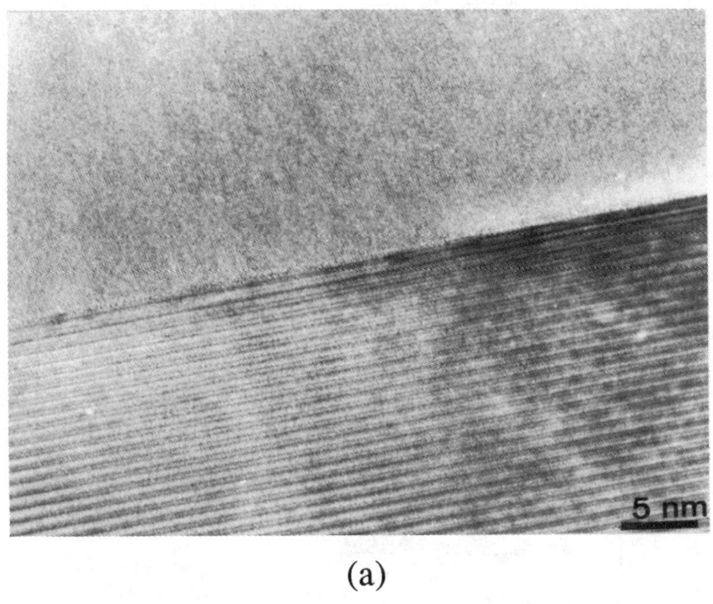

(a)

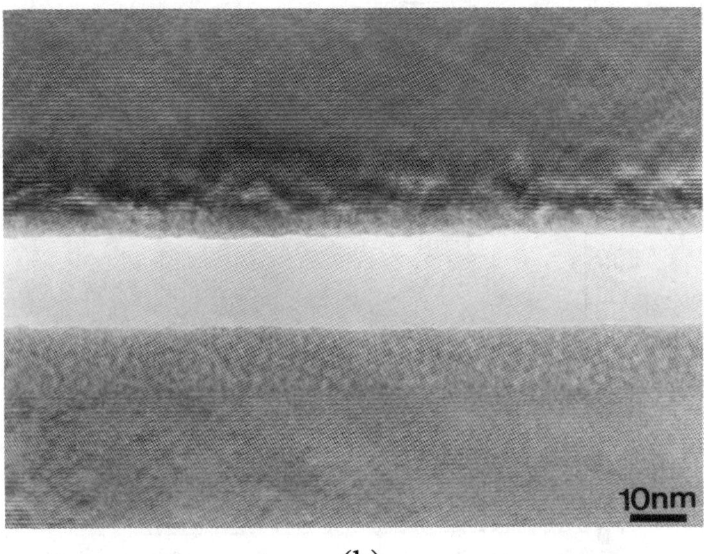

(b)

Figure 3   High resolution TEM images for wires sintered at (a) 910°C and (b) 960°C. The wires were prepared from 500°C calcined powder.

higher values compared to the data shown in Fig. 1 may be explained by the nature of the binder. Since HPMC is water based, the grains of 800°C calcined powder which are already Y123, are attacked by the water during mixing of the binder. Phase decomposition occurs at the surface of these grains as noted in a number of studies [5]. Thus, the surface of these grains must recover during a sintering step. This recovery can occur more rapidly at higher temperature and can improve grain boundaries but the higher temperature also initiates liquid formation, which suppresses the J_C.

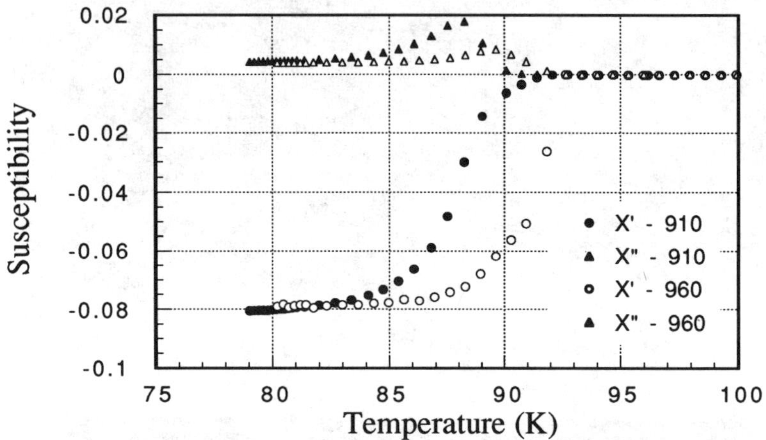

Figure 4 - Temperature dependence of a.c. magnetic susceptibility for the wires sintered at 910 and 960°C.

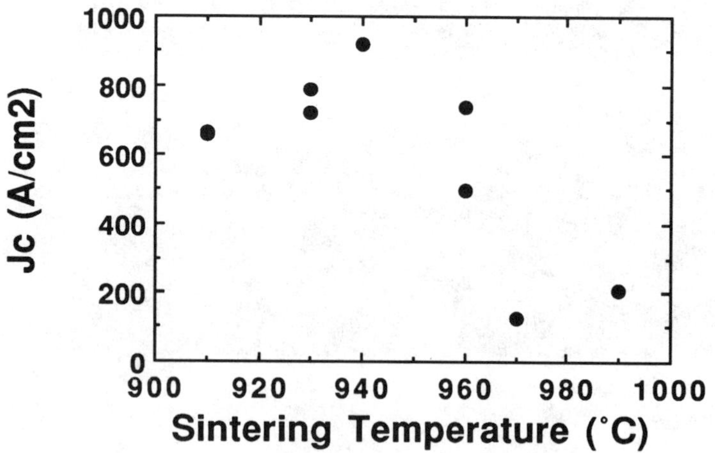

Figure 5 - Temperature dependence of Jc for the wires prepared from 800°C calcined powder.

800°C Calcined Powders (Y123) with Y211 powders and HPMC

To improve J_c of these wires, a Y211 phase, which is expected assist with pinning centers [2], was extruded with Y123. The 800°C calcined Y123 powders and fully converted Y211 powders were mixed with HPMC and a small amount of water and then extruded. The extruded wires were sintered at incremental temperatures between 930 and 1010°C for 10 hours followed by an oxygenation step. Though it is expected that Y211 phase would improve J_c, the highest J_c obtained was 200 A/cm^2 for the sample which contained 25wt% Y211 sintered at 990°C.

Figure 6 Schematic figure for the sintered wire. A-F indicate
the positions of electrodes.

J_c of Long Length Y123 Wires

A long length Y123 wire was extruded from the mixture of 500°C calcined precursor powder and binder. The green wires had a length of about 75cm. The wire was sintered at 890°C for 10 hours followed by an oxygenation step. A schematic figure of this final sintered long length of wire is shown in Figure 6. After sintering, the length of this wire reduced to about 60 cm. The positions of electrodes for the voltage measurements are also indicated in Figure 6. The J_c for the entire length of this wire (distance between the electrodes for potential difference measurement was 56cm) was 120 A/cm^2 while the J_c for a short length (1.3 mm) was 340 A/cm^2. Although the low J_c of 140 A/cm^2 for the section A-B is observed, the J_c's for the sections B-C, C-D and E-F which include a sharp curve are almost equivalent as shown in Table I. Essentially, no monotonic reduction of the J_c with variation wire length was observed.

Table I J_c vs distance for a long length wire. The alphabets, A - F,
indicate the electrode positions illustrated in figure 6.

	Distance (cm)	Ic (A)	J_c (A/cm^2)
A - F	56.0	0.6	120
A - B	10.2	0.7	140
C - D	12.5	1.5	300
E - F	10.8	1.4	280
B - C	3.2	1.3	260

A monotonic reduction of J_C may be expected when grain boundaries influence current deterioration i.e., the accumulation of weak link problems. Data from this long length of wire indicates that the low overall J_C value is due to the section A-B where large pores are visible. Therefore, if dense wires can be extruded without pores, it will be possible to manufacture long lengths of wires without significant deterioration of J_C's over the length of the wire.

Conclusions

Wires have been extruded from mixtures of calcined Y123 powder, 2 wt% of HPMC and water. The highest J_C obtained for sintered wires was 1100 A/cm^2. For wires extruded from 500°C calcined powders, a plot of J_C vs sintering temperature showed that an optimum sintering temperature for high J_C occurs between 900 and 920°C. This optimum temperature can be strongly correlated with microstructure. For example, when grain growth occurs, J_C decreases. On the other hand, similar analysis of wires prepared from 800°C calcined powders showed an optimum temperature of 940°C and a lower J_C value of 920 A/cm^2. This shift of optimum temperature and loss of J_C may be attributed to poor grain boundaries induced mixing of the starting materials. A long length of wire was also extruded in an "S"-shape and sintered. The J_C of this wire is strongly dependant on the presence of macro-scale defects such as pores. However, in regions of wire without macro-pores, there appears to be minimal reduction of J_C over long distances which include sharp curvature.

Acknowledgements

The authors would like to thank Professor Godon L. Dunlop of the Department of Mining and Metallurgical Engineering, The University of Queensland for helpful discussion. This work was supported by Syndicated R&D investment funds through a partnership between Bankers Trust (Australia) and Comquest Pty. Ltd.. These funds were awarded to I. D. R. Mackinnon, J.C. Barry and Professor G. L. Dunlop as chief investigators.

References

1. T. J. Doi, M. Okada, K. Higashiyama, T. Kamo and S. Matsuda, "Flux Pinning in High Tc Superconductors," Extended Abstracts, International Workshop on Superconductivity, Co-sponsored by ISTEC and MRS, (1992), 315-316.

2. M. Murakami, S. Gotoh, N. Koshizuka, S. Tanaka, T. Matsushita, S. Kambe and K. Kitazawa, "Critical Currents and Flux Creep in Melt Processed High Tc Oxide Superconductors," Cryogenics, 30 (1990), 390-396.

3. S. J. Golden, T. Yamashita, A. Bhargava, J. C. Barry, I. D. R. Mackinnon and D. Page, "The Extrusion of Continuous Lengths of Wire Using Hydroxypropylmethylcellulose," Physica C in press, (1993).

4. A.P. Malozemoff, L. Krusin-Elbaum and J. R. Clem, "Interlayer Coupling in High Temperature Superconductors," Physica C, 162-164 (1989), 353.

5. R. Zhao, M. J. Goringe, S. Myhara and P. S. Turner, "Transmission Electron Microscopy and High-resolution Transmission Electron Microscopy Studies of the Early Stages in the Degradation of YBa$_2$Cu$_3$O$_{7-\partial}$ Superconductor in Water Vapour," Phil. Mag., 66 (1992), 491-506.

CONSIDERATIONS IN THE PROCESSING OF LONG LENGTHS OF MELT TEXTURED

$YBa_2Cu_3O_{6+x}$ WIRES

by P. J. McGinn, C. Varanasi, D. Balkin
Center for Materials Science and Engineering
Dept. of Electrical Engr.
Univ. of Notre Dame
Notre Dame, IN 46556

Abstract

Processing of long lengths of textured $YBa_2Cu_3O_{7-x}$ wire is hindered by the slow rates required by the solidification process. Factors influencing the microstructural development of textured wires are discussed. Theories describing the rate limiting steps of the solidification process are reviewed, and potential means for increasing process speeds are discussed.

Processing of Long Lengths of Superconductors
Edited by U. Balachandran, E.W. Collings and A. Goyal
The Minerals, Metals & Materials Society, 1994

<u>Introduction</u>

One of the most serious obstacles to the practical application of $YBa_2Cu_3O_{6+x}$ (Y-123) bulk high temperature superconductors involves the weak link nature of most high angle grain boundaries. Although this problem has been minimized by using various melt texturing techniques on pellets and small lengths of wire, only limited success has been achieved in applying these methods to complicated shapes and longer lengths of wire. The main limitation of many of the melt texturing processes for fabricating long wires is due to the requirement that the entire specimen be heated above the peritectic temperature. The poor structural integrity of the partially melted Y-123 specimens require that they be supported by a substrate. Unfortunately, however, very few substrates are non-reactive with the barium rich liquid that forms when Y-123 decomposes, and this reactivity severely limits the extendability of the technology.

The zone melting process, however, requires neither the complete heating of an entire sample nor the presence of a supporting substrate to fabricate textured wires. Here, sintered wires are moved at controlled rates (which depend on the wire cross-section) relative to the localized hot zone of a vertical tube furnace which has a controlled thermal gradient. Although some significant successes have been achieved, the extendability of this technology to the texturing of long wires has been hindered by an inadequate understanding of the inter-related variables involved in the fabrication and zone-melt processing of Y-123 based materials. This work summarizes several significant aspects of melt texturing in general, and zone melting in particular, which affect the processing of long lengths of $YBa_2Cu_3O_{6+x}$.

<u>Melt Texturing Techniques</u>

While there are a number of variations of the texturing techniques, they all involve heating Y-123 above its peritectic decomposition temperature (1010^oC in air) and slow cooling to form aligned, generally large, grains of Y-123. Upon heating $YBa_2Cu_3O_{6+x}$ above 1010^oC, it will decompose to Y_2BaCuO_5 (211) and a liquid phase consisting of $BaCuO_2+CuO$. Upon slow cooling (eg. $1^oC/hr.$) from 1010 to approximately 900^oC the Y-123 will nucleate and grow.

A variety of solidification techniques have been reported, but they can be classified into essentially three types:

a) slow cooling without a temperature gradient

b) slow cooling in a defined temperature gradient without sample transport

c) slow cooling through a temperature gradient by sample transport

Some other variations exist, such as texturing by slow cooling in a magnetic field, but they will not be discussed here for the sake of brevity.

Growth of Y-123 single crystals in a flux generally produces thin platelets because of the more rapid growth in the a- and b-directions than in the c-direction of the crystal. This tendency can be used to produce regions of aligned grains (or "domains" of texture) by slow cooling through the peritectic, even in the absence of a temperature gradient. However, without a temperature gradient, domains tend to nucleate and grow at multiple sites throughout a sample. This approach was used by Salama, et al.[1] and has now been used by many others. Large textured regions can be produced by this technique. They can then be sectioned from a sample for characterization and testing. However, because of multiple nucleation sites, there will be high angle boundaries and perhaps $BaCuO_2$-CuO rich phases between domains. This limits the transport of current between domains, although the transport properties of individual domains are extremely good.

The use of a temperature gradient during slow cooling helps to minimize the possibility of multiple nucleation sites for domains along the length of the sample. The size of the imposed temperature gradient is important in determining the resulting microstructure, as is described below. Because no sample (or furnace) transport is involved, only samples of relatively modest dimension can be textured by this technique.

Further modifications to this technique have been made by introducing seed crystals. For example, a Sm-123 crystal, chosen because of its elevated melting point, can be positioned at the hot end of the sample. Its orientation can be selected to promote basal plane growth in the desired direction. Thus, for flux trapping in a cylinder for bearing applications, the crystal can be positioned so that the c-axis is parallel to the longitudinal axis of the cylinder.

The most versatile melt texturing processes for long lengths employs the motion of a sample relative to a temperature gradient to achieve texture. This is accomplished either by moving a heating source (eg. a focussed IR line heater) along a sample, or moving a sample past a heating source. The movement of the sample is preferred from a commercial standpoint, because it permits long samples to be processed in a semi-continuous fashion.

<u>The Horizontal Bridgman method</u>

This directional solidification process involves horizontally transporting a sintered YBCO sample through a pre-set temperature profile to texture relatively long sections of material. The use of a temperature gradient helps to eliminate random nucleation and growth of Y-123 grains throughout the length of the specimen and thereby facilitates the formation of a single, unified planar growth front. The horizontal furnace geometry used with this method, however, requires that the precursor material be supported on a substrate (which may react with the Y-123 (often YSZ is employed)). Although the substrate/reaction layer can contribute to random nucleation and growth, usually precursor cross-sectional dimensions are great enough to restrict interaction effects to a relatively small, localized volume. The primary advantage of the Bridgman method is that it permits the independent control of temperature gradient (G) and crystal growth rate (R) which facilitates the study of the effect of G/R and cooling gradient (GR) on the peritectic growth front morphology. This has been employed both with standard Y-123[2] and also with PMP processing,[3] utilizing 211 and the liquid phase as the precursors.

<u>The Floating Zone Melting Process</u>

This directional solidification process involves passing a vertically suspended sintered YBCO wire through a temperature gradient to produce a textured microstructure. The process requires that a sintered wire or bar be centered in and mechanically supported above a furnace containing a well defined hot zone region. Depending on the length of the wire support both from the top and bottom may be necessary. The wire is mounted above (and usually outside) the furnace to preclude any chemical interaction between the support and the the highly reactive molten precurser material. The process is unique in that only a small section of the wire is melted, and the molten zone is moved from one end of the wire to the other. The furnace hot zone (which defines the wire molten zone), has been produced by a variety of methods which include resistive elements,[4] halogen lamps, infrared lamps, Nd-YAG lasers[5] and CO_2 lasers[6] (Cima et al., 1992). The equipment can be constructed in several different ways to permit the relative movement of the wire through the hot zone, but the simplest and most mechanically stable geometry involves rigidly mounting the wire to a fixed support and moving the furnace up and over the sample to be textured. When support is only provided at the top of the wire as the hot zone moves up the wire length, the textured length (below the hot zone region) is supported by the surface tension of the partial melt. For a given wire diameter and processing rate, the surface tension of the partial melt is affected by the hot zone temperature, the furnace design (hot zone width and cooling gradient), the atmosphere, and the stoichiometry (i.e. 211 content) of the sintered wire. Together, these factors influence the extent of liquid phase segregation which occurs from the partial melt during zone melt processing. When the YBCO precursor material is heated into the 211 + Liquid region, some liquid migrates towards the cooler section of the wire (i.e. above the hot zone region), where it ultimately demixes and solidifies.

For a given furnace design, the primary processing variables involved in this technique are, again, the temperature gradient (G) and the growth rate (R). The temperature gradient can be adjusted by changing the hot zone design, the maximum hot zone temperature and by installing controllable cooling plates at the bottom (exit end) of the furnace. The growth rate is adjusted by changing the speed at which the furnace moves vertically upward. This processing technique offers the only viable means of fabricating long, well textured lengths of YBCO superconductor (free of substrate contamination) in a continuous or semi-continuous manner.

<u>Solidification Conditions</u>

The practical application of long YBCO wires critically depends on the ability to reproducibly form an aligned microstructure from a planar (or faceted, for Y-123) growth front. According to constitutional supercooling theory developed to describe the solidification of metals, during directional solidification a planar growth front becomes stable when the actual temperature gradient in the liquid at the interface (G_L) is larger than the liquidus gradient. This condition, which is represented by equation (1) results in the immediate dissolution and melting of any small protuberance in the growth front since T actual is greater than T liquidus at any point in the liquid (i.e. ahead of the growth front). The condition is

$$G_L/R > - (m_L C_o/D_L)((1-k_o)/k_o) \qquad (1)$$

where R is the growth rate (or pull rate), m_L is the slope of liquidus line, C_o is the alloy composition, D_L is the diffusion coefficient of the solute (211) in the liquid and k_o is the equilibrium distribution coefficicient. Conversely, if G/R is smaller than the limit defined by equation (1), the actual temperature becomes lower than the equilibrium liquidus temperature and a constitutional supercooling region forms in front of the planar interface. The presence of this region results in the preferential growth of plane front perturbations (discontinuous grain growth) and a denditic or celluar growth front. The product of G and R (GR) describes the cooling rate at the interface, which determines the scale of the microstructures formed. Maintaining G/R at a constant value, while progressively increasing G and R can be expected to lead to a refinement of the structure (ie. finer grains) without altering the morphology of the solidified structure.

However, Cima et al.[6] developed a Y-123 solidification model which suggests that the 211 particles dissolved in the liquid become supercooled close to the growing crystal interface to create the concentration gradient required to drive mass transport and thereby sustain planar growth. This supposition seemingly contradicts classical constitutional supercooling theory which predicts that the presence of a supercooled region will result in an unstable interface, with perturbations growing and leading to the formation of a cellular or dendritic growth front. However, they resolve the apparent contradiction by suggesting that the surface energy relationships associated with the highly faceted Y-123 plates permit interface stability with a limited amount of undercooling.   An example of such a faceted interface is shown in Figure 1, which shows the quenched growth front in a zone melted Y-123 wire. Consequently, it is proposed that the planar growth is limited by the undercooling that will cause the nucleation of new crystals in front of the interface. A quantitative development of this model yielded the following:

$$R_{max} = \left\{ D_L \, / \, l(c_{s\gamma} - c_{Lp}) \right\} \left\{ [(\Delta T_s)_{max} + Gl]/m_{L\gamma} \right\} \qquad (2)$$

which suggests that the maximum YBCO growth rate (R_{max}) at which a planar interface could be sustained is directly proportional to the diffusion coefficient (D_L) of the solute in the liquid and inversely proportional to diffusion distance to the interface (l), which is assumed as half the 211 particle separation (2l), where the Y211 particle size in the partial melt is l. $(\Delta T_s)_{max}$ is the maximum constitutional supercooling at a distance from the interface $x=l$, c_{Lp} is the peritectic liquid composition, $c_{s\gamma}$ is the 123 solid composition, and $m_{L\gamma}$ is the slpe of the equilibrium 123 liquidus. Cima et al also performed quench studies on comparable Y211 rich wires that were processed at different growth rates (R) through different thermal gradients (G) via a floating zone melt process. The results of the studies highlighted the fact that the growth front was very sensitive to processing speed and relatively *insensitive* to the thermal gradient. At low processing speeds, the growth front was planar. At increasing speeds, the growth front became cellular/dendritic and finally blocky in nature.

Izumi et al.[7] developed an alternate Y-123 solidification model, also based on the assumption that the rate controlling step in Y-123 growth was the diffusion of solute to the interface. However, they assumed that the concentration gradient in the liquid was due to the change in the chemical potential at the 211/liquid interface caused by the curvature of the 211 particles. The change in the chemical potential was related to the undercooling (ΔT_r) by the Gibbs-Thomson relationship:

$$\Delta T_r = 2\Gamma \, /r \qquad (3)$$

where Γ is the Gibbs-Thomson Coefficient ($\sigma/\Delta s$) where σ is the interface energy between the 211 and the liquid, Δs is the volumetric entropy of fusion, and r is the radius of the 211 particle. This relationship was used in conjunction with the Y-123 phase diagram to determine the resulting composition difference (i.e yttrium concentration gradient) ΔC between the liquid at the 211 particle and that at the Y-123 interface. This yielded the following:

$$\Delta C = C^r_L - C^\infty_L = 2\,\Gamma/rm_L \qquad (4)$$

where C^r_L is the rare earth concentration at the 211 particle, C^∞_L is the rare earth concentration in the liquid at the Y-123 interface, and m_L is the liquidus slope of the 211 phase. The model was based on a mass balance which equated the flux required for Y-123 growth with that provided from the 211 particles. Given the initial size distribution of 211 particles, the initial volume fraction of 211 phase and several thermo-physical constants (diffusivity, Gibbs-Thomson coefficient, maximum 211 decomposition rate, etc.), the balanced velocity for steady state directional solidification was estimated. It was found that the theoretical limit of the continuous growth rate (a few mm/hr) and the size distribution of 211 particles in the Y-123 crystal agreed well with experimental results obtained from samples prepared via the Bridgman method. Using the same experimental set-up to study stoichiometric 211 rich samples Izumi et al.[7] held the cooling rate (GR) constant and found an enhancement in magnetic J_c properties with increasing G/R values. They attributed their results to an associated improvement in Y-123 grain alignment.

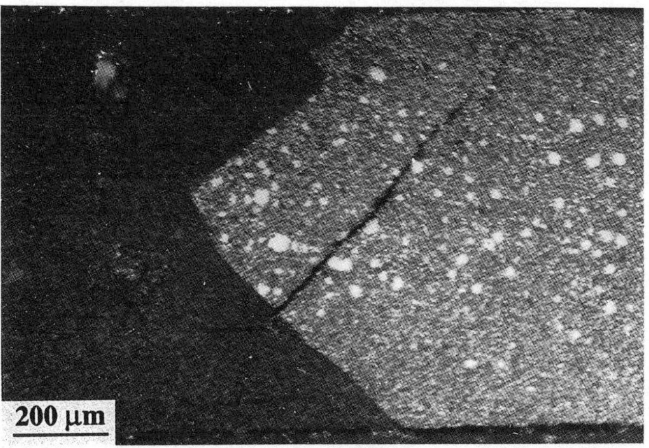

Figure 1: Faceted growth front in Y-123 wire quenched during zone melt texturing

Related to this model was the finding by Izumi, et al[7] that the final 211 particle size was independent of growth rate, R, which was attributed to only particles greater than a certain critical radius, r_c, being entrapped in the growth front. If the surface energy between a 211 particle (P) and the solid Y-123 (S) ,σ_{PS}, is greater than the sum of the surface energy σ_{PL} between the particle and the liquid(L) and the surface energy between the liquid and the Y-123, σ_{LS}, then the material in the fluid surrounding the particle will be transferred to the solidification front and the particle will be pushed ahead into the liquid. In fact, evidence of such particle pushing has been observed by Varanasi, et al.[8] Izumi, et al suggest that the small 211 particles which are pushed by the solidification front will coalesce with other small particles in the liquid and coarsen until a critical radius is reached at which point the particle is no longer pushed, but becomes entrapped in the growth front. This then yields the growth rate independent 211 size.

Control of Lateral Gradient

When texturing relatively thick sections in a gradient one of the concerns is whether or not the lateral temperature gradient in the sample is uniform. This aspect of texture processing was investigated by Selvamanickam, et al.[9] They used an auxialiary heater to maintain the outside of the sample hotter than the inside, with a lateral temperature gradient of -15 C/cm. This minimizes the growth of grains with non-axial alignment, permitting improved texture to be achieved in thick wires. Similar work by Meng et al [10] showed that by changing the ratio of the longitudinal gradient to the lateral gradient ((dT/dx)/(dT/dy) in their terminology), the grain orientation could be varied from being aligned along the wire axis to being off the wire axis. In thin samples this is less of a problem as geometrical constraints encourage growth parallel to the axis of the wire.

Liquid Flow in a Temperature Gradient

One complicating feature of processing YBCO in a thermal gradient that arises is demixing. Migration of liquid occurs in the thermal gradient, producing a concentration gradient along the texturing direction.[11] This liquid (BaCuO-rich) migrates down the temperature gradient and freezes, producing a "hump" in a wire. The temperature corresponding to the freezing point of the liquid has differed between reports in the literature. In zone melting, the liquid will initially migrate both up and down the wire. Thus the material in the hot zone will become deficient in Ba and Cu and will not yield stoichiometric YBCO on freezing. If zone melting proceeds by the movement of the furnace up the wire, the "hump" on the top end of the wire will continue to be pushed out ahead of the hot zone as the wire slowly moves into the hot zone. Thus a steady state condition will be achieved. However, the occurrence of this migration is important in a process such as zone melting because the Y-123 sintered "feed" material coming into the hot zone experiences this excess liquid phase in the "hump" prior to entering the hot zone. This hump can be 1-2 mm. in length, so that, assuming a 2 mm/hr travel speed, the sintered Y-123 will be in the liquid for up to one hour. This liquid will promote grain growth of the Y-123 at the least, and perhaps decomposition of the Y-123, depending on the composition of the liquid. If grain growth of the Y-123 results, for pure Y-123, the final textured microstructure will have larger 211 than a sample that experienced no grain growth.[12] If decomposition occurs, for example, by the reaction of $123 + CuO \rightarrow 211 + L$, then excess 211 will be produced in a liquid which can then begin to coarsen. In fact, a change in 211 size with distance along the wire has been observed.[13] This size change might also be due to Y enrichment along the wire, as observed by Meng, et al. They observed the Y content to increase along the wire during zone melting, which was attributed to the zone refining that occurs during solidification.[10] However, it is not clear if this is, in fact, the cause, as zone refining is usually associated with the transport of low melting point material along with the hot zone, not high melting point compounds such as 211. Research by Jiang et al has shown that the contact angle of the liquid phase on the Y-123 varies with atmosphere.[14] In their work O_2 pressures above 1 atm. resulted in decreased liquid flow away from the hot zone.

Microstructural Development

To further enhance the J_c of textured Y-123, it is necessary to provide additional flux pinning centers. Y_2BaCuO_5 (211) particles have been noted to lead to improvements in the critical current density as long as they are below a critical size.[15] This critical size has not been clearly identified, but it has been shown that large 211 precipitates tend to degrade Y-123 properties, while small 211 particles improve the properties.[15-17] Thus, it is of great importance to be able to control the size of 211 particles, which is to some extent a more difficult task in zone melting as the length of time above the peritectic is controlled by the width of the hot zone and the travel speed. To minimize coarsening it is desirable to have as narrow a hot zone as possible.

A number of different strategies have been employed in bulk Y-123 processing to produce fine 211. These include the following:

1) additions of Pt and PtO_2 have been found to lead to decreases in 211 size and also to change the morphology of the 211 into particles with a very high aspect ratio by reducing the coarsening rate by

decreasing the energy of the 211/liquid phase interface.[8,18-23] This has been widely used to reduce 211 size in bulk melt texturing techniques, but has not been reported for wire fabrication. In techniques such as QMG, the Pt additions can result from corrosion of the Pt crucible by the melt, whereas in other techniques discrete additions are required.

2) rapid heating of Y-123 during processing above the peritectic has been found to result in refined 211. The faster the Y-123 is heated above the peritectic (to 1050, for example) the more unstable the Y-123 becomes. This in turn, provides a greater free energy change with decomposition, promoting greater nucleation of 211, resulting in the production of finer 211.[24] Thus, in normal melt texture growth, a more rapid heating rate will yield finer 211 in the textured microstructure, assuming the cooling rate is kept constant.

3) the addition of heterogeneous nucleation sites, either in the form of excess 211, or other second phase additions, to act as nucleation sites for the peritectically produced 211, thereby decreasing the average 211 particle size.[25,26] 211 particles are very effective in this manner. Jin, et al added aerosol processed 211 particles that were 0.2 μm in size, and after texturing had a 0.7 μm 211 size in the Y-123 matrix.[27] Figure 2 shows a zone melted Y-123 wire with Er211 additions. The fact that most of the 211 particles in Figure 2 contain Er-rich cores (the bright centers) indicates that the added 211 acts as a heterogeneous nucleation site for the Y211 during Y-123 decomposition. The inability to resolve bright cores in all the 211 particles is due to polishing angle effects.

Other additions have been employed to reduce the 211, but it is not clear that any of them actually act as heterogeneous nucleation sites. Some promote 211 dissolution, resulting in finer 211 such as $BaSnO_3$, which reacts with the melt and ties up Y in a Y-Ba-Sn-O compound, resulting in refined 211, along with excess liquid phase.[28] CeO_2 additions appear to act similar to Pt in that they refine the 211 size, and change the particle morphology to being more acicular.[29] No clear evidence of any particle acting as a heterogeneous nucleation site has been shown (with the exception of 211), although many authors have offered it as an explanation for observations of 211 refinement.

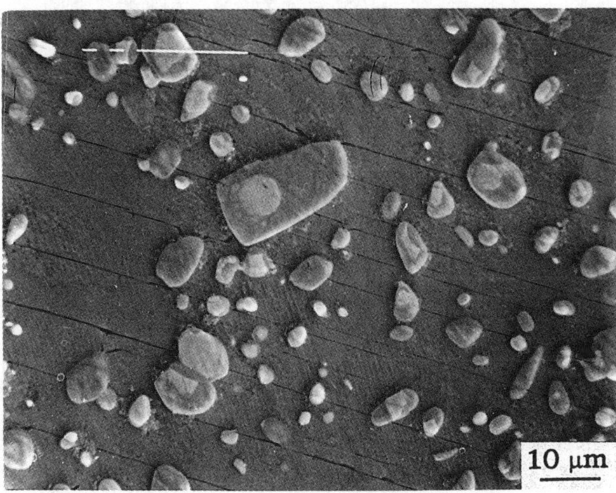

Figure 2: Melt textured Y-123 with Er-211 additions showing Er-211 cores in mixed rare earth 211 particles

4) the use of alternate precursors, such as beginning with 211 + liquid phase (PMP),[3] Y_2O_3 + liquid,

produced as a result of a quench process (QMG or MPMG)[30,31] or well dispersed Y_2O_3 + liquid by a solid state route (SLMG),[32,33] all of which result in the formation of refined 211 particles in the Y-123 matrix.

Since much of the melt texture processing relies on the use of sintered Y-123 as the precursor, research has also been performed to determine whether the grain size of the Y-123 affects the size of the 211 that forms during decomposition.[34] This work indicates that the nucleation and growth of 211, in the absence of any second phase additions which might act as heterogeneous nucleation sites, depends on the Y-123 grain size. However, this effect is expected to be less important when, for example, different heating rates are employed or there are significiant second phase additions. From the perspective of zone melting, the sharpness of the gradient on the entrance side of the hot zone, along with the travel speed, will control the heating rate. However, because the speed is a greater concern from the point of controlling the solidification, altering the speed to increase the heating rate is not a practical option. The only option is to select a furnace design or heating technique that will yield a sharp gradient. Just as for generating a sharp gradient on cooling, a laser source is probably the best option in this regard.

Wire Preparation

For producing the precursor wires to be used in melt texturing, several techniques have been employed. Both spinning techniques and extrusion have been reported. As long as the wire is relatively pore free the wire quality does not appear to be a major factor in limiting successful production of textured wire. Large pores can lead to liquid "pooling" and an inhomogeneous microstructure containing regions that are denuded of 211. Pore evolution can be controlled by proper binder burnout during sintering. Density gradients in the extruded wires can also cause problems by making the production of straight wires more difficult, but this can be minimized through constrained sintering (e.g. sintering under a slight applied load).

The addition of externally prepared Y211 powder to YBCO, (or the use of Y211-rich YBCO powders prepared from the oxides), greatly facilitates zone melt processing by providing more of the reactive "residual" solid phase in the partial melt. The presence of this excess 211 not only enhances the viscosity of the partial melt, but it also helps to effectively tie up the peritectic liquid which has a tendency to migrate and "de-mix" in a thermal gradient.

Future Directions

Although a desirable microstructure has been identified for producing high J_c values, the difficulty in practice is achieving highly aligned microstructures over extended distances. The longest reported distances over which "single domain" types of structures have been generated are on the order of one meter. Unfortunately, generating such structures so far has necessitated extremely slow travel speeds (1 - 3 mm/hr.) This is the greatest disadvantage of the directional solidification processes as currently practiced. However, it is hoped that this will not prove to be a limitation as understanding of the solidification process improves.

If melt texturing is to become a commercially significant process, means will have to be developed to speed up the process beyond the rates currently being employed. For example, some work has been reported involving attempts to solidify directly from the melt (completely liquid) to form Y-123, avoiding the peritectic reaction altogether, by combining a very high gradient with a rapid sample movement (20 cm/sec).[5] Although not successful so far, such developments will be necessary in order to texture long sections of wire.

In addition, it is clear that future work on texturing will also have to involve studies of metal interactions with molten Y-123, since the need to include metals for mechanical strength almost a foregone conclusion. To date most work on melt texturing has concentrated on Y-123 without metal additions, which results (if it is well textured) in an easily cleavable sample with low mechanical strength. For practical applications, greater strength through appropriate additions of metal will be required. The compatibility of such additions with melt texturing will have to be shown. Only preliminary work on the compatibility of a range of metals (other than silver) with molten Y-123 have been performed to date.[35] Some preliminary work has been performed on texturing of composite wires, with again, most of the work only considering silver additions. Moreover, these studies have tended to consider pellets rather than wires.   Silver additions do not alter the textured microstructure significantly, but do tend to prevent crack formation and propagation, and improve the mechanical strength of the Y-123. Ag does not appear to enhance the J_c in textured Y-123, as it does in sintered Y-123, but mixed results have been reported on whether it actually degrades J_c as compared to textured Y-123 with no Ag additions.

One consideration in using silver is that the melting point is below that of the peritectic of Y-123 in air. Two potential solutions to this difficulty have been offered. By alloying Ag with approximately 10% Pd, the melting point of the alloy is raised above the peritectic of the Y-123. At this doping level the amount of Pd is still low enough to avoid any substantial degradation of the Y-123, although there is some reaction of the melt with the Pd, leading to the formation of some intermetallic compounds. The other solution is to texture in a reduced pO_2 atmosphere to lower the peritectic temperature of the Y-123 below the silver melting point. This is presently the subject of several investigations.

Conclusions

Zone melting and Bridgman processing are both effective means of producing high critical current density Y-123 wires. The limitation of these processes for manufacturing use in producing textured Y-123 is the slow travel speeds that are necessary. Theories developed to describe the factors controlling the growth rate indicate that the rate limiting factor is Y diffusion to the solidification front. Processing techniques that can yield a very fine distribution of 211 in the liquid phase can be expected to allow somewhat faster processing rates.

Acknowledgements

The authors wish to acknowledge the support of this work by the Midwest Superconductivity Consortium (DOE Contract DE-FG02-90ER45427).

References

1   K. Salama, V. Selvamanickam, L. Gao, K. Sun, Appl. Phys. Lett., 54, 2352 (1989)

2   T.Izumi, Y. Shiohara, J. Mater.Res.,7 (1992) 16.

3   Z. Lian, Z. Pingxiang, J. Ping, W. Keguang, W. Jingrong, and W. Xiaozu, Supercond. Sci. Tech.,   3 (1990) 490

4   P. McGinn, W. Chen, N. Zhu, U. Balachandran, M. Lanagan, Physica C, 165 (1990) 480

5   S. Nagaya, M. Miyajima, I Hirabayashi, Y. Shiohara, S. Tanaka, IEEE Trans. Mag., 27 (1991) 1487.

6   M. Cima, M. Flemings, A. Figueredo, M. Nakade, H. Ishii, H.Brody, J. Haggerty, J. Appl. Phys., 72 (1992) 179

7   T. Izumi, , Y. Nakamura, and Y. Shiohara, J. Mater. Res., 7 (1992) 1621

8   C. Varanasi, P. J. McGinn, Physica C, 207 (1993) 79

9   V. Selvamanickam, C. Partsinevelos, A. McGuire, K. Salama, Appl. Phys. Lett. 60 (1992) 3313

10   R. L. Meng, Y. Sun, P. Hor, C.W. Chu, Physica C 179 (1991) 149

11   D. Dube, B. Arsenault, C. Gelinas, P. Lambert, Matls. Lett. 15 (1992) 1

12   C. Varanasi. P. McGinn, unpublished

13   D. Balkin, unpublished

14   X. P. Jiang, M. Cima, H. Brody, J. Haggerty, M. Flemings, Proc. of Intl. Workshop on Supercond. (MRS, Pittsburgh, 1992) 259

15   D.F. Lee, V. Selvamanickam, K. Salama, Physica C 202 (1992) 83

16   P. McGinn, N. Zhu, W. Chen, S. Sengupta, T. Li, Physica C, 176 (1991) 203

17   M. Murakami, S. Gotoh, N. Koshizuka, S. Tanaka, T. Matsushita, S. Kambe, K Kitazawa, Cryogenics 30 (1990) 390

18   N. Ogawa, I. Hirabayashi, S. Tanaka, Physica C, 177 (1991) 101

19   N. Ogawa, M. Yoshida, I. Hirabayashi, S. Tanaka, Supercond. Sci. Technol. 5 (1992) S89

20   S. Gauss, S. Elschner, H. Bestgen, Cryogenics, 32 (1992) 965

21   M. Morita, M. Tanaka, S. Takebayashi, K. Kimura, K. Miyamoto, K. Sawano,   Jap. J. Appl. Phys. 30 (1991) L813

22   C. Varanasi, P. J. McGinn, V. Pavate, E. Kvam, submitted to Physica C

23   T. Izumi, Y. Nakamura, T.H. Sung, Y. Shiohara, J. Mater. Res. 7 (1992) 801

24   J. Rignalda, X. Yao, D.G. McCartney, C.J. Kiely, G.J. Tatlock, Materials Letters, 13 (1992) 357

25   C. Varanasi, D. Balkin, P. McGinn Applied Superconductivity, 1 (1993) 71-80

26   D. Balkin, C. Varanasi, P. McGinn, in *Layered Superconductors: Fabrication,   Properties and Applications*, ed. by D.T. Shaw, C.C. Tsuei, T.R. Schneider, Y. Shiohara, (Materials Research Society Proceedings, Vol. 275, 1992) p. 207

27   S. Jin, G. Kammlott, T. Tiefel, T. Kodas, T. Ward, D. Kroeger, Physica C, 181 (1991)57

28   C. Varanasi, P. McGinn, submitted Supercond. Sci. Tech

29   C.J. Kim, et al , J. Mater. Res., in press

30   M. Murakami, M. Morita, N. Koyama, Jap. J. Appl. Phys., 28 (1989) L 1125

31   M. Murakami, M. Morita, K. Doi, K. Miyamoto, Jap. J. Appl. Phys., 28 (1989) 1189

32   C. Varanasi, S. Sengupta, P. J. McGinn, D. Shi, Applied Superconductivity, in press

33   D. Shi, S. Sengupta, J.S. Lou, C. Varanasi, P. McGinn, Physica C 213 (1993) 179

34   C. Varanasi, P.J. McGinn, Materials Letters, 17 (1993) 205

35   G. Kammlott, T. Tiefel, S. Jin, Appl. Phys. Lett., 56 (1990) 2459

Effect of mechanical deformation on the evolution of c-axis texture of

$Bi_{1.6}Pb_{0.4}Sr_2Ca_2Cu_3O_z$ superconductor HIP cladded on Ag substrate

J. M. Yoo and K. Mukherjee

Department of Materials Science and Mechanics
Composite Materials and Structure Center
Michigan State University, East Lansing, MI 48824

Abstract

The c-axis texture evolution of 2223 BSCCO superconductor has been studied by x-ray polefigure method, using "popLA" software package, which can elucidate the evolution of texture associated with mechanical processing of BSCCO compound. The specimens were fabricated on Ag substrate by a HIP cladding technique to induce randomly oriented state as the starting material, and to investigate the role of non (001) grain during deformation. As the amount of cold rolling reduction (%) increased, a tighter clustering of the (00$\underline{14}$) poles around the surface normal, indicated that randomly oriented c-axes' grains from initial HIP cladded surface rotated towards the normal direction. In addition (105), (109), and (110) experimental polefigure studies indicated that under deformation, those plane normals tend to form a fibre texture around the compression direction of rolling (normal direction). After achieving certain degrees of texturing, saturation due to textural hardening was also observed. Detailed texture analysis, along with inverse polefigure, and microstructure analysis are reported.

Processing of Long Lengths of Superconductors
Edited by U. Balachandran, E.W. Collings and A. Goyal
The Minerals, Metals & Materials Society, 1994

I. Introduction

For practical engineering applications, the superconducting compounds should have sufficient current carrying capability and should be fabricated into desired shapes such as wires and tapes for high current and high field applications [1-3]. The fabrication of tapes or wires from these ceramics is an extremely challenging task, since the processing must lead to chemically and mechanically optimised microstructure if these tapes are to support large current densities in high magnetic fields.

The need for a strong microstructural control is due to the intrinsic anisotropic properties of these materials. For example, they tend to be strongly superconducting in the basal (001) plane [4]. Microcracking tend to occur preferentially on the (001) plane, as well as on the (100) and (010) planes due to the weak van der Waals bonding between adjacent BiO layers [5]. Prefered orientations, *i.e.* "texture" appears to be a key parameter in enhancing critical current density (J_c). The overall implications of a second phase component are unknown, but it seems probable that they induce damage in the form of microcracks during deformation as well as affect the development of texture. Therefore, it is necessary that the processing must achieve : i) a crystallographic texture characterized by a high degree of alignment of superconducting crystal planes lying parallel to the conducting direction, ii) a high degree of densification, and iii) a minimal volume % of the second phase.

Our research is focused on Bi-Sr-Ca-Cu-O system, with a major emphasis on the $Bi_{1.6}Pb_{0.4}Sr_2Ca_2Cu_3O_Z$ compound, because of its high critical temperature (T_c). For these compounds, there are number of experimental evidence which show that a strong crystallographic texture is essential to minimize weak links and achieve a high critical current density. N. Enomoto et al. [6] found that the critical current density increased sharply as the degree of c-axis texture of their tapes increased. A similar result has been reported for 2223 (BSCCO) tapes by Jin et al. [7]. The c-plane grain boundaries which have large misorientations are also reported to decrease critical current density [8].

An important, and as yet little understood area, is the relationship between thermomechanical treatment and grain alignment in this class of superconductor materials. The exact mechanism for texture formation in the silver-clad BSCCO tape, during the rolling process, is not clearly described in the literature. Consequently, a systematic study has been undertaken by us to investigate the effect of mechanical deformation and annealing on the c-axis texture of $Bi_{1.6}Pb_{0.4}Sr_2Ca_2Cu_3O_Z$/Ag composite compacted by a HIP-cladding technique. An understanding of these effects would be useful in controlling the microstructure that is essential to achieve high critical current density (J_c). Our objective of this research is to develop a basic understanding of their mechanical behavior, and to utilize this understanding to develop synthesis and processing methods (rolling, hot extrusion and pressing) including prediction of the microstructures that result from processing.

Using a conventional process it is very difficult to clad high T_c BSCCO superconductor on Ag substarte (due to a very narrow sintering temperature range [9]), without decomposition of the high T_c phase. In order to obtain a good BSCCO/Ag interface bonding, and high density BSCCO layers prior to rolling, we used hot isostatic press (HIP) technique to clad BSCCO superconductor on Ag. In addition our previous results [10] show good stability of BSCCO superconductor with respect to oxygen stoichiometry during HIP processing.

II. Experimental

High purity(>99.9%) oxides or carbonates of Bi, Pb, Sr, Ca, and Cu, with the cation ratio of 1.6:0.4:2.0:2.0:3.0, are carefully mixed, calcined, and reground. The high T_c phase powder thus prepared, was first compacted into a cylindrical shape by uniaxial compression. The next step was HIP consolidation. Multi-layer stacks consisting of alternate layers of 2223 (BSCCO) and silver, were coated with boron nitride and encapsulated in a pyrex tube under vacuum, prior to HIP consolidation.

As shown in the schematic diagram in Figure 1, the sealed samples were HIP processed in an Ar atmosphere at a temperature of 850 $^{\circ}$C and a pressure of 138 MPa for a duration of 3 hours to obtain high density BSCCO layers. These samples also exhibited good BSCCO/Ag interface bonding strength prior to rolling. The Hipped samples were cooled slowly at a rate of 100 $^{\circ}$C/h. Preforms for the cold rolling process were cut from these multi-layer stacks.

The phase identification and crystal structure determination were performed using a Scintag-XDS-200 x-ray diffractometer, equipped with a computerized data collection system. The texture analysis was also performed by using a Scintag polefigure goniometer, equipped with popLA software package. The Preferred Orientation Package-Los Alamos (popLA) is a comprehensive coherent menu-driven set of utility programs that is independent of the texture measurement hardware [11].

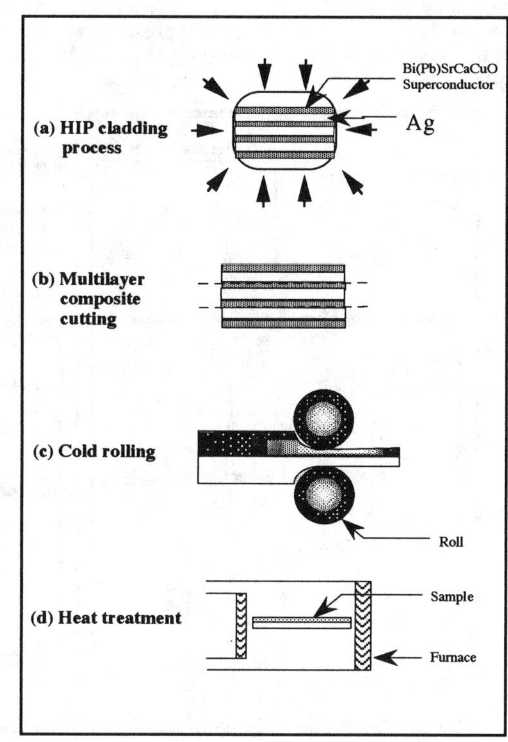

Fig. 1   Schematic illustration of processing steps for BSCCO/Ag superconductor tape

For microstructural and compositional studies, polished longitudinal cross-section samples were etched for 20-25 second with 1 part 65% perchloric acid mixed with 99 parts 2-butoxy-ethanol to expose the grain structure, with special emphasis on revealing interior alignment.
Fracture surfaces, produced by fracturing the tapes parallel to the rolling direction in the thin-longitudinal cross-section, were also studied. Both etched and fracture surface of longitudinal cross-section were examined by using a Hitachi S-2500C scanning electron microscope (SEM) equipped with a Link energy dispersive spectrometer (EDS).

III. Results and discussion

A. X-ray θ-2θ diffraction

Figure 2. shows the measured X-ray diffraction data for a successively cold rolled sample as a function of cold rolling thickness reduction(%). A thickness reduction ratio R(%) was used to describe the deformation extent, as:
$$R = (t_1 - t_2)/t_1$$

where t_1 and t_2 are the thickness of the BSCCO/Ag before and after deformation. After the HIP cladding process, the surface X-ray diffraction contains numerous peaks such as (105), (109), (110), and (10$\underline{11}$) along with the (00l)peak. The (00$\underline{10}$), (00$\underline{14}$) peak intensities increased while (105), (110), (109), and (10$\underline{11}$) peak intensities decreased after 20 % reduction. After 95 % reduction, the (105) peak intensity almost disappeared, and a very little trace of (109), (10$\underline{11}$) remained. There is, however, a slight peak broadening, which means smaller particle size resulting from severe mechanical deformation, which is often found for low temperature and high pressure synthesis [12]. This comparison of the X-ray data reveals that the (00<u>10</u>) and (00<u>14</u>) peak intensities increased with the amount of cold rolling reduction.

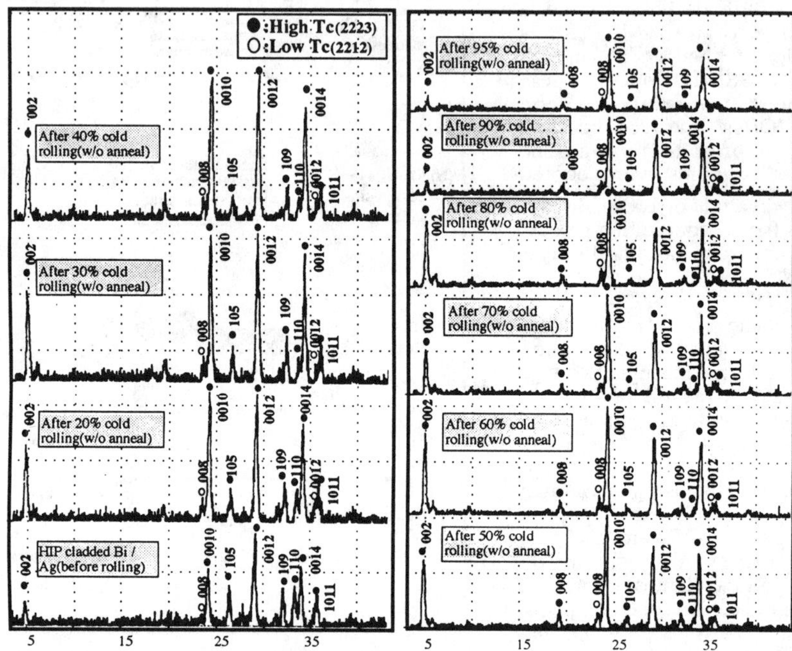

Fig. 2 The measured X-ray diffraction data of successive cold rolled sample with respect to cold rolling reduction(%)

Lotgering introduced a simple method to quantify the degree of c-axis orientation in polycrystalline ceramics. The ratio of the intensities of the (00l) to the (hkl) reflections increases with the improving c-axis orientation, and this fact can be used to quantify the extent of texturing. A factor F, called the Lotgering factor, is defined as :

$$F = (P - P_0)/(1 - P_0), \quad \text{where } P = \sum_l I(00l) \Big/ \sum_{hkl} I(hkl)$$

where I refers to the X-ray peak intensity [13]. P is the sum of integrated intensities for all (00l) reflections divided by the sum of all intensities of (hkl) in the textured specimen. P_0 is an equivalent parameter for a random specimen.

In the present experiments our experimental results from ground powders of the BSCCO 2223 phase gave $P_0 = 0.29$. Figure 3 shows a plot of Lotgering factor F versus deformation extent R. The four points in the plot, with rapidly increasing F, correspond to specimens after 20, 30, 40, and

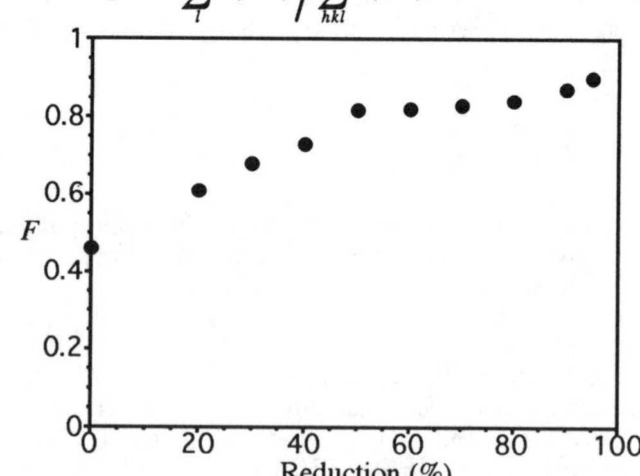

Fig. 3. A plot of Lotgering factor vs. deformation extent R(%), calculated from fig. 2.

134

50% R of cold rolling deformation. It is observed that the F factor reached 0.83 when a sample was 50 % cold rolled, increasing from an initial F value of 0.465, and steadily increased with respect to further rolling. After 95 % reduction, the F factor was 0.901.

During the superconducting tape processing, the annealing steps help to heal the fractured particles which result from cold rolling deformation. In order to study the effect of annealing temperature on the c-axis texture, cold rolled 2223 BSCCO/Ag tape was annealed at various temperatures as shown in fig. 4. It can be observed that the peak intensities, corresponding to the high T_c phase, are maintained up to an annealing temperature of 850 °C, and the strong (00l) reflection is dominant. As the annealing temperature increased to 860 °C, the high T_c(2223) phase abruptly decomposed into the low T_c(2212) phase. Normally, the high T_c(2223) phase is reported to decompose around 870 °C [14].

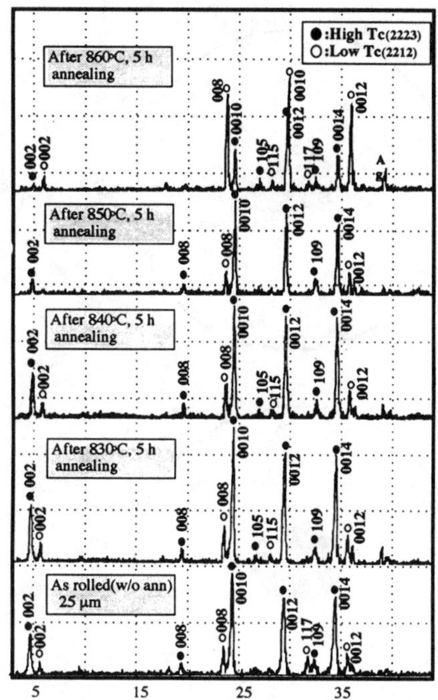

Fig. 4 The comparisonal X-ray diffraction data BSCCO/Ag tapes are annealed at 830, 840, 850, and 860oC for 5 H

This change is presumably due to the lowering of the superconductor melting point by silver [15].

<u>B. X-ray polefigure and texture analyses</u>

<u>1. Experimental polefigure</u>   X-ray diffraction studies, however, do not provide enough information about the preferential orientation, *ie* "texture" and the orientational perfection of the crystal with respect to a certain direction. Such properties are better determined by X-ray pole figure analysis, since the X-ray pole figure goniometers are designed to measure the diffracted intensity from a sample as it is tilted and rotated to different orientations with respect to the X-ray beam. In brief, a polefigure is a map of the statistical distribution of the atomic plane normals of a polycrystalline sample, and therefore, it provides a complete picture of the texture of the sample.

Experimental (0014) pole figures were measured from the initial HIP cladded surface to different amounts of cold rolled surface of samples as shown in fig. 5. As the amount of cold rolling reduction increased, a tighter clustering of the (0014) poles indicated that randomly oriented c axis of grains, from the initial HIP cladded surface, rotate nearly parallel to the compression direction of rolling as shown in fig. 5 (a)-(h). This evolution of c-axis texture during deformation is also confirmed by the distribution of (109) poles as shown in fig. 6 (a)-(h). At low deformation (~15-40% R), it appears that a weak fibre-like texture is beginning to appear at this stage of deformation as shown in fig. 6 (a)-(d). The angle between the compression direction of rolling (normal direction) and the fiber is the angle between the c axis and the (109) plane normals of the orthorhombic unit cell. The observed tilt angle of (109) fiber texture from normal direction is 45 to 50 degrees, which is in good agreement with the theoretical calculation of 47.16 degrees. From the group of (109) polefigure, it exhibited a basal texture of the (00l)[hk0] type that we could distinguish as orientations of the (001)[010], (001)[110], and (001)[210] types. The group of the experimental (0014) and (109) polefigures show that under deformation, the c-plane

135

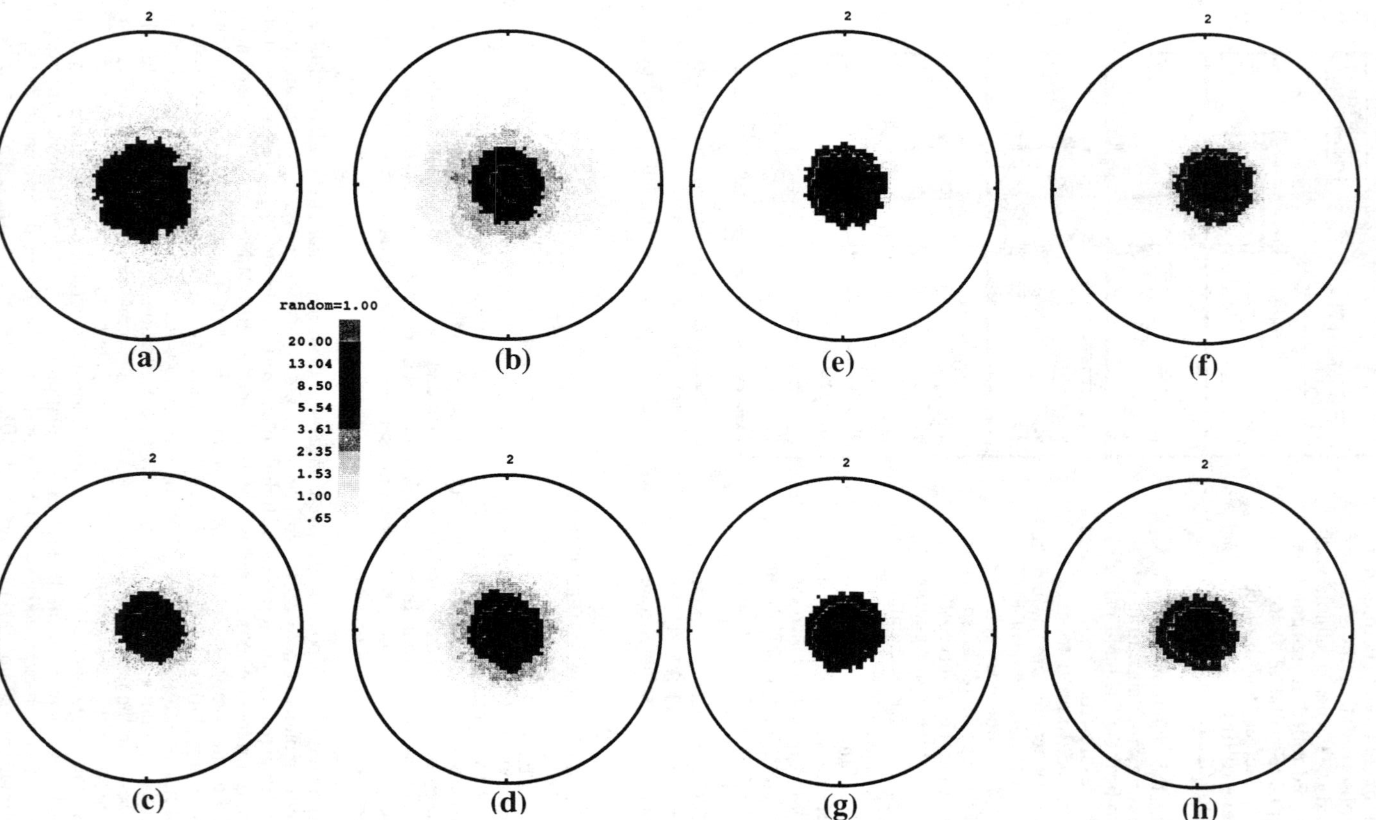

Fig. 5. Measured (00$\underline{14}$) pole figures from successively rolled and annealed sample. a). HIP cladded sample, b). 15%, c). 30%, d). 40%, e). 50%, f). 70%, g). 90%, and h). 98% cold rolled sample.

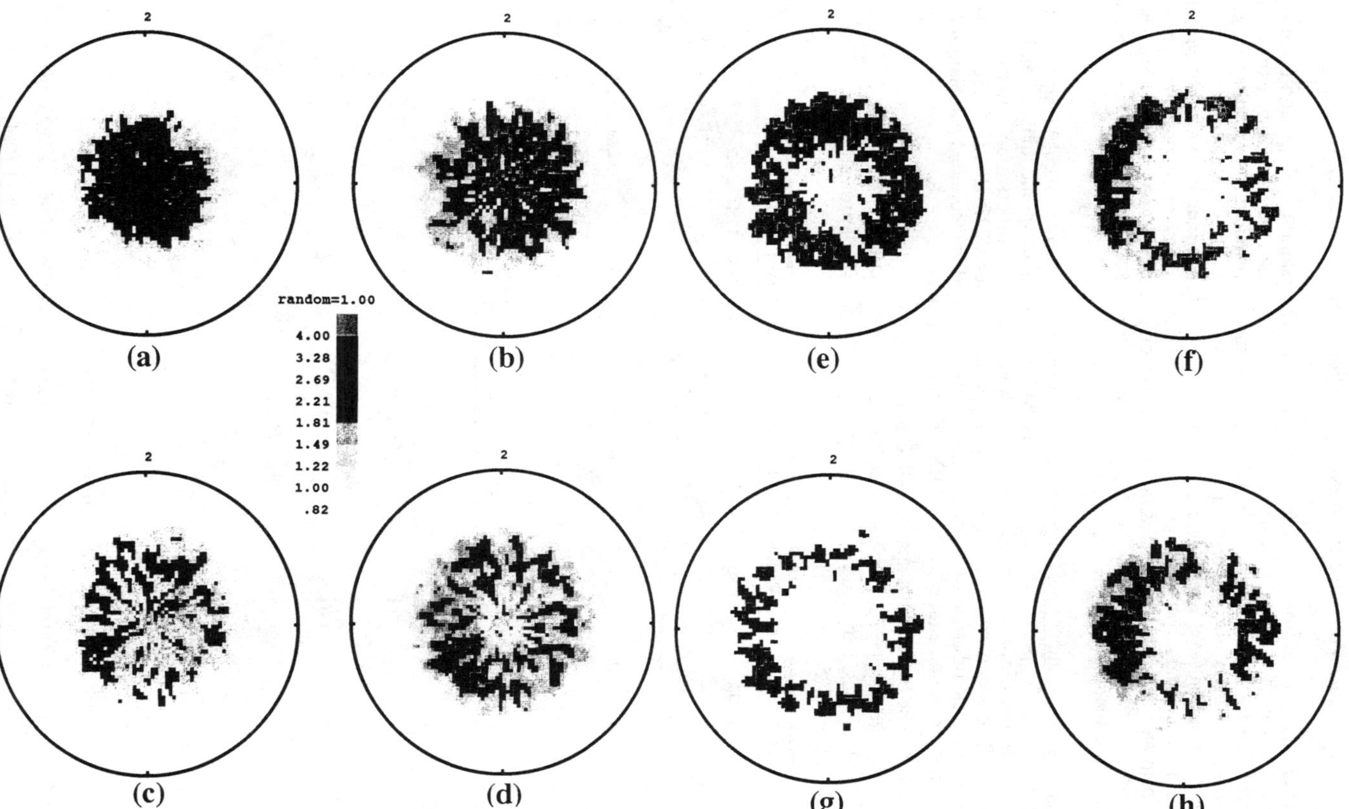

Fig. 6. Measured (109) pole figures from successively rolled and annealed sample. a). HIP cladded sample, b). 15%, c). 30%, d). 40%, e). 50%, f). 70%, g). 90%, and h). 98% cold rolled sample.

normals rotate toward the compression direction of rolling and the (109) plane normals tend to form a fibre texture around the compression direction (normal direction).

After achieving a certain degree of texturing (at ~50% R), saturation of the evolution of deformation texture was observed with respect to further deformation. This can be explained by the fact that the number of non (00l) oriented crystal, responsible for accommodating deformation, rotate toward (00l) orientations that do not support continued deformation. For a higher deformation (~98% R), some of the c-plane normals continue to rotate, but with a spread of these normals toward the constrained direction (transverse direction). The spread of the c-plane nornals is also indicated in the distribution of the (109) poles (fig. 6 (h)). The strong fibre-like distribution of the (109) poles that is observed at ~70% R is degraded with increasing deformation (~98% R).

<u>2. c-axis-oriented grains: [001] projection</u> From the experimental (105), (109), (00<u>14</u>) polefigures, using "popLA" software package, which employs harmonic analysis and the WIMV (Williams-Imhoff-Matties-Vinell) method [11], the inverse polefigure was calculated.   The standard (001) stereographic projection for 2223 BSCCO (fig. 7) is constructed with the information given in table 1 which shows the theoretical calculation of stereographic projection angles for 2223 BSCCO superconductor.

Table 1. The calculated d-spacing value and stereographic projection angles of the Bi(Pb)SrCaCuO 2223 phase.(The latitudinal angles Φ [001], Φ [010], and Φ [100] are designated as those for the [001], [010], and [100] projections. a = 3.818, b = 3.825, c = 37.070 Å)

hkl	d-spacing	Φ [001]	Φ [010]	Φ [100]	hkl	d-spacing	Φ [001]	Φ [010]	Φ [100]
002	18.53		90	90	017	3.10	54.16	35.84	90
004	9.27		90	90	107	3.10	54.21	90	35.79
006	6.18		90	90	00<u>12</u>	3.09		90	90
008	4.63		90	90	019	2.80	47.09	42.89	90
00<u>10</u>	3.71		90	90	109	2.80	47.16	90	42.83
013	3.65	72.78	17.22	90	110	2.70	90	45.07	44.93
103	3.64	72.82	90	17.17	112	2.67	81.69	45.67	45.52
015	3.39	62.68	27.31	90	00<u>14</u>	2.65		90	90
105	3.39	62.74	90	27.25					

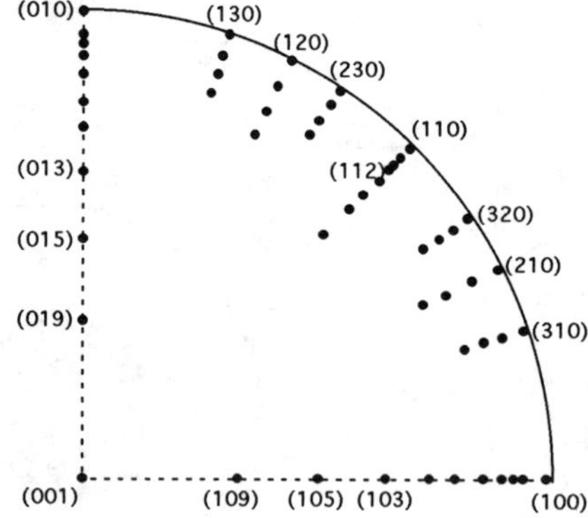

Fig. 7. Calculated [001]stereographic projection of the Bi(Pb)SrCaCuO 2223 phase

Our preliminary inverse pole figure data(fig. 8) reveals that: i) scattered (00l) grains are oriented nearly perpendicular to the plane of the tape, ii) (019), (109), (015), and (105) grain

138

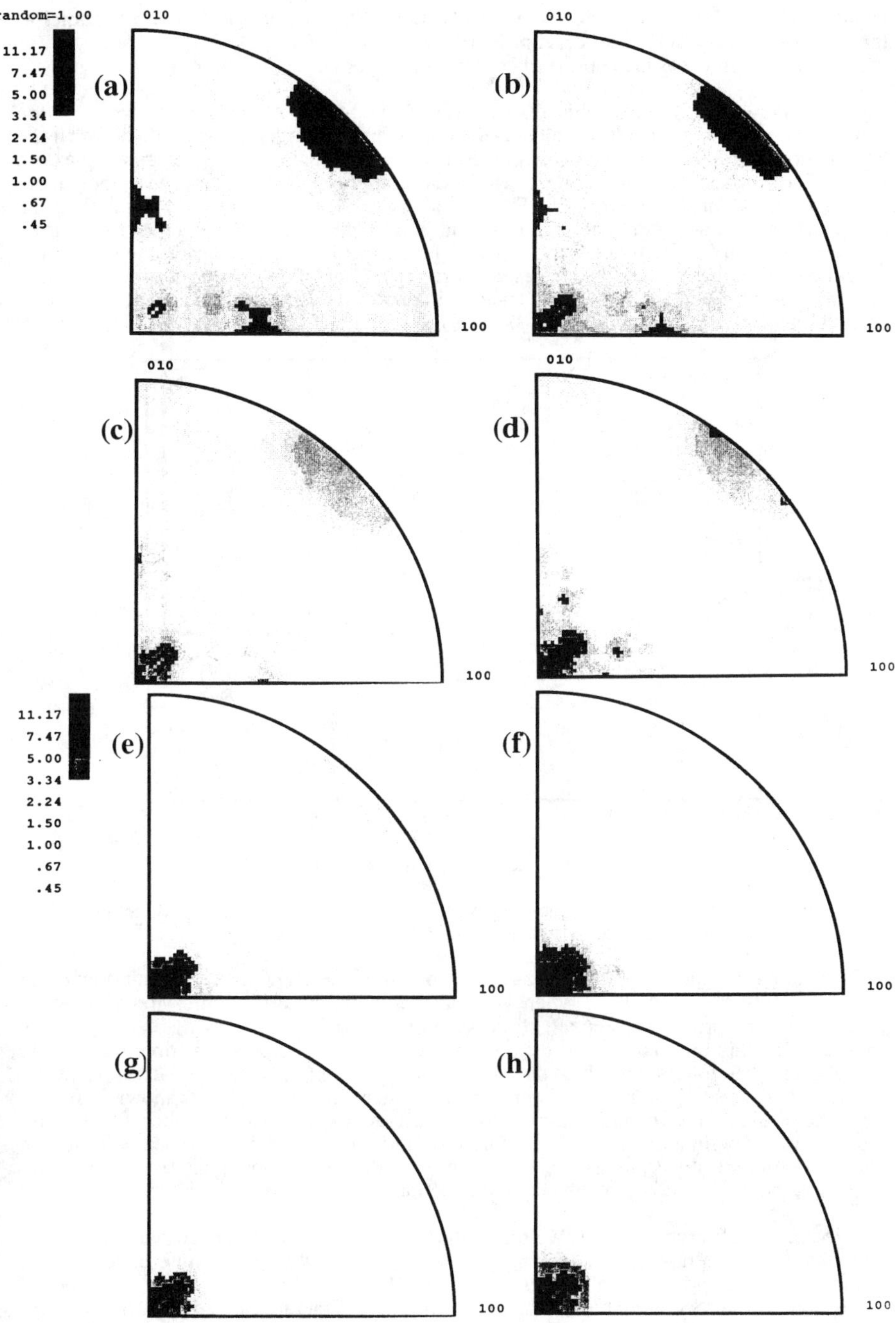

Fig. 8. Inverse pole figures for the normal direction calculated from the SOD of the samples after a). HIP cladded sample, b). 15%, c). 30%, d). 40%, e). 50%, f). 70%, g). 90%, and h). 98% cold rolled sample.

could rotate towards compression direction of rolling (normal direction) that is parallel to the c-axis, iii) after achieving certain degree of texturing (at ~50% R), the saturation of deformation texture was also observed with respect to further deformation.

Plots are shown in fig. 9. for the distribution of the c-plane normals with respect to the compression direction of rolling calculated from inverse polefigure. The angle between the c-plane normals and the compression direction decreases with increasing deformation extent R(%). The c-plane normals rotate towards the compression direction under increasing deformation extent R(%). At ~50% R, most of the c axes become nearly parallel to the compression direction with a negligible spread in other direction. However, as the amount of non (00l) grain such as (109), (105), (112), and (110) that is responsible for accommodating deformation is decreased, increase in orientation density of c-plane normal become restricted and saturated. This result is in good agreement with the result of F factor calculations and measurement of (00<u>14</u>), (109) experimental polefigures.

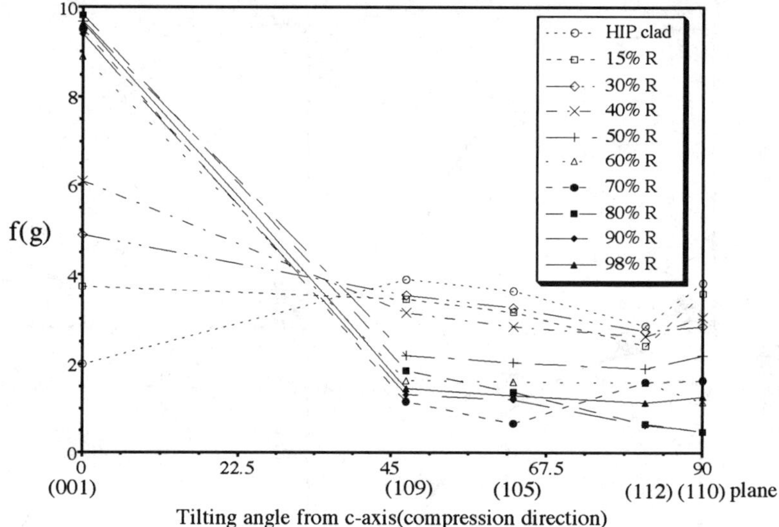

Fig. 9. Plots for orientation density f(g) vs. tilting angle from compression direction, calculated from the inverse polefigure. Random = 1.0 in f(g).

3. <u>Textural hardening</u> Fig. 10 (a). shows how these non (00l) grains play an important role in accomodating deformation, when we consider plane strain compression is used to approximate the primary stress and strain state of rolling. After achieving a strong c-axis texture, with this approximation, c-axis is parallel to the compression direction for the crystals, and therefore, the resolved shear stress on the basal (c-planes) and lateral planes (a, b-planes) vanishes and no more deformation can be accommodated [16] as shown in fig. 10 (b). The c-axis texture aggregate becomes elastically stiff with respect to further compression of rolling due to the lack of non (00l) grains to support continued deformation (the predominant slip systems are no longer oriented to support slip), so the textural hardening induces a locking of the textured material at high strains.

This locking is observed in Figs. 10 (b). by the rapid increase of the equivalent stress level [16]. These explain how plastic locking occurs as a result of strong texture development; this results in a rapid increase in stress level since further deformation along the imposed strain path occurs by essentially elastic deformation. Fracture, as a result of the high stresses, is inevitable.

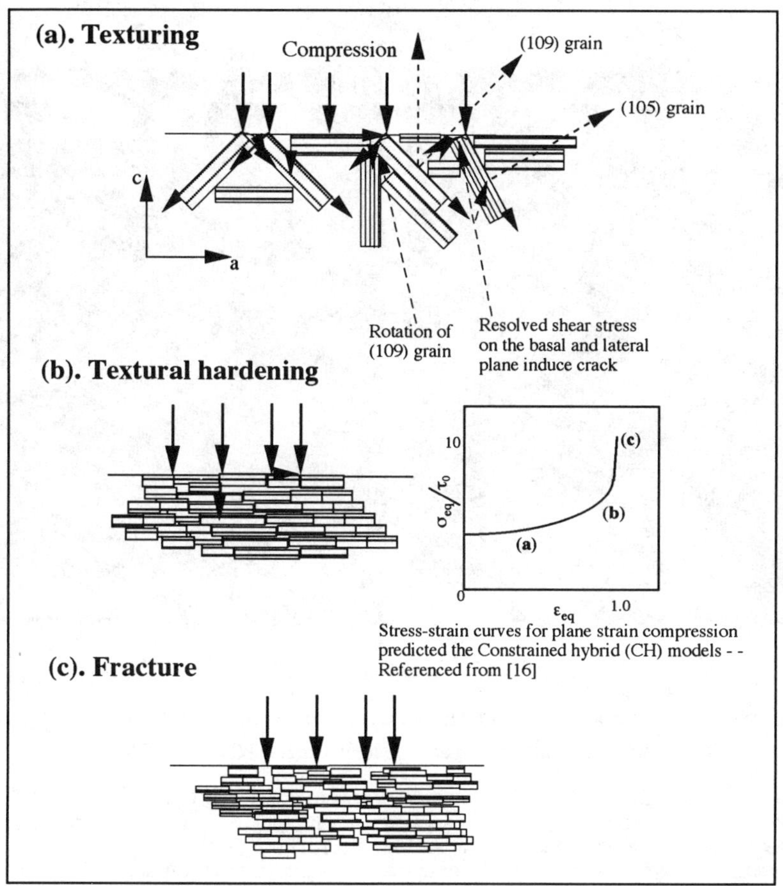

Fig. 10. Schematic illustration of the development of texture, textural hardening, and fracture.

C. Microstructure analysis

We investigated fracture surfaces that were produced by fracturing the tapes parallel to the rolling direction in the thin longitudinal cross section. Fig. 11. compares SEM micrographs of a longitudinal cross section of a cold rolled sample and an annealed sample. Well aligned grains along the rolling direction are reasonably clear. The 2223 BSCCO grains are small and fragmented, which is in agreement with the results of the X-ray data after cold rolling. The fracture morphology of the annealed tape suggests a plate-like grain morphology.

A more representative method for examining the c-axis texture along the rolling direction is to prepare polished sections. Until recently such sections have not been very informative, because the BSCCO grain structure (unlike the twinned microstructure of $YBa_2Cu_3O_X$ superconductor) is not cleary evident on a polished surface. However, it can be effectively revealed by the perchloric acid etch, although it should be noted that excessive etching results in destruction of 2223 BSCCO structure. Polished longitudinal cross-section samples were etched with 1 part 65% perchloric acid mixed with 99 parts 2-butoxy-ethanol to expose the grain structure, with special emphasis on revealing interior alignment.

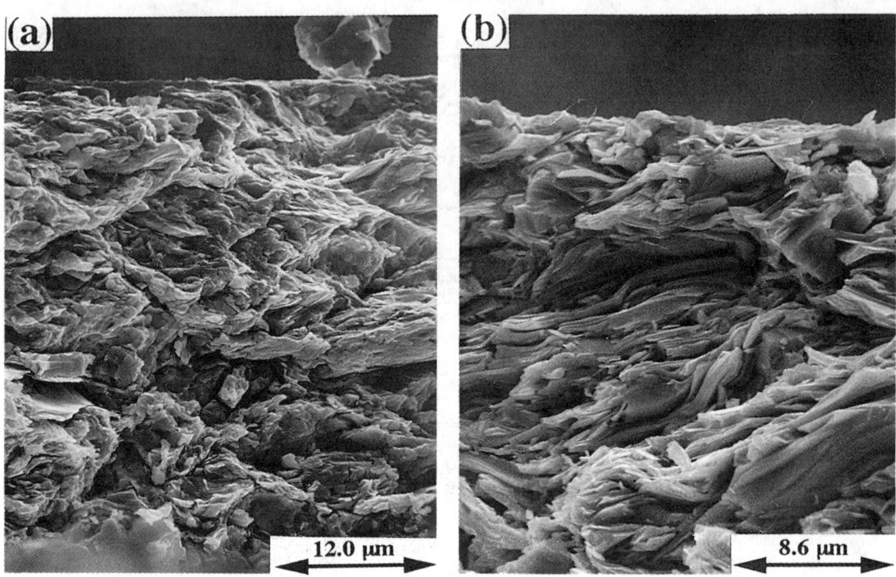

Fig. 11. SEM micrographs of fractured longitudinal cross sections of 2223 BSCCO/Ag tape:
a). as rolled(without annealing) and b). annealed at 850 °C for 5 hours.

Fig. 12. shows a group of SEM secondary electron images for HIP cladded and 30% R sample. These are longitudinal cross-section micrographs with the horizontal direction parallel to the long axis of the sample. At low deformation (~30% R), it appears that non (00l) grain such as (109) and (105) begin to rotate toward c-axis at this stage of deformation. During this mechanical processing, texturing, cracking, and fracturing (resulting from rotation) are intimately related. The second phase appears to be an important parameter of interior alignment at this low deformation stage. It appears that the second phase induces damage in the form of fragmented particles during deformation as well as interrupt the local grain alignment, as shown in fig.12. (d). At higher deformation (~70% R) such as repeated rolling process, however, second phase is broken up due to the severe mechanical deformation and dispersed as small size particles throughout the microstructure as shown in fig. 13. (a).

Fig. 13. compares longitudinal cross section of 70% and 98% cold rolled samples . As deformation extent R(%) increases to 98% R, the micrograph (fig.13.(d)) shows propagation of cracks throughout the microstructure, and very small fractured basal plane aligned particles. As mentioned above, after developing strong c-axis texture, it leads to elastically stiff response with respect to further compression of rolling due to the lack of non (00l) grains to accommodate deformation. Further deformation along the imposed strain path will produce fracture, as a result of the high stresses. Eventhough producing strong crystallographic textures has the tendency to induce cracking, large cracks have an obviously deleterious effect on conductivity especially if the crack planes are perpendicular, instead of parallel, to the current path. This implies that the method of cold rolling process must be chosen to mitigate mechanical states that tend to promote fracture as well as produce strong crystallographic textures.

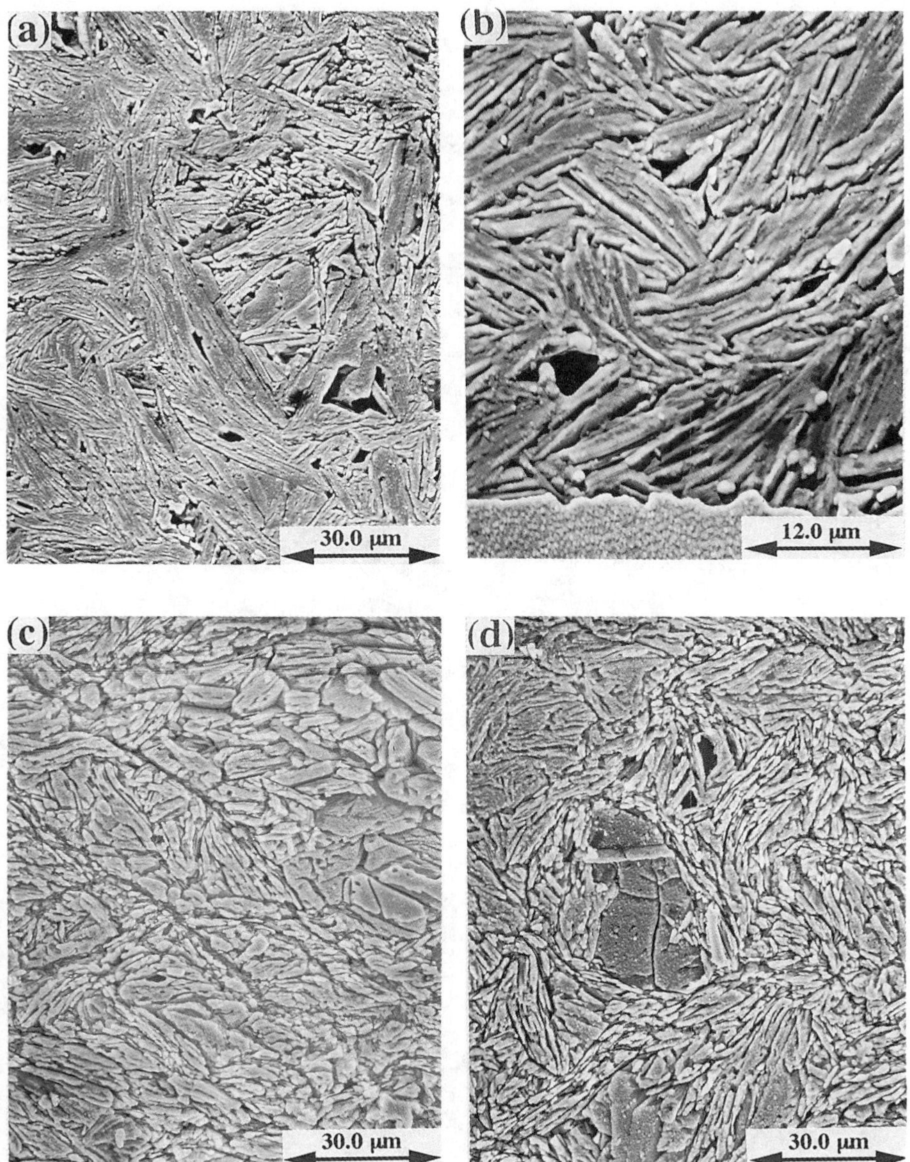

Fig. 12. SEM secondary electron micrographs from polished and etched longitudinal cross-section samples: (a) HIP cladded sample (interior area), (b) HIP cladded sample (near Ag interface area), (c) 30% cold rolled sample, (d) 30% cold rolled sample with the presence of second phase particles- - - The local grain alignment interrupted by second phase particles can be seen cleary.

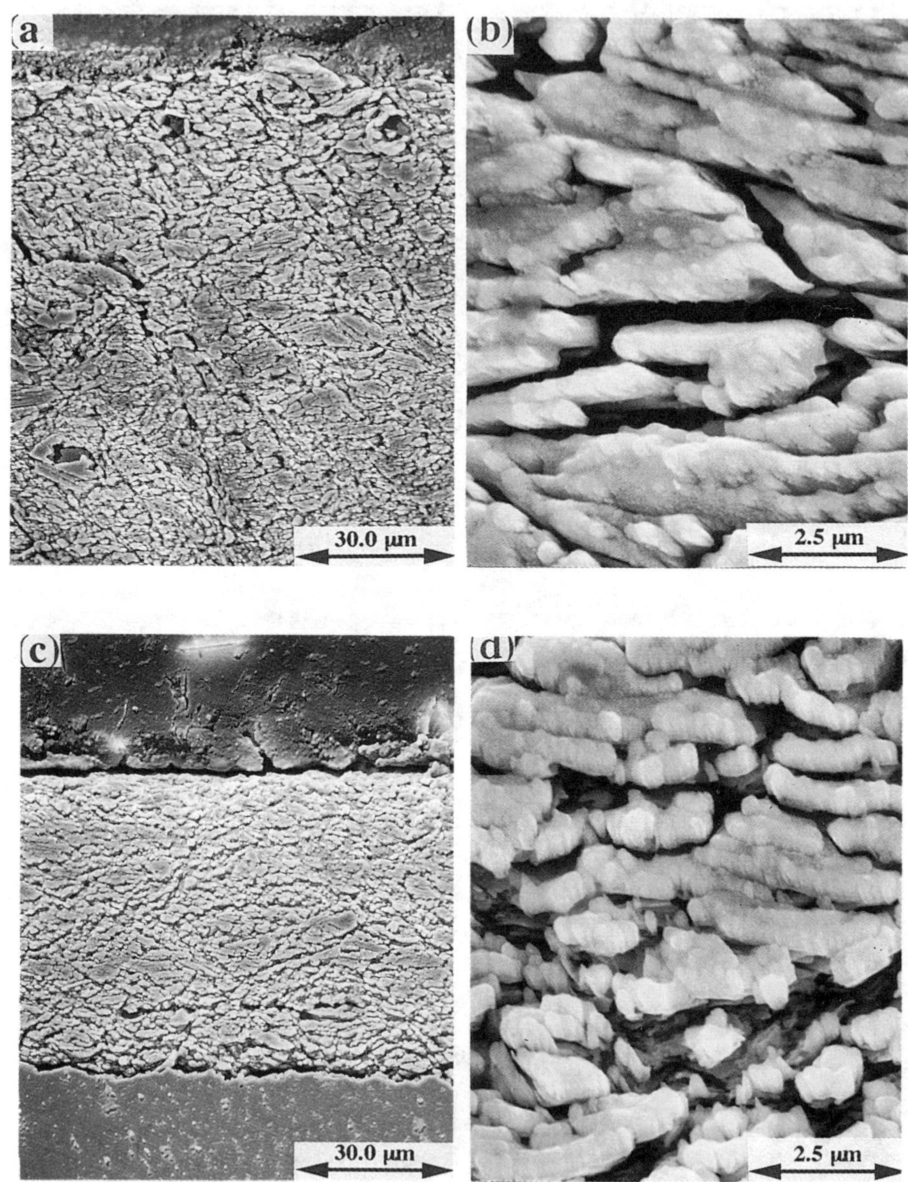

Fig. 13. SEM secondary electron micrographs from polished and etched longitudinal cross-section samples: (a) 70% cold rolled sample, (b) 70% cold rolled sample with higher magnification, (c) 98% cold rolled sample, (b) 98% cold rolled sample with higher magnification

Plane strain compression is usually used to approximate the primary stress and strain state of rolling. However, based on the microstructure observation, the propagation of shear crack throughout the microstructure propose that the additional strain states achieved in rolling are characterized by superimposed shear strains on the compressive strains [17]. In practice, it is necessary to assume a superimposed shear strain (due to friction) on the compressive strains to explain the propagation of shear crack and larger amount of inelastic deformation before such strong textures are obtained.

<u>Acknowledgment</u>

The authors would like to thank The State of Michigan Research Excellence Fund (REF) for supporting this research through The Composite Materials and Structures Center of The Michigan State University.

<u>References</u>

1. T. Hikata, K. Sato, and H. Hitotsuyanagi, <u>Jpn. J. Appl. Phys</u>, 28 (1989) L82.

2. M. Ueyama, T. Hikata, T. Kato, and K. Sato, <u>Jpn. J. Appl. Phys</u>, 30 (1991) L1384.

3. J. Kase, K. Togano, H. Kumakura, D. R. Dietderich, N. Irisawa, T. Morimoto, and H. Maeda, <u>Jpn. J. Appl. Phys</u>, 29 (1990) L1096.

4. T. R. Dinger, T. K. Worthington, W. J. Gallagher, and R.L. Sandstrom, <u>Phys. Rev. Lett</u>, 58 (1987) 2687.

5. H. K. Liu, Y. C. Guo, and S. X. Dou, <u>Supercond. Sci. Technol</u>, 5 (1992) 591.

6. N. Enomoto, H. Kikuchi, N. Uno, H. Kumakura, K. Togano, and N. Watanabe, <u>Jpn. J. Appl. Phys</u>, 29 (1990) L447.

7. G. Jin, J. E. Graebner, T. H. Tiefel, R. B. Van Dover, A. E. White, and G. W. Kammlott, <u>Phisica C</u>, 177 (1991) 189.

8. Y. Takehashi and T. Suga, *Jpn. J. Appl. Phys.* **29** (1990) L2006.

9. D. W. Johnson, W. W. Rhodes, <u>J. Am. Ceram. Soc</u>, 72 [12] (1989) 2346.

10. J. M. Yoo and K. Mukherjee, <u>J. Mater. Sci</u>, 28 (1993) 2361.

11. J. S. Kallend, U. F. Kocks, A. D. Rollet, and H. R. Wenk, <u>Mater. Sci. and Eng</u>, A132 (1991) 1.

12. H. Enami, N. Kawahara, T. shinohara, S. Kawabata, H. Hoshizaki, A. Matsumuro, and T. Imura, <u>Jpn. J. Appl. Phys</u>, 28 (1989) L377

13. Y. Feng, K.E. Hautanen, Y.E. High, D.C. Larbalestier, R. Ray II, E.E. Hellstrom, and S.E. Babcock, <u>Phisica C</u>, 192 (1992) 293.

14. Y. Suzuki, T. Inove, S. Hayashi, and H. Komatsu, <u>Jpn. J. Appl. Phys</u>, 28 (1989) L1382.

15. G. Kozlowsky, I. Maartense, R. Spyker, R. Leese, and C. E. Oberly, <u>Phisica C</u>, 173 (1991) 195.

16. S. Ahzi, R. J. Asaro, and D. M. Parks, <u>Report of research in mechanics of materials,</u> Report No. 2 (1992).

17. C. S. Lee and B. J. Dugan, <u>Metall. Trans</u>, 22A (1991) 2637.

INVESTIGATION INTO THE STRAIN TOLERANCE

OF BSCCO COMPOSITE TAPES

Gherardi L., Caracino P., <u>Metra P.</u>, Vellego G.

Pirelli Cavi Spa Div. Italia, v. Sarca 202, 20126 Milano, Italy.

Abstract

Intense research activity carried out worldwide has led to the development of high Tc superconducting tapes with electrical performances closer and closer to those required for power cables applications. In this perspective, the mechanical behaviour of these tapes, which is known to be intrinsically rather poor, can turn out to be a critical factor. Mono and multifilamentary composite (silver sheathed) BSCCO tapes have been tested using a specially designed apparatus. The dependence of critical current on stress and strain has been investigated for different configurations, and possible models have been considered taking into account the structural parameters as well as the thermal history of the composite tapes. In particular, the effect of the pre-compression which is assumed to be imposed to the superconductor by the silver sheath during cool down has been analysed. The comparison between experimentally determined maximum tolerable stress and strain of present tapes, and values typically required during cable manufacturing, handling and service conditions, is discussed. Possible ways to improve the strain tolerance of superconducting tapes are analysed.

Processing of Long Lengths of Superconductors
Edited by U. Balachandran, E.W. Collings and A. Goyal
The Minerals, Metals & Materials Society, 1994

<u>Introduction</u>

Worldwide efforts on materials and materials processing has led to the development of High T_c composite tapes with superconducting properties closer and closer (at least as far as "record" performances are concerned) to the targets required for power transmission applications. Topics more related to the industrial development of long tapes and to their actual use in the construction of power cables are therefore more and more important.

Among these, the mechanical performances of high T_c tapes, which are characterised by a ceramic core of extreme intrinsic brittleness, are easily recognized to be critical.

Available literature on BSCCO 2223 "Powder in tube" tapes reports about samples and measurements that are often not directly comparable, but it is widely recognised that the transport properties (namely, J_c) of tapes can be affected (irreversibly !) by strains well below 1%, that is, lower than those generally associated with typical cable technologies.

Although some further efforts can be done in designing cable structures and processes to minimize mechanical deformation of elementary tapes, improvements in the mechanical behaviour are generally considered necessary.

To achieve this, efforts are required both in the direction of empirical optimization, and for a better understanding of the basic mechanisms governing such mechanical behaviour.

<u>Requirements</u>

The required performances for HTS tapes to be used in power cable conductors have been considered in details by several studies oriented to this application.

In addition to the fundamental electrical properties (current capability, a.c. losses) the cable application requires in particular the crucial availability of continuous long lengths of already reacted HTS tapes (fig. 1a), having a flexibility sufficient for the manufacturing of the cable, and for its installation.

Much less demanding is the stress performance, due to the fact that cable conductors are to be operated at moderate current densities and in weak magnetic fields, as opposed to the typical case of conductors for SC magnets.

An overview of the required levels for strain, as indicated by the design parameters considered in the most advanced projects on HTS power cables (1,2,3), is given in fig. 1b, which reports also the state of the art of HTS tapes in this respect (4,5,6).

It is apparent from this comparison that HTS tapes having a suitable strain tolerance for cable conductors are already marginally feasible: however the remaining challenge is to make this performance systematically repeatable and to combine it with the best levels of the other (electrical) properties.

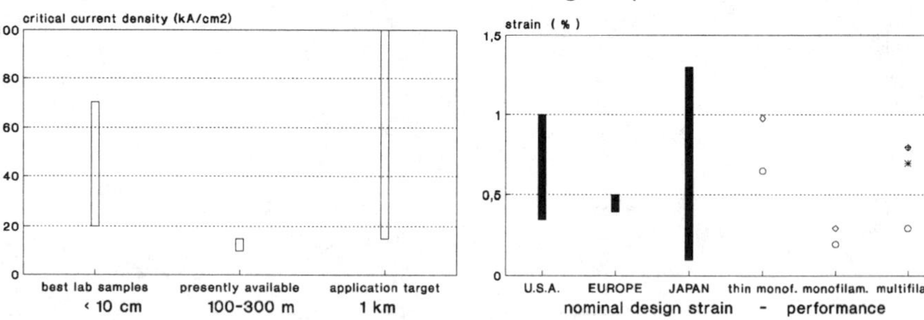

Figure 1 - Targets for cable application and achieved performances for HTS tapes:
a) current-length capability   b) strain tolerance.

<u>Samples</u>

The tapes characterised in this work were all fabricated by the Powder in Tube method, by filling the superconducting powder (or s.c. precursor) into a pure silver tube, drawing the composite to reduce its diameter, and then either rolling to obtain a monofilamentary ribbon structure, or cutting pieces of the wire so obtained, and packing them into a bundle inside a new silver tube, to be again drawn (possibly repeating such step to increase the number of filaments) and finally rolled, for multifilamentary structure. The starting powders were generally home made, by a proprietary method (7), but commercially supplied raw materials were also used in some cases, with results which were generally poorer in terms of critical current density, but substantially equivalent in terms of mechanical behaviour.
Initial diameters of the silver tube were in the range $4 \div 6$ mm and $8 \div 10$ mm for the inner and outer diameter, respectively, allowing for fill factors spanning from below 20% to about 40%. Final tapes were generally thin, as is recognized to be needed for promoting texturing of the material, with overall thickness from about 200 μm down to 70 μm.
In addition to tapes, a number of screen printed specimens, made of 2212 BSCCO film on Ag substrate, were prepared for comparison purposes (see discussion).
Thermal treatments given to the samples are described in ref. (8).

<u>Experimental</u>

Tensile tests in liquid nitrogen bath were carried out using a specially designed apparatus allowing for stress-strain and transport critical current versus stress (and strain) analysis.
A load cell was used to measure the applied force, whereas high sensitivity cryogenic strain gauges were applied to the sample to monitor the deformation. Care was taken in selecting such gauges with very low thickness and tensile strength, in order to avoid any significant reinforcement of the tapes. Transport I_c was measured by standard four probe technique, with the voltage taps placed so as to include almost exactly the region where strain was measured.
The effect of bending (at room temperature) on J_c was also investigated by measuring J_c after bending the tapes on given diameters. Single bending (on one side only), reverse bending, and bending cycles were carried out and compared.
Other physical properties, relevant for the interpretation of the mechanical behaviour of the composite tapes, as e.g. thermal expansion coefficients of silver and superconductor, were also investigated by standard techniques.

<u>Results and discussion</u>

<u>J_c vs stress and strain - Tensile tests.</u> A large number of tests on mono - and multi - filamentary composite samples with 2212 and, more extensively, with 2223 BSCCO, were carried out to qualify the dependence of the transport J_c of such tapes on applied mechanical stress and strain.
Figures 2a and 2b show the relative J_c versus average stress characteristics for typical 2212 and 2223 monofilamentary tapes, and for 2223 BSCCO tapes with different structures, respectively.
As appears from fig. 2a, the two materials have a similar response to mechanical stress, with an initial stress range where J_c is not affected by stress, followed by a sharp drop (apparently sharper for 2223) starting around 50 MPa.
The initial "flat" region appears somewhat extended (fig. 2b), for multifilamentary specimens, so confirming (3,4) that multiple structure with thinner filaments provides some advantage, presumably due also to increased area of the interface between superconductor and silver, and the reduced depth of initial flaws, both factors being supposed to improve the fracture resistance of the oxide superconductor.
As a matter of fact, partial fracture of the superconductor is supposed to be the only cause for J_c decay; this is confirmed also by the measurements (see the same fig. 2a) done after stress release, which show that J_c is absolutely unaffected by unloading, that is, on one hand J_c decay is totally irreversible, and, on the other hand, stresses applied in the elastic range do not affect J_c.

149

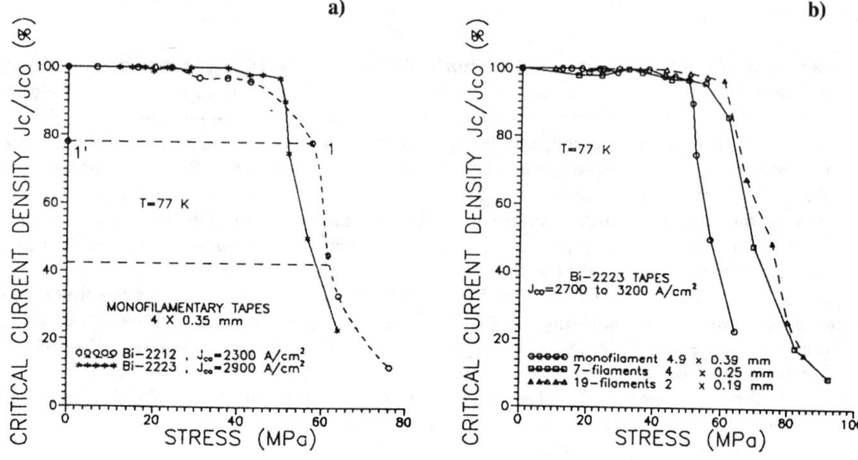

Figure 2 - **Effect of mechanical stress on the critical current for different HTS tapes:**

        a) comparison of Bi-2212 and Bi-2223 monofilamentary tapes with similar current performance and fill factor (25 to 30 %)

        b) comparison of Bi-2223 tapes with different number of filaments (fill factor 28% for mono, 18% for 7-fil. and for 19-fil.

Bending tests. Figure 3 illustrates the J_c vs strain behaviour of BSCCO 2223 multifilamentary tapes tested in different conditions.

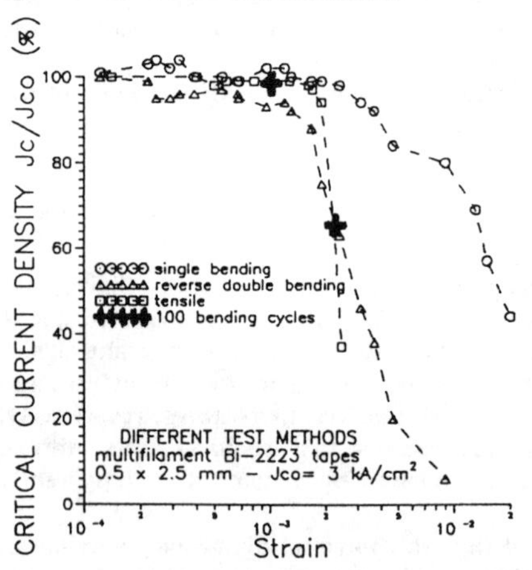

Figure 3 - Comparison of the effect of strain on critical current, when evaluated with different test methods: single and double reverse bending at room temperature, bending fatigue at room temperature and tensile test at 77 K.

Results on bent samples (bending at room temperature) are very well consistent with ones from tensile tests: smaller but well comparable degradation for the same maximum strain is found on the "single" bent samples, which experience a non-uniform deformation, whereas samples "double reverse" bent showed very similar degradation. The effect of bending cycles is also shown in the same figure: bending 100 times on the same diameter did not affect J_c for curvatures giving strains below the "irreversible damage" threshold (that is the "flat" region was confirmed to correspond to an elastic regime), whereas the J_c decay consequent to repeated bending was shown to approach, for the same maximum strain, that induced by traction.

<u>Comparison with Nb$_3$Sn</u>. The flat "safe" region shown by these BSCCO composite specimens in their J_c vs strain (and stress) characteristic demonstrates, as pointed out above, that there is at least one range of deformations the s.c. material is (electrically) indifferent to. This is known not to be the case with Nb$_3$Sn composites, which have a similar (metal sheathed) structure, but show a typical J_c vs strain behaviour characterised by a non monotonical dependence of J_c upon applied strain. This is understood as due to the remarkable dependence of J_c of the s.c. material on both compression and traction (see in fig. 4), that manifests itself in a structure where the superconductor is in a state of initial pre-compression.

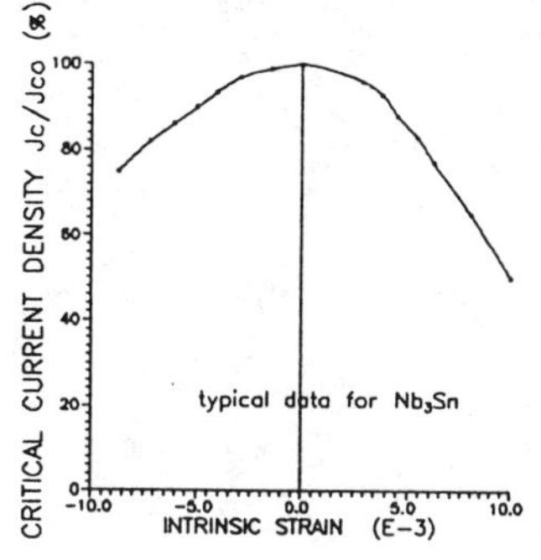

Figure 4 - Effect of strain on the current performance of NB$_3$Sn conductors, represented as the actual strain for the SC material.

Such initial pre-compression arises from the difference in the thermal expansion coefficients of superconductor and metal sheath when the composite is cooled down from the annealing temperature.

As a matter of fact, the same situation is expected to be found in BSCCO composites after thermal treatment, since BSCCO and silver do have different α_T.

<u>Thermomechanical beaviour of the composites.</u> To try to quantitatively study such presumed pre-stress (and pre-deformation), and better interpret the J_c vs ϵ curves (is the flat region more extended than it would be according to the intrinsic material behaviour, because of pre-strain? to which extent?), in view of possibly exploiting this effect by optimizing the composite structure, the following investigation was carried out.

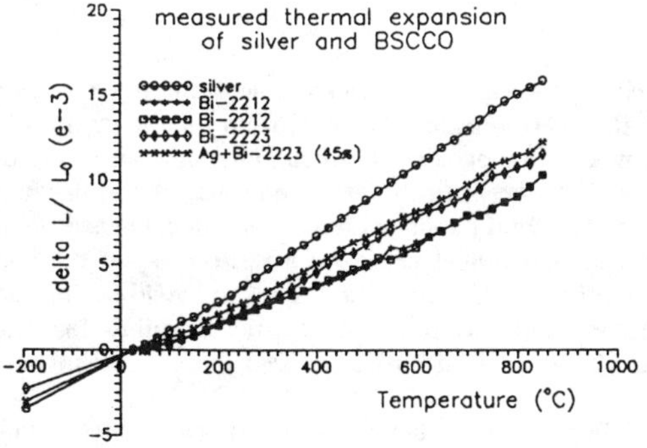

Figure 5 - Thermal expansion measured on samples made of pure silver, on Bi-2212 and Bi-2223 sintered rods and on a composite structure.

Thermal expansion of BSCCO and silver was accurately determined as a function of temperature by means of TMA analysis (above room temperature up to 850°C), and A.S.T.M. method E 228-71 down to 77 K. It is well evident from fig. 5 that the expansion of BSCCO 2212 is slightly lower than that of 2223, and that both are substantially lower than that of silver. The results for BSCCO+silver are for rods with a 2 mm diameter and about 40% fill factor, and seem to indicate that, at least for such geometry and fill factor, the expansion of the composite is closer to that of BSCCO, rather than that of silver.

Direct qualitative evidence of such effect is given in figure 6a) for BSCCO 2212 film deposited (by screen printing) or a pure silver substrate. Here the "asymmetric" specimen appears bent, after thermal treatment, as a consequence of differential thermal contraction during cooling down, with a curvature corresponding to a bending deformation of about 3×10^{-3} for silver (note that the ends of the substrate, where there is no BSCCO deposit, are practically straight).

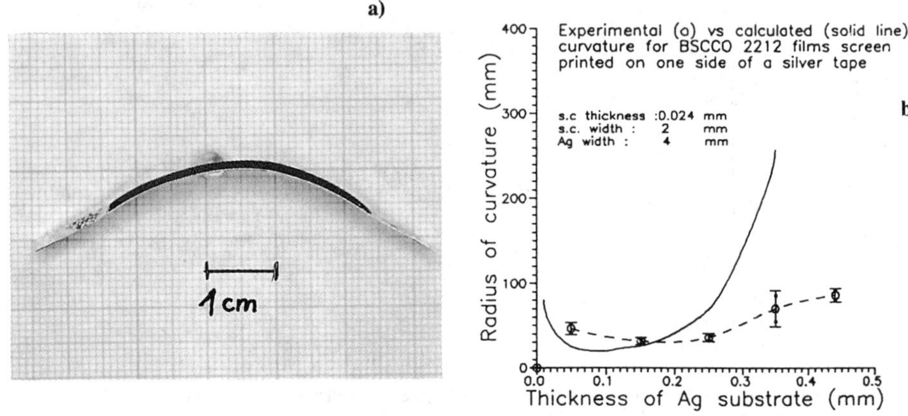

Figure 6 - Samples of screen printed Bi-2212 on silver tape, showing the bending originated by the thermomechanical history during the thermal treatment:

    a) typical appearance of a bent sample
       silver tape 60 x 4 x 0.2 mm, SC layer 43 x 4 x 0.024 mm
    b) experimental data for the curvature radius of the samples, versus the silver
       substrate thickness, compared with the values calculated from the mechanical model
       (elastic + yield) of the structure.

To take advantage of the favourable geometric configuration of such specimens and try to gather some more quantitative information about the thermomechanical behaviour of the composite (easily transferrable to the case of the "symmetric" tape), a series of samples with an identical amount of superconductor on Ag substrates with different thickness were prepared and thermal treated. The resulting curvature radii, see fig. 6b, shows a surprisingly weak dependence upon the silver thickness. In fact, a simple mathematical model based on measured α_T of the two materials, on the elastic modulus and on a tentative "apparent" modulus beyond yield (estimated from a tipical stress-strain curve) for silver and assuming an elastic modulus for the superconductor of the same order than that of Ag, as it had been deduced from σ-ϵ behaviour of specimens of fig. 2, failed to describe such behaviour.

As a matter of fact, quantitative modelling of the mechanical behaviour of silver after yield turns out to be a very difficult task, especially when dealing with annealed Ag, since the yield zone is very extended, and extremely sensitive to a variety of factors, from thermal and mechanical history, to even testing conditions, so that a "typical" behaviour is difficult or practically impossible to define.

Direct empirical evaluation of the predeformation of silver in thin standard "symmetric" tapes with 2223 BSCCO was then carried out by preparing a series of 10 specimens and very carefully measuring (under 200 X magnification) their lengths before and after each of the three thermal treatments our tapes are normally given (we followed for these specimens the exact procedure as usual, including the rollings between thermal treatments, which reduced the thickness by 10 μm each time).

For comparison, two pure silver ribbons of comparable thickness, produced by rolling in the same way as the composites, were also annealed, together with the s.c. tapes.

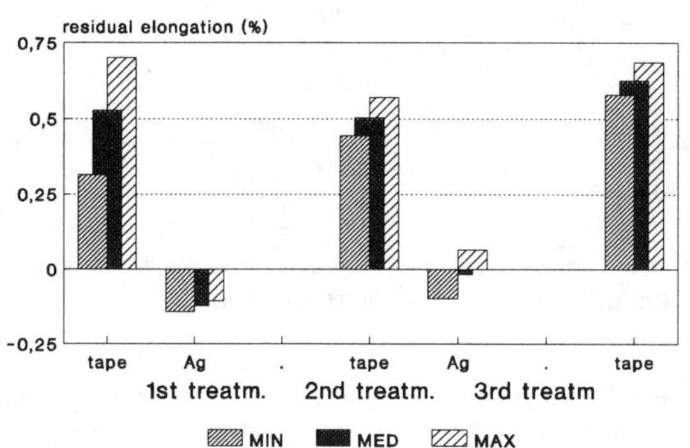

Figure 7 - Residual elongations of Bi-2223 tapes, measured after each thermal treatment, compared to the residual elongation of silver alone:

thickness: 90 to 100 μm
length: 50 mm
silver cross sec-
tion: 0.13 to 0.15 mm^2
fill factor: 35 to 40 %

The results are shown in fig. 7. An average permanent elongation of about 5×10^{-3} was measured after each thermal treatment, to be compared to the slight contraction shown by the silver specimens after the first annealing only, probably due to the relaxation of axial stress "frozen-in" during the initial rolling deformation (the rolling given between annealings is much "milder"). Among the possible ways to quantitatively interpret these results, which, by the way, seem to agree well with the ones from films, we tried to follow a very simple semiquantitative model based on the assumptions that the superconductor being not sintered, but just a "compacted powder" during heating, it offered no resistance to silver expansion, so that mutual interaction between s.c. and Ag occurred only during cooling. This means that the permanent elongation we measured on annealed tapes represents exactly the deformation applied to silver due to the presence of BSCCO; this, in principle, should allow to determine exactly also the residual deformation of BSCCO, provided that the elastic modulus of BSCCO itself, and the σ-ϵ experimental curve of Ag, both at room temperature, are known (of course, also the sections must be known). As for the former, direct measurement is very difficult to realize, especially because "significant" BSCCO specimens, that is ones formed in an actual tape, (what is known to make a substantial difference compared to larger rods, at least for the s.c. properties), are practically impossible to handle, and certainly not suitable for standard techniques. It is however reasonable to assume that the mechanical behaviour of BSCCO is elastic up to fracture (as is typical for oxide ceramics), and that its modulus is in a range around 80 GPa (in literature values from 40 to over 100 GPa, measured with different methods, are reported). On the opposite, the behaviour of annealed silver should be assumed to be perfectly known, but we decided anyway to systematically measure a series of annealed specimens with thicknesses comparable to the ones of the tapes and higher, which gave interesting results.

σ-ϵ **of thin annealed silver and composite tapes.** The first important result was the confirmation that the elastic limit for these specimens is at deformations generally one order of magnitude lower than the residual ϵ measured on the composite tapes (see for example figs. 8a,b).

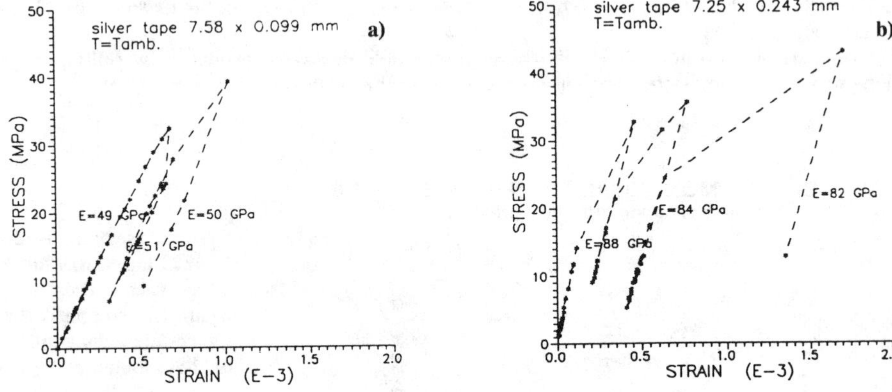

Figure 8 - Stress-strain characteristics at room temperature for silver tapes of different thickness, showing the yield behaviour and different values of the elastic modulus.

That is, during cooling the silver sheath is deformed mainly in the plastic regime, which explains very well the weak dependence of the curvature of the films upon the substrate thickness.

The other interesting result came from the comparison among the σ-ϵ curves for specimens with different thickness. It was found (see again fig. 8a) that measuring a very thin (100 μm in that instance) specimen can provide a value for the elastic modulus that is different (and much lower) from the "intrinsic" value which is well known and independent of annealing.

This unexpected result (see fig. 10) for the empirical dependence of such apparent modulus upon thickness) may have different explanations: very thin specimens with few very large crystals (because of annealing), and an extremely high surface to volume ratio (because of shape and size) are not conceptually the best ones for measuring a bulk property like the "intrinsic" elastic modulus. Measurements on composite specimens with different thickness gave very similar results (see fig. 9a,b), with the thinner specimens (70 μm) showing an elastic modulus about half that of the thick (650 μm) one.

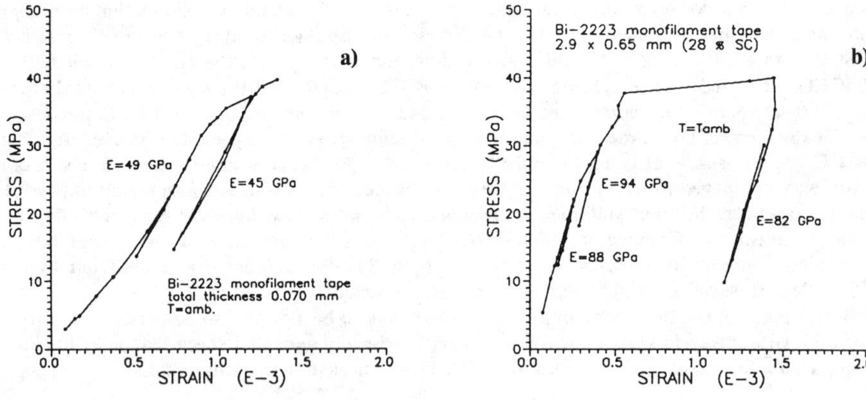

Figure 9 - Stress-strain characteristics at room temperature for HTS tapes of different thickness, showing the yield behaviour and significantly different values of the elastic modulus.

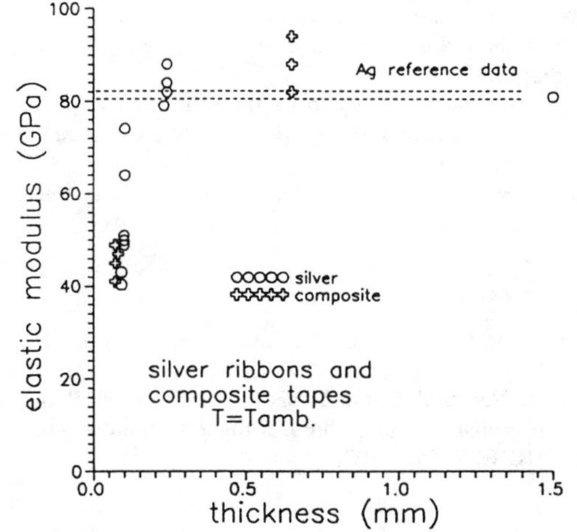

Figure 10 - Variation of the elastic modulus of silver and of composite tapes, as measured on samples of different thickness, compared to literature data.

Compared to those measured on pure silver (see again fig. 10), these moduli appear slightly higher: although not very precise, these results support therefore the indication that the elastic modulus of the superconductor alone is about the same or higher than that of silver.

As an experimental remark, it may be worth pointing out that such moduli of the composite were determined by taking the slope of the <u>unloading</u> characteristics, which reflects the true elastic behaviour of the specimen; instead, the <u>initial</u> slope of the curve results from the contribution of plastic (supposed pre-strained beyond yield) silver, and pre-compressed BSCCO.

<u>Estimated pre-compression of superconductor</u>. As for our specific purpose, that is for calculating the pre-deformation imparted by silver to the superconductor, this anomalous behaviour brings about one more degree of uncertainty, because in actual tapes the thickness of silver is of the order of $20 \div 30$ μm, that is even lower than for the Ag specimens we tested, and its interface surface may also have a more "corrugated" geometry.

Bearing all this in mind, our estimate for the pre-deformation of the superconductor in such a composite is in the order of $1 \div 1.5 \times 10^{-3}$ if we assume a modulus around 80 GPa for BSCCO. Assuming a higher value around 100 GPa would give a slightly lower pre-deformation, whereas a modulus of 40 GPa would give a very high, probably unrealistic, pre-deformation.

In any case, this pre-deformation seems to represent a significant fraction of the deformation up to which the superconducting performance of the tapes remains unaffected, that is its contribution to the strain tolerance of the tapes can be very important.

The main factors controlling such pre-deformation are the fill factor and the silver plastic behaviour: a fill factor typically lower for the multifilamentary samples could be one additional explanation for their better mechanical performances.

Improvement of the sheath plastic behaviour could be achieved by doping or by the addition of another structural element (for example an outer sheath of a stronger metal), but both ways may bring about also some drawbacks, so that a compromise solution should be carefully studied for each particular application.

<u>Conclusions</u>

The strain tolerance of BSCCO silver sheathed composite tapes is generally not yet fully adequate compared to the requirements of power cables applications.

The pre-compression imposed by silver to the superconductor during cooling from the annealing temperature was found to contribute significantly to the mechanical response of tapes.

Ways to improve the mechanical behaviour of tapes should include the adoption of a multifilamentary structure, and the reinforcement of the silver sheath.

For a given fill factor, improvement of silver itself by doping, or partial replacement by another stronger metal, provided that they do not affect the s.c. performances of the tapes or other design parameters, could provide significant advantages.

The limit for the improvement achievable by this way is likely to be the ultimate compression yield of the superconductor, as there is at present no evidence that the J_c of BSCCO is affected by compression in the elastic regime. To check this assumption, and in general to better describe the behaviour of the superconductor under compression, as well as under traction, further experimental investigation is needed.

<u>Aknowledgement</u>

Thank are due to Edoardo Schiavi and Roberto Giuliacci for their contribution to the experimental work.

This work was partially supported by Italian National Council for Research under Progetto Finalizzato "Tecnologie Superconduttive e Criogeniche", and by the European Community under BRITE/EURAM Projects BREU-0095-C and BRE2-CT92-0207.

<u>References</u>

1. S.P.Ashworth et al., <u>The technical and economic feasibility of HTS power transmission cables</u> (EUCAS '93, Gottingen, Oct. 1993).

2. D.Von Dollen et al., <u>Design concept of a room temperature dielectric HTS cable</u> (American Power Conference, Chicago, Apr. 1993).

3. T.Nakahara, <u>Review of Japanese R&D on SC</u> (Sumitomo Review, N°35, Jan. 1993).

4. S.X.Dou, H.K.Liu, <u>Ag-sheathed Bi(Pb)SrCaCuO SC tapes</u> (Superc. Sci. Technology, 6, 1993).

5. J.W.Ekin et al., <u>Effect of axial strain on the critical current of Ag-sheathed Bi-based superconductors in magnetic field up to 25 T</u> (Appl. Phys. Lett. 61, p. 858, 1992).

6. M.J.Minot, <u>Advances in long length BSCCO HTS multifilamentary composite wire development</u> (ICMC 93, Albuquerque, July 1993).

7. G.Szabo Miszenti et al., <u>Preparation of high quality HTSC powders: a breakthrough approach in the citrate route</u> (Europ. Conf. Appl. Superc., Gottingen, Oct. 1993).

8. P. Metra et al., <u>Transport properties of different BSCCO wires</u> (ICTPS '90, Rio de Janeiro, Apr. 29 - May 4, 1990).

SYNTHESIS AND CHARACTERIZATION

Synthesis and Microstructure of the 2223 Phase ($Bi_{2.0}Sr_{2.0}Ca_{2.0}Cu_{3.0}O_x$)

in the (Bi,Pb)-Sr-Ca-Cu-O System Obtained from Freeze Dried Precursors

P. Krishnaraj, M. Lelovic, T.A. Deis, N.G. Eror, U. Balachandran[*]

Dept. of Materials Science and Engineering
University of Pittsburgh, PA

[*]Energy Technology Division
Argonne National Laboratory

Abstract

Amorphous precursor powders for $Bi_{1.8}Pb_{0.4}Sr_2Ca_{2.2}Cu_3O_x$ and $Bi_{1.8}Pb_{0.4}Sr_2Ca_{2.0}Cu_3O_x$ (2223) were prepared by a freeze-drying process incorporating a splat-freezing step. The critical parameters affecting the drying process are discussed. The freeze-dried powders have high reactivity and led to essentially phase pure 2223 phase formation in 12 hr. The microstructure of the 2223 shows fine coherent particles of 2201 in the matrix. The grain boundaries of the 2223 grains appear to be clean, leading to good intergrain contact between 2223 grains. It was shown that both lead content and oxygen partial pressure affected liquid-phase formation. Lead content affected dissolution of the 2212 phase into the liquid phase from which the 2223 high-T_c superconducting phase is formed.

Processing of Long Lengths of Superconductors
Edited by U. Balachandran, E.W. Collings and A. Goyal
The Minerals, Metals & Materials Society, 1994

Introduction

Since Maeda et al[1] discovered superconductivity in the Bi-Sr-Ca-Cu-O (BSCCO) system, much effort has been focused on the synthesis of the superconducting phases of the BSCCO system, in particular the $Bi_2Sr_2Ca_2Cu_3O_x$ (2223) phase with a superconducting transition temperature (T_c) of 110K. It is difficult to produce the 2223 phase without the partial substitution of Pb for Bi to promote the formation and stabilization of the 2223 phase.[2] It is well known that partial substitution of lead for bismuth has a stabilizing effect on the 2223 phase.[3-5] However, the exact role of the lead ions and the nature of the stabilization are not clear. One approach considers lead ions as a liquid-phase-former via decomposition of Ca_2PbO_4.[5,6] Idemoto et al.[6] studied annealing effects on the lead-doped superconductor at various temperatures and oxygen pressures. Ca_2PbO_4 was shown to form over a significant range of oxygen partial pressure. Another way of looking at the effect of lead is to consider lead ions as a bismuth substitute, i.e., lead is only present in bismuth-containing phases. A systematic study of 21 compositions on a compositional line connecting all three superconducting phases for the temperature range of 825 -1100°C in static air has been reported.[8] The reported observation was that no specific lead-containing phase (Ca_2PbO_4 or Sr_2PbO_4) was detected, and lead was only present in bismuth-containing phases. Hong and Mason[9] studied solid-solution ranges in the $Bi_2Sr_2CaCu_2O_x$ (2212) and 2223 superconducting phases at 800°C in air. They found the Ca_2PbO_4 phase to be present in most of the lead-doped samples. Both arguments[8,9] seem to agree with Idemoto's results.[6]

The importance of Ca_2PbO_4 in the system has been carefully studied at Argonne National Laboratory.[10,11] A two-step powder-in-tube process has produced a transport critical current density (J_c) as high as 50,000 A/cm^2 at 77K in zero applied field. In this process, lead-doped 2212 was mixed with Ca_2PbO_4, Ca_2CuO_3, and CuO. DTA results indicated involvement of the Ca_2PbO_4 phase in the formation of liquid phase during heat treatment. Lead content in the 2212 phase was shown to be important. The highest I_c and J_c values were obtained when all of the lead was incorporated into the 2212 phase.

The formation of the 2223 phase is promoted by synthesis in reduced oxygen partial pressure[12,13] and modification of the initial composition with excess Ca and Cu.[14-16] Production of a material with a high volume fraction of the 2223 phase by conventional solid state processing requires very long synthesis times.[17] Solid state reaction techniques commonly employed in the synthesis of these compounds have several disadvantages for multicomponent oxide systems such as $(BiPb)_2Sr_2Ca_2Cu_3O_x$. Such techniques do not yield chemically homogeneous precursors, and the rate of 2223 phase formation is slow due to transport limitations in the solid state. A further disadvantage arises from segregation due to the slow decomposition of $SrCO_3$ and $CaCO_3$ when these are used as starting materials.

The actual formation mechanism of the high-T_c phase has not yet been established due to the lack of an appropriate phase diagram. Chen and Stevens[18] used high-resolution electron microscopy to study the formation mechanism of the 2223 phase. They concluded that the actual mechanism, at 845°C, is a three-step process: dissolution of the 2212 phase into the liquid phase (liquid phase was formed from decomposition of Ca_2PbO_4), precipitation of the $Bi_2Sr_2CuO_x$ (2201) phase from the liquid, and nucleation and growth of the 2223 phase from the 2201 phase. The final state was a mixture of the 2223 phase and a 2201 grain-boundary phase. Different results were reported by Umezawa et al.[19] who showed the presence of the 2212 phase at grain boundaries. Even in the sample showing only X-ray diffraction peaks for the 2223 phase, one unit cell of the 2212 phase was present at the grain boundary. Luo et al.[20] carried out a detailed kinetics and mechanistic analysis, by an isothermal equilibration method, of the growth of the 2223 phase in Ag-sheathed wire. Their study showed a two-dimensional growth geometry of the 2223 phase and evidence of an amorphous phase at grain boundaries.

Chemical methods such as coprecipitation often result in a sequential precipitation of the elemental compounds serving only to reduce the particle size of the same components used in the solid state techniques. Such processes typically employ oxalic acid, the presence of which leads to the formation of $SrCO_3$ and $CaCO_3$. Sol-gel processing leads to similar problems of carbonate formation due to the presence of citric acid or acetic acid.

To overcome problems of chemical inhomogeneities in the precursor material, as well as the presence of carbonaceous anions, freeze-drying has been applied to the synthesis of $YBa_2Cu_3O_x$[21-24] and has recently been applied to the synthesis of $(BiPb)_2Sr_2Ca_2Cu_3O_x$.[25,26] In this study a freeze-drying technique utilizing nitrate precursors for the synthesis of $(BiPb)_2Sr_2Ca_2Cu_3O_x$ was adopted. The critical parameters for obtaining highly reactive powders leading to rapid formation of the 2223 phase, and the resultant microstructure, will be discussed.

Experimental Methods

Solutions of Bi, Sr, Ca and Cu nitrates were prepared using distilled-deionized water. These solutions were characterized by thermogravimetric analysis (TGA) to determine the cation content. The solutions were mixed in the cation ratio Bi/Pb/Sr/Ca/Cu = 1.8:0.4:2:2.2:3, with the Pb added in the form of PbO. The solution was diluted to a concentration of 0.1M in Bi. The pH of the solution was adjusted to fall in the range of 0.3-0.7 by adding HNO_3. Powders with varying lead content were also prepared via freeze-drying: $Bi_{2.0}Sr_{2.0}Ca_{2.0}Cu_{3.0}Oy$, $Bi_{1.8}Pb_{0.2}Sr_{2.0}Ca_{2.0}Cu_{3.0}O_y$ and $Bi_{1.8}Pb_{0.4}Sr_{2.0}Ca_{2.0}Cu_{3.0}O_y$.

The nitrate solution was injected onto a teflon coated stainless steel block cooled to liquid nitrogen temperature. The resulting flakes of frozen material were ground using a high shear homogenizing tool (IKA Works W45 MA) to a particle size of less than 20 μm. A schematic of the experimental set-up used to freeze the powders is shown in Fig. 1.

The frozen powder was then transferred to a commercial freeze drier (Edwards High Vacuum, Supermodulo 45) preset to a temperature of -40°C and operated at a pressure of 0.1 mbar. Sample temperature and chamber pressure were continuously monitored and the temperature was raised to 20°C over a period of 2 days. The dried powder was then transferred to an oven and heated to 125°C at a rate of 5°C/hr under flowing argon.

The dehydrated precursor was then transferred to a furnace preheated to temperatures between 600°C and 730°C and flash heated in a 22% O_2 (balance Ar) gas flow. The denitrated powders were then ground, pressed into pellets and introduced into a furnace preheated to 845° C and sintered for varying times under a 7% O_2 atmosphere. Advantages of this process over conventional solid-state methods for $YBa_2Cu_3O_y$ and $(Bi,Pb)_2Sr_2Ca_2Cu_3O_y$ have been described elsewhere.[21-26]

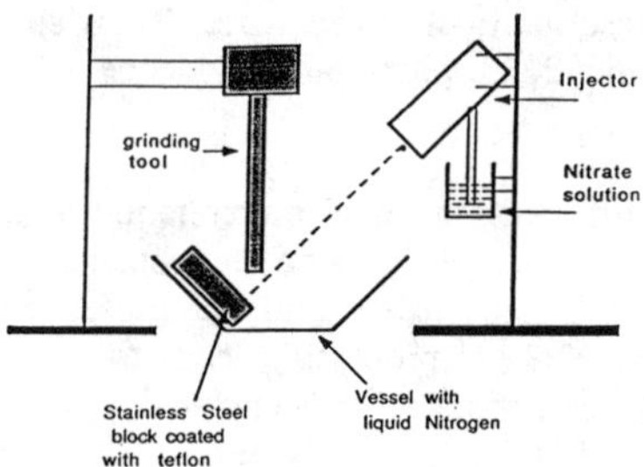

Fig. 1 Schematic of the apparatus used in the splat freezing process.

Fig. 2 XRD plot of the freeze dried powder dried to 120°C indicating amorphous nature of the powders obtained in the splat freezing process. Peaks corresponding to $Sr(NO_3)_2 \cdot xH_2O$ resulting from incorporation of waters of crystallization by the sample.

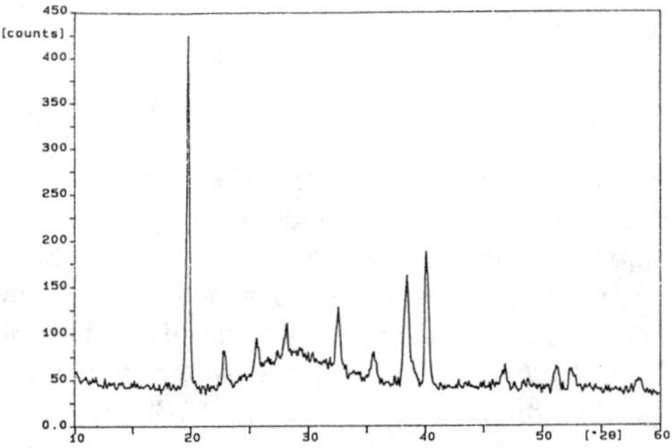

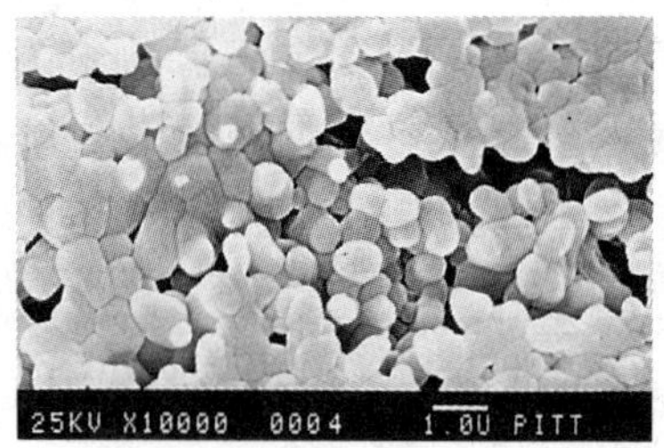

Fig. 3 SEM micrograph of the freeze dried powder dried to 120°C shows the particle size be $\approx 1 \ \mu$m.

Differential thermal analysis (DTA) and thermogravimetric analysis (TGA) were used to study the dehydration and denitration behavior, and the onset of liquid phase formation. X-ray diffraction (XRD) was used to characterize the phase content of the samples and determine the extent of the 2223 phase formation. Scanning electron microscopy (SEM) and transmission electron microscopy (TEM) were used to characterize the microstructure of the samples, while energy dispersive spectroscopy (EDS) was used for compositional analysis.

Results

Freeze Drying of the Nitrate Solutions

The freeze-drying process is very sensitive to the NO_3^- concentration and the pH of the solutions. The solutions used in this study which had a pH outside of a narrow range (0.3-0.7) melted during the drying step.

The splat freezing process adopted in this study led to the formation of amorphous precursors. Fig. 2 shows an XRD plot of the powders dried at 120°C. The only peaks seen are for $Sr(NO_3) \bullet xH_2O$ which could be due to the sample incorporating waters of crystallization when exposed to air. The particle morphology of the powder dried to room temperature in shown in Fig. 3. The individual particle size is seen to be of the order of 1 μm. DTA and TGA traces, shown in Fig. 4a, on the same powder, indicate the presence of a substantial amount of residual water, associated with the waters of hydration of the nitrate salts. The endotherms seen below 150°C on the DTA trace correspond to the removal of waters of hydration and melting of the sample. Dehydration is accomplished by heating the sample at the rate of 5°C/hr under flowing argon. The DTA trace in Fig. 4b on the dehydrated powder indicates the absence of endotherms associated with melting and water removal. A TGA trace on the same powder indicates that the residual water content is small, of the order of 1.5 weight % of the sample. Further, both TGA and DTA traces indicate that denitration of the sample is complete by 600°C.

Zheng et al.[27] showed that copper ions in Bi-Sr-Ca-Cu-O glasses exist in the +1 ionic valence state. Oxygen loss is associated with transition between CuO and Cu_2O. To detect weight loss due to the formation of a liquid phase[27], the scale of the TGA run is enlarged between 600 and 900°C in Fig. 5. Onset of liquid-phase formation was associated with a change in slope at $\approx$790°C. Between 830 and 860°C, the TGA curve shows no transformation, due to the slow transformation kinetics relative to the TGA heating rate.

Because we are able to detect the initiation of liquid-phase formation, a plot was constructed of onset temperature for liquid phase-vs-lead content at different Po_2 levels (Fig. 6). An increase in lead content and decrease in Po_2 decreased the initiation temperature by $\approx$40°C.

Synthesis of 2223 from Freeze-Dried Nitrate Precursors

The dehydrated powders were denitrated by introducing them into a furnace preheated to temperatures ranging from 600°C to 730°C in an atmosphere of 22% O_2 and holding for 1 hr. TGA was used to determine the onset temperature of the melting reaction, involving Ca_2PbO_4, as a function of oxygen partial pressure. Denitration temperatures were chosen to be below the liquid phase formation temperature. The denitrated powders were then crushed and pelletized at a pressure of 500 MPa and introduced into a furnace preheated to 845°C and an atmosphere of 7% O_2. XRD patterns of samples held for varying times are

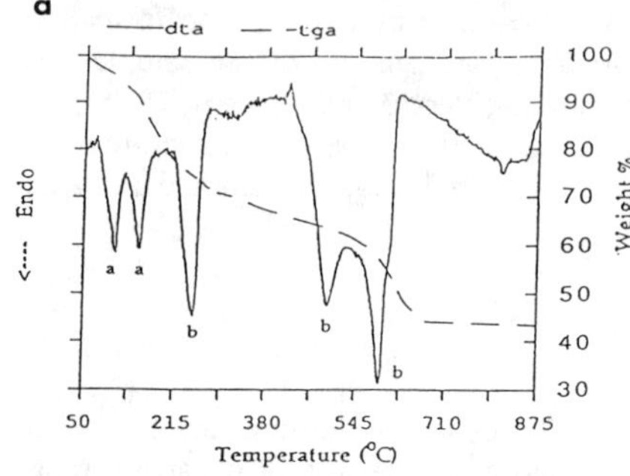
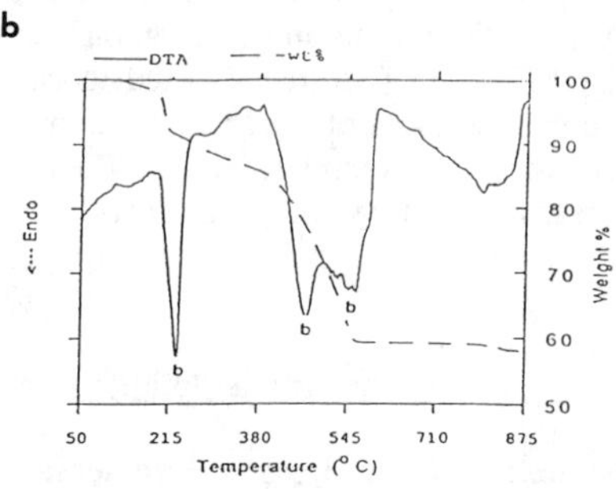

Fig. 4 Combined DTA and TGA trace of the freeze dried powder. The peaks marked "a" correspond to melting and evolution of water, peaks marked "b" correspond to decomposition of the nitrates. 4a) For the freeze dried powder dried to room temperature the presence of substantial water is seen from both traces. 4b) For the freeze dried powder dried to 120°C the DTA endotherms associated with water removal and melting are absent. The TGA data indicate the residual moisture to be 1.5% of the sample weight.

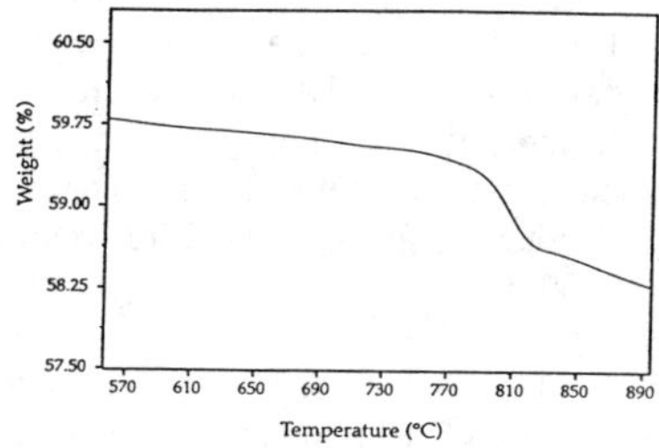

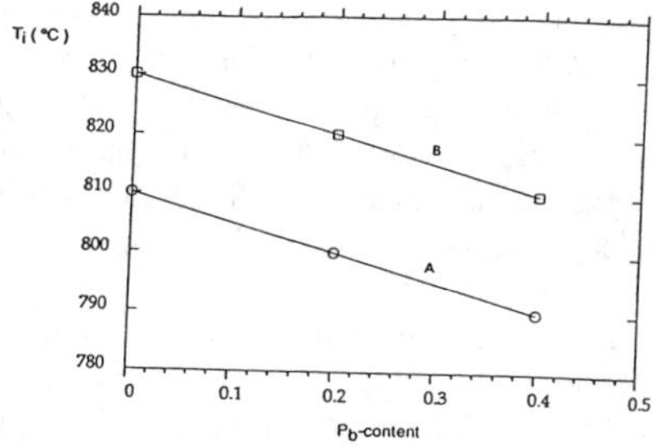

Fig. 5 Enlarged TGA trace of Fig. 4b to show weight loss due to formation of liquid phase.

Fig. 6 Onset temperature for liquid phase vs. lead content at different Po_2 levels.

shown in Fig. 7, each step is accompanied by regrinding and pelletizing. A sample held for 2 hrs. shows the primary phase to be $Bi_2Sr_2CaCu_2O_x$ (2212). Sintering for an additional 5 hrs. indicates the major phase to be 2223; however the presence of 2212 is seen from the XRD data. A further sintering of 5 hrs. indicates the sample to be 2223 (Sample A). The low angle peaks for Sample A are shown as an insert in Fig. 7.

Fig. 8 shows the XRD pattern for a sample without lead which was sintered for (1) 2 hours, (2) 2+5 hrs., and (3) 2+5+5 hrs. at 845°C in 7% O_2. Each step was accompanied by grinding and pelletizing. After 7 hrs., the 2212 phase becomes stable, with no indication of the 2223 high-T_c phase peaks. Low-angle peaks shown in the inset indicate that only the (002) peak for the 2212 phase (m) was present.

In a specimen with composition $Bi_{1.8}Pb_{0.2}Sr_{2.0}Ca_{2.0}Cu_{3.0}O_y$, a small volume fraction of the 2223 high-T_c phase was detected after 7 hrs. (Fig. 9). This is shown by the diffraction peak at 4.7° 2Θ corresponding to the (002) plane for the 2223 high-T_c phase. The 2212 phase was the major phase after 12 hrs., as can be seen from curve (3).

Microstructure of the 2223 Phase Obtained from Freeze-Dried Precursors

Microstructural analysis was carried out for the $Bi_{1.8}Pb_{0.4}Sr_{2.0}Ca_{2.0}Cu_{3.0}O_y$ samples used in the XRD analysis. Fig. 10 is a high-resolution TEM micrograph taken in the dark field (DF) image mode for a sample heat treated for 2 hrs. at 845°C (XRD pattern 1 in Fig. 7) showing the coexistence of a crystalline phase (A) and an amorphous phase (B). From the diffraction patterns and spacing between lattice fringes of ≈ 3.0 nm, the crystalline phase was established to be 2212. Energy dispersive spectroscopy (EDS) on both regions had identical chemical compositions close to the 2212 stoichiometric ratio.

Fig. 11 shows an SEM micrograph of the cross section of a pellet sintered for 12 hrs. in 7% O_2 at 845°C, displaying the micaceous morphology typical of the 2223 phase. TEM micrographs of a $Bi_{1.8}Pb_{0.4}Sr_{2.0}Ca_{2.2}Cu_{3.0}O_y$ sample, shown in Fig. 12a, indicate fine particles within the grains of the 2223 phase. It should be noted that the micrograph shown is atypical in that it had an unusually large number of such particles within the 2223 grain. High resolution microscopy establishes these to be the 2201 phase as seen from the lattice fringes, shown in Fig. 12b, which indicate a "c" axis of 25 Å. EDS analysis shows these particles to be close compositionally to the 2201 phase. Lattice fringes from the matrix indicate a "c" axis of 37 Å corresponding to the 2223 phase. The grain to grain contact between 2223 grains is seen to be good, with no observable grain boundary phase present, as illustrated in Fig. 13.

Evidence of Amorphous Phases in 2223 Synthesized From Solid State Reaction Precursors in Standard Powder in Tube Processing

Silver sheathed $(Bi,Pb)_2Sr_2Ca_2Cu_3O_x$ (2223) tapes were fabricated by Argonne National Laboratory using the powder-in-tube (PIT) technique. Fig. 14a and b show the same area of a heat treated cross section before and after etching. A web-like network that corresponded one-to-one with the large dark areas between regions of 2223 grains was revealed and was shown to be Bi-rich.

XRD was employed to initially investigate the crystallinity of the Bi-rich phase revealed by etching the heat treated sample. The before and after etch XRD spectra are shown in Fig.

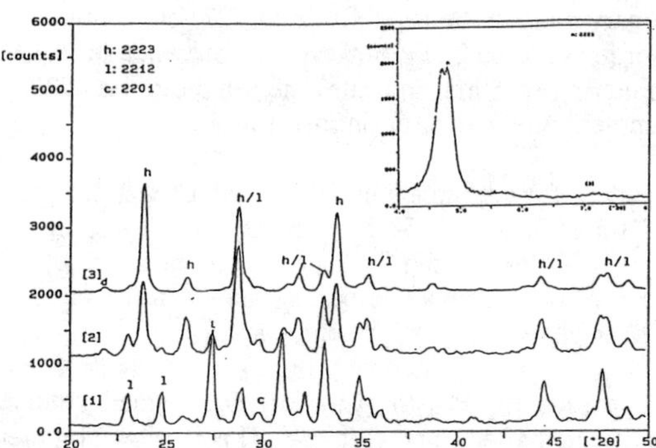

Fig. 7 XRD of the sample sintered at 845°C in 7% O_2 indicating the evolution of the 2223 phase. Each step is accompanied by grinding and pelletizing. Shown in insert, is the detail of the low angle 2223 phase (002) reflection for the sample sintered for 12 hrs. [1]: 2hrs.; [2]: 2+5 hrs.; [3]: 2+5+5 hrs.

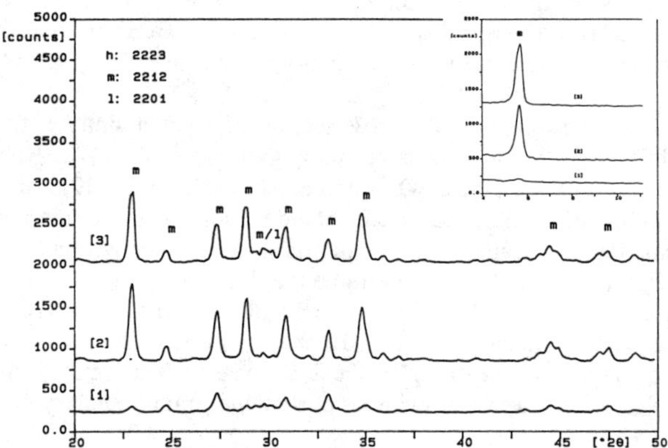

Fig. 8 XRD of samples without lead sintered at 845°C in 7% O_2 with no indication of 2223 phase. Each step is accompanied by grinding and pelletizing. Shown in insert, is the detail of the low angle 2212 phase (002) reflections. [1]: 2hrs.; [2]: 2+5 hrs.; [3]: 2+5+5 hrs.

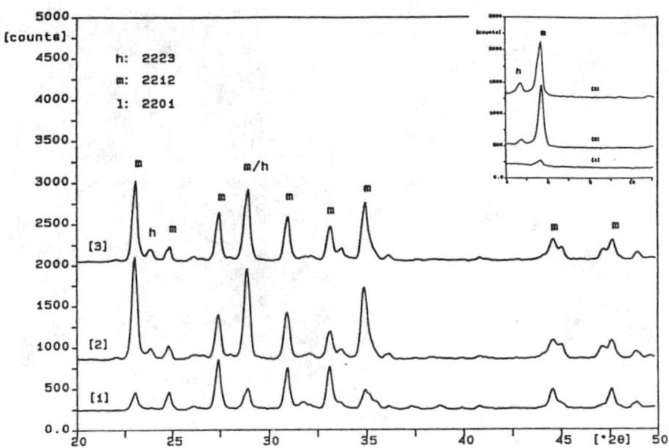

Fig. 9 XRD of $Bi_{1.8}Pb_{0.2}Sr_{2.0}Ca_{2.0}Cu_{3.0}O_y$ samples sintered at 845°C in 7% O_2 with small fraction of 2223 phase detected after 7 hrs. Each step is accompanied by grinding and pelletizing. Shown in insert, is the detail of the low angle 2212 and 2223 phase (002) reflections. [1]: 2hrs.; [2]: 2+5 hrs.; [3]: 2+5+5 hrs.

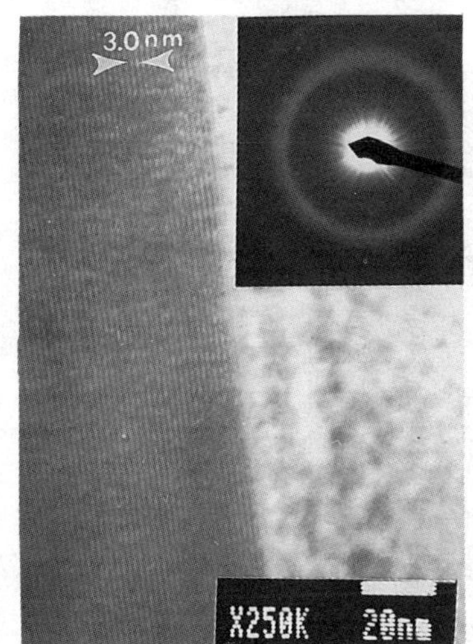

Fig. 10 High-resolution TEM micrograph taken in the dark field (DF) image mode of sample heat treated for 2 hrs. at 845°C (XRD pattern 1 in Fig. 7) showing the co-existence of a crystalline phase (A) and an amorphous phase (B).

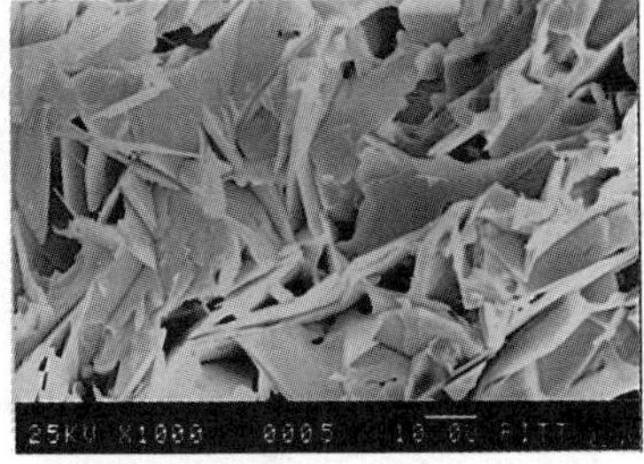

Fig. 11 SEM micrograph showing the micaceous morphology typical of the 2223 phase.

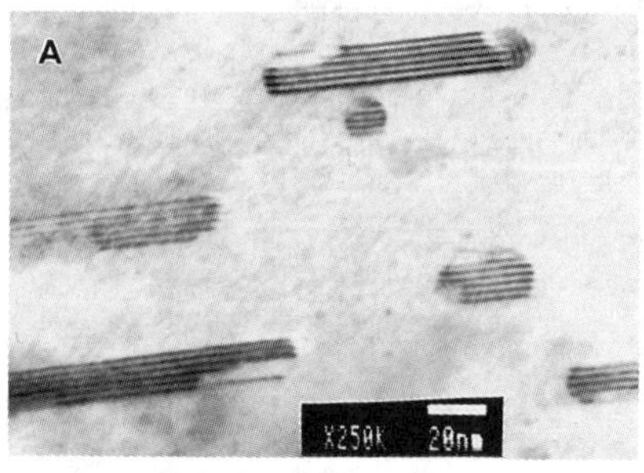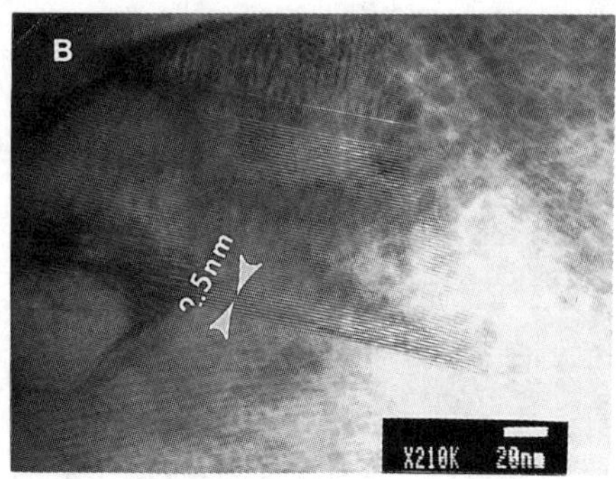

Fig. 12 TEM micrographs showing a) Fine particles of 2201 within the 2223 grains. b) Lattice fringes from particles within the matrix indicate a "c" axis of 25 Å corresponding to the 2201 phase.

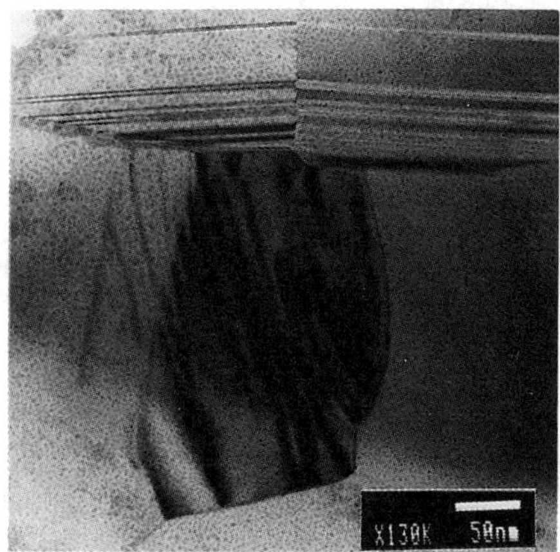

Fig. 13 TEM micrographs showing of contact between 2223 grains is seen to be good, with no evidence of the presence of a grain boundary phase.

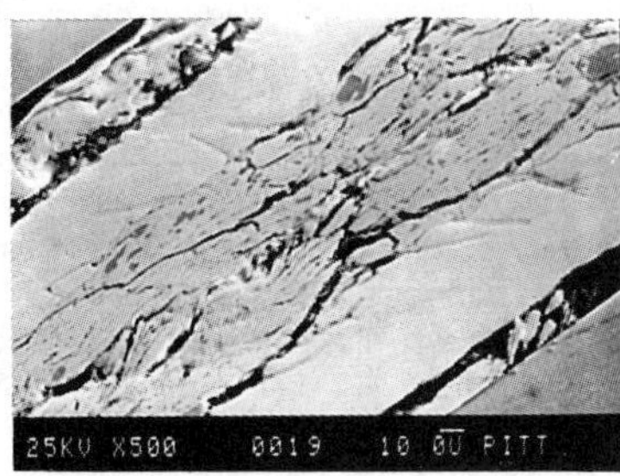
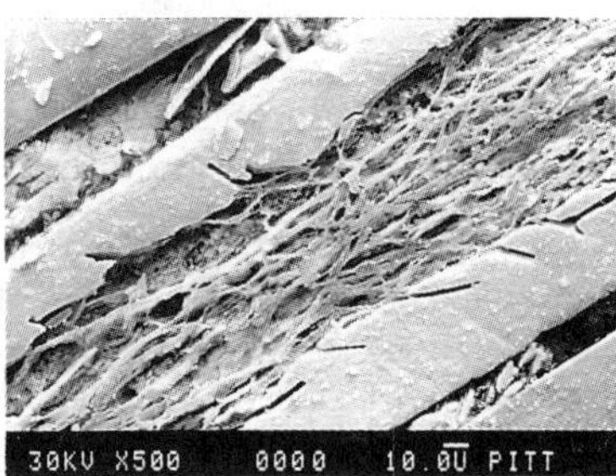

Fig. 14 SEM micrographs showing a longitudinal cross section of a silver sheathed 2223 tape a) before and b) after etching, revealing a predominantly amorphous Bi-rich phase.

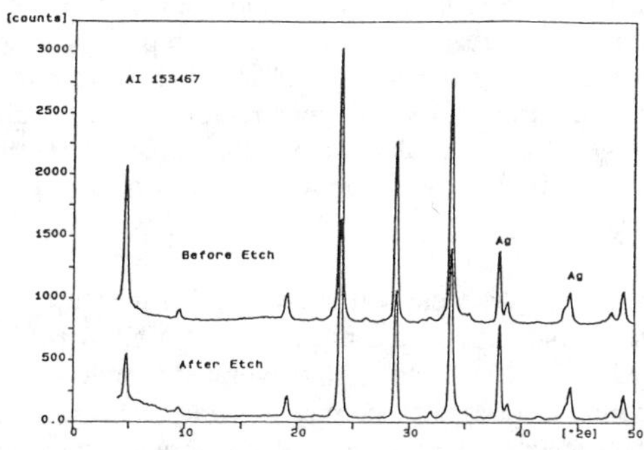

Fig. 15 XRD spectra a) before and b) after etching silver sheathed 2223 tapes.

15. The after etching XRD spectrum was identical to the before etching spectrum except for a slightly lower intensity which is expected from the decrease in material due to etching. TEM was then employed to further investigate this Bi-rich phase.

TEM samples were prepared by depositing the remains of powder samples, etched for 24 hrs., onto carbon coated TEM grids. Particles observed were mainly amorphous, as determined by electron diffraction, but most had small regions near the edges where electron diffraction patterns (EDP's) could be obtained (i.e. the presence of partial crystallinity was observed). Dark field imaging revealed a crystallite size in the range 5-20 nm. Also observed, while less frequent than the polycrystalline ring diffraction patterns, were small portions of particles where single crystal EDP's could be obtained. The spot EDP and ring patterns were indexed to be the $Bi_{24}Pb_2O_{40}$ phase. Many such particles were analyzed by electron diffraction and all could be indexed to the $Bi_{24}Pb_2O_{40}$ phase except one which indexed to lead free bismuth oxide phase.

Discussion

Prior studies on the freeze-drying of superconductors[21-26] have used atomization of solutions into liquid nitrogen as a means of achieving freezing. There are two disadvantages to this process. One is that the rates of freezing achieved by atomization into liquid nitrogen are modest. Lower freezing rates allow for the growth of ice crystals and a subsequent segregation of the solute. The second limitation, inherent in any atomization process, is the low output of the process in terms of product quantity. In the splat freezing process adopted in this study, both of these limitations are overcome. The heat transfer rates obtained by splat freezing the solutions against a metal surface cooled to liquid nitrogen temperature are sufficient to generate amorphous precursors. Such high cooling rates ensure that segregation of the solute in the freezing process is minimized and the atomic scale mixing obtained in the liquid is approached in the frozen material. We propose that the melting of these amorphous samples outside a narrow range of pH and NO_3^- concentration is the result of a phase separation into solute rich and solute poor regions. Such phase separation results in loss of chemical homogeneity and a subsequent reduction in the reactivity of the powders leading to a much slower rate of 2223 formation. Thus, control of pH and $[NO_3^-]$ are critical in the freeze drying process.

A further occurrence of melting is seen in the freeze-dried nitrate samples, which contain hydrated metal nitrates. This is due to the presence of $Ca(NO_3)_2 \bullet 4H_2O$ (m.p. 42°C) and $Cu(NO_3)_2 \bullet 3H_2O$ (m.p. 112°C). Flash heating the hydrated sample to above the temperature of decomposition of the nitrates does not completely avoid melting. This is particularly true of a packed bed, such as a powder in a crucible. A dehydration step with low heating rates (5°C/hr) is required to remove the waters of crystallization from the sample without causing melting.

It has been proposed that the 2223 phase grows from a partially molten phase containing Ca_2PbO_4.[18,28,29] Therefore, in the initial calcination step for the freeze-dried precursor the temperature was chosen to be below the temperature of liquid phase formation. The high reactivity of the freeze-dried precursor is seen from formation of a significant fraction of 2223 after holding for 7 hrs. and nearly phase clean 2223 is seen after 12 hrs. The displacement fringe contrast[30] seen in Fig. 12a is due to the partially coherent 2201 particles in the 2223 matrix. Due to the similar values of the "a" axis of the 2223 and the 2201 phases (a = 5.4 Å), it is energetically favourable to establish coherency along the basal planes, implying that the c-axes are parallel for both the matrix and the particles. The grain boundaries are clean, with good contact between adjacent 2223 grains. The absence of a grain boundary phase in these materials is particularly encouraging from the standpoint of maximizing critical current density (J_c) values in these materials. The presence of small amounts of impurity phases need not lead to a significant decrease in the transport properties of the superconducting material. It is the presence of a continuous transport path for supercurrent through the material that is of importance.

Onset temperature for liquid-phase formation can be lowered with the addition of lead and/or a decrease in Po_2. The weight loss associated with formation of the liquid phase can be detected by TGA with a heating rate of 5°C/min. As shown in Fig. 6, a decrease in Po_2 for the sample without lead lowers the onset temperature by ≈ 20°C. Interpretation of weight loss results is complicated by other concurrent effects such as exchange of oxygen between the sample and environment, variation in oxygen content, and oxygen diffusion in the 2212 system (without lead).[31,32] If the reported Arrhenius-type behavior for diffusion of oxygen into the lead-free 2212 phase is used[32], a diffusivity of $D \approx 4 \times 10^{-5}$ (cm^2/s) is obtained at 800°C. This indicates that the weight loss measured by TGA in a sample without lead is actually the difference between oxygen gain by the 2212 phase[31,32] and oxygen loss due to formation of the liquid phase.[27] An increase in lead content up to 0.4 lowers the onset temperature for formation of liquid phase from 810 to 790°C (7% O_2 atmosphere). This seems to favor arguments about Ca_2PbO_4 phase formation, and therefore the role of lead as a liquid-phase former.

XRD results indicated that increase in lead content affected not only formation of the liquid phase but also formation of the high-T_c 2223 phase. From the XRD data for a $Bi_{1.8}Pb_{0.4}Sr_{2.0}Ca_{2.0}Cu_{3.0}O_y$ a significant amount of 2223 phase was detected after heat treating for 7 hrs. at 845°C (Fig. 7, curve 2). However, XRD data for the sample without lead for the same heat treatment (Fig. 8, curve 2), showed 2212 as the major phase, with no indication of peaks characteristic of the 2223 high-T_c phase. This experimental observation indicates that the lead content actually affected the dissolution of the 2212 phase into the liquid phase. If the lead content in the initial composition becomes greater than about 20% of the bismuth content, the 2212 phase will dissolve. If the ratio of lead to bismuth is < 20%, the 2212 phase will not dissolve completely (Fig. 15, curve 2) and will remain even after further heat treatments.

To verify observations from TGA, DTA, and XRD measurements, high-resolution microscopy, coupled with EDS analysis, was performed. As shown in Fig. 10a, coexistence of the 2212 phase and amorphous phase was verified. The MBD diffraction pattern from the amorphous phase exhibited only one diffuse ring. The dissolution interface lies along the c-plane, indicating that dissolution occurs preferentially along the a and b planes due to the short diffusion paths. EDS analysis from both regions showed the presence of lead and bismuth (Fig. 10b). This was at least a qualitative indication that the presence of lead at a certain level in the 2212 phase was affecting its dissolution behavior. This observation correlates well with the study by Dorris et al.[10] The highest I_c and J_c values were obtained when all of the lead was incorporated into the 2212 phase.[10]

Our observation is in general agreement with the report by Chen and Stevens.[18] The actual formation mechanism of the 2223 high-T_c phase consists of the dissolution of the 2212 phase into the liquid phase and subsequent nucleation and growth of the high-Tc phase out of the liquid phase. We have previously reported particles of the 2201 phase inside the grains of the 2223 high-T_c phase.[33] Morgan et al.[34] indicated that 2223 is slowly formed from precursors by small, bismuth and lead-rich, liquid droplets that migrate over growing platelets and deposit ledges of product in their wakes.

The application of these precursors for wire fabrication through a powder in tube process is being investigated. Sintering times with freeze-dried powders are an order of magnitude shorter than for powders obtained from conventional solid state processing, which typically requires 200-300 hrs. of sintering to achieve near single phase 2223. Therefore long length conductor fabrication using powders derived by the freeze drying technique is commercially and economically viable. Further, the possibility of sufficiently small coherent particles of 2201 in the matrix may provide sites for flux pinning.

Conclusions

A freeze-drying approach incorporating a splat freezing process was adopted for the synthesis of the 2223 phase from nitrate precursors. The freeze-dried nitrate precursor is seen to be highly reactive, leading to single phase 2223 material after 12 hr of sintering. TEM coupled with EDS analysis indicates that the 2201 particles are present within the 2223 grains. The presence of these coherent particles in the matrix may provide sites for flux pinning. The grain boundaries of the 2223 grains are free from secondary phases, allowing for good contact between the 2223 grains.

The onset temperature for liquid-phase formation in the Bi-Sr-Ca-Cu-O system was lowered by both lead doping and a decrease in oxygen partial pressure. The increase in lead content also increased the rate of formation and increased the stability of the 2223 phase. If the lead concentration was less than $\approx 20\%$ of the bismuth content, the 2212 phase did not dissolve completely.

High-resolution TEM demonstrated that the 2212 phase dissolves along a and b planes. The 2223 phase then grows from the liquid phase by nucleation and growth.

Evidence for a noncrystalline phase is evident from XRD results showing the only the presence of the 2212 phase during the early stages of synthesis of the 2223 phase, when the overall composition is 2223.

Further evidence of a noncrystalline phase is also evident from etched "2223" synthesized by standard solid state powder in tube processing. The combined analyses of SEM, XRD, and TEM indicate this phase to be a largely amorphous Bi-rich phase with a significant ($\approx 10\%$) amount of Pb. The amorphous nature of this phase is consistent with the existence of a precursor liquid phase.

Acknowledgments

We thank Mr. C. Van Ormer, from the University of Pittsburgh, for his assistance with the microscopy. This work was supported by the U.S. Department of Energy, Energy Efficiency and Renewable Energy, as part of a program to develop electric power technology, under Contract W-31-109-Eng-38.

References

1. H. Maeda, T. Tanaka, M. Fukotomi, and T. Asano, Jpn. J. Appl. Phys. 27 (1988) L209.

2. S.A. Sunshine, T. Siegrist, L.F. Shneemeyer, D.W. Murphy, R.J. Cava, B. Batlogg, R.B. van Dover, R.M. Fleming, S.H. Glarum, S. Nakahara, R. Farrow, J.J. Krajewski, S.M. Zahurak, J.V. Wasczak, J.H. Marshall, P. Marsh, L.W. Rupp, and W.F. Peck, Phys. Rev. B. Condens. Mater., 38 [1] (1988) 893.

3. R.J. Cava, B. Batlogg, S.A. Sunshine, T. Siegrist, R.M. Fleming, K. Rabe, L.F.Schneemeyer, D.W. Murphy, R.V.van Dover, P.K. Gallagher, S.H. Glarum, S. Nakahara, R.C. Farrow, J.J. Krajewski, S.M. Zahurak, J.V. Waszczak, J.H.Marshall, P. Marsh, L.W. Rupp Jr., W.F. Peck and E.A. Rietman, Physica C 153-155 (1988) 560.

4. B.W. Staff, Z. Wang, M.J. Lee, J.V. Yakhmi, P.C.De Camargo, J.F. Major and J.W.Ruter, Physica C 156 (1988) 251.

5. U. Balachandran, D. Shi, D.I. Dos Santos, S.W. Graham, M.A. Patel, B. Tani, K. Vandervoort, H. Claus and R.B. Poeppel, Physica C 156 (1988) 649.

6. Y. Idemoto, S. Ichikawa and K. Fueki, Physica C 181 (1991) 171.

7. M. Xie, L.W. Zhang, T.G. Chen, X.T. Li and J. Cai, Physica C 206 (1993) 251.

8. P. Strobel, J.C. Toledano, D. Morin, J. Schneck, G. Vacquier, O. Monnereau, J. Primot and T. Fournier, Physica C 201 (1992) 27.

9. B. Hong and T.O. Mason, J. Amer. Cer. Soc., 74 [5] 1045 (1991).

10. S.E. Dorris, B.C. Prorok, M.T. Lanagan, S. Sinha and R.B. Poeppel, Physica C 212 (1993) 66.

11. U. Balachandran, Argonne National Laboratory, private communication (1993).

12. U. Endo, S. Koyama and T. Kawai, Jpn. J. App. Phys., 27 (1988) L1476.

13. U. Endo, S. Koyama and T. Kawai, Jpn. J. App. Phys., 27 (1988) L1861.

14. A. Sumiyama, T. Yoshitomi, H. Endo, J. Tsuchiya, N. Kijima, M. Mizuno and Y. Oguri, Jpn. J. App. Phys., 27 (1988) L542.

15. N. Kijima, H. Endo, J. Tsuchiya, A. Sumiyama, M.Mizuno and Y. Oguri, Jpn. J. App. Phys., 27 (1988) L821.

16. D. Shi, M. Tang, K. Vandervoot and H. Claus, Phys. Rev. B39 (1989) 9091.

17. J.M. Taracon, W.R. McKinnon, P. Barboux, D.M. Hwang, B.G. Bagley, L.H. Greene, G.W. Hull, Y. LePage, N. Stoffel and M. Girgoud, Phys. Rev. B 38 (1988) 8885.

18. Y.L. Chen and R. Stevens, J. Am. Ceram. Soc., 75 [5] (1992) 1142.

19. A. Umezawa, Y. Feng, H. Edelman, Y. High, D. Larbalestier, Y. Sung and S. Flesher, Physica C, 198 (1992) 261.

20. J.S. Luo, N. Merchant, V.A. Maroni, D.M. Gruen, B.S. Tani, W.L. Carter and G.N. Riley Jr., Applied Superconductivity 1 (1993) 101.

21. S.M. Johnson, M.I. Gusman, D.J. Rowcliff, T.H. Geballe and J.Z Sun, Adv. Ceram. Mater., 2[B] (1987) 337.

22. M. Strasik and N.G. Eror, "A Novel Preparation Technique for High-Temperature Superconductor $YBa_2Cu_3O_7$", Proc. of Materials Research Society Symp. on Atomic and Molecular Processing of Electronic and Ceramic Materials, (Sept 1987), 189.

23. N.V. Coppa, G.H. Myer, R.E. Salomon, A. Bura, J.W. O'Reilly, J.E. Crow and P.K. Davis, J. Mater. Res., 7[8] (1992) 2071.

24. Y. Kimura, T. Ito, H. Yoshikawa, T. Tachiwaki and A. Hiraki, Jpn. J. App. Phys., 29 (1990), L1409.

25. K.H. Song, H.K. Liu, S.X. Dou and C.C. Sorrel, J. Am. Ceram. Soc., 73[6] (1990) 1771.

26. V. Primo, F. Sapina, M.J. Sanchis, R. Ibanez, A. Beltran and D. Beltran, Mat. Lett., 15 (1992) 149.

27. H. Zheng, M.W. Colby and J.D. Mackenzie, Journal of Non-Crystalline Solids 127 (1991) 143.

28. T. Hatano, K. Aota, S. Ikeda, K. Nakamura and K. Ogawa, Jpn. J. App. Phys., 27 (1988) L2055.

29. T. Uzumaki, K. Yamanaka, N. Kamehara and K. Niwa, Jpn. J. App. Phys., 28 (1989) L75.

30. R.B. Nicholson "Precipitation Studies in Electron Microscopy and Microanalysis of Metals" eds J.A. Belk and A.L. Davies, Elsevier Publ. Co. (1968).

31. G. Triscone, J.Y. Genoud, T. Graf, A. Junod and J. Muller, Physica C 176 (1991) 247.

32. S. McKernan and A. Zettl , Physica C 209 (1993) 585.

33. P. Krishnaraj, M.Lelovic, N. Eror and U. Balachandran, Physica C 215 (1993) 305-312.

34. P.E. Morgan, R.M. Housley, J.R. Porter and J.J. Ratto, Physica C 176 (1991) 279.

PRECURSOR PREPARATION BY A MODIFIED OXALATE COPRECIPITATION AND ITS SUITABILITY FOR MANUFACTURING BI(PB)SRCACUO-CONDUCTORS

Ch.Lang,[1,2] S. Gauss [1], W. Becker [1], M. Rupich[2] and W. Carter[2]

[1]Corporate Research II, Hoechst AG, Frankfurt/Main, FRG

[2]American Superconductor Corporation, Westborough, MA

Abstract

Two of the most promising methods for the preparation of conductors for energy applications of HTSC are the "Powder-in-tube-method" and the Doctor-Blade process using Ag as substrate material. We compared the suitability of precalcined B(PSCCO) precursor powders made by different routes, especially by a modified coprecipitation of oxalates. The material was tested in wire-form and as coated Ag-substrates. Phase purity, stoichiometry, homogeneity, reactivity, C-content and particle sizes were investigated by XRD, SEM, EDX, ICP-AES, AC-susceptibility and particle-size measurements. The precursor powders prepared by the modified coprecipitation exhibit improved properties in comparision to conventional routes. For "Powder-in-tube" tapes the precursor could be converted in monofilamentary samples in less than 100h with only one intermediate pressing step to j_c-values up to $1.5 * 10^4$ A/cm^2 (0T/77K) at this time. In the case of the Doctor Blade-process large surfaces could be homogeneously coated with a thickness of 15μm and reacted to dense and highly textured microstructure. Optimized j_c-values are up to $1.0 * 10^4$ A/cm^2 (0T/77K).

Processing of Long Lengths of Superconductors
Edited by U. Balachandran, E.W. Collings and A. Goyal
The Minerals, Metals & Materials Society, 1994

<h1 align="center">1. Introduction</h1>

The superconductors of the Bi-family, the 2212 low-T_c phase and the 2223 high-T_c phase, are the most promising ceramic materials for the preparation of conductors for energy applications. The Doctor-Blade-process with Ag as a substrate (1-3) and the Powder-in-tube method (4-6) have been successfully developed for conductors with high critical current density. Both preparation routes start from a precursor powder instead of phase pure material, which is processed to the wire or tape and is then finally reacted to the superconducting phase. In the literature authors often maintained that an important factor in achieving improved electrical properties is having a homogeneous precursor material, that is achieving a uniform distribution of phases within the precursorpowder. Normally the conventional mixed-oxides process is chosen for the preparation of the precursor material with iterational treatments to the powders including grinding (milling) steps and annealings. In our contribution we show the advantages of a precursor powder made by a modified coprecipitation of Bi(Pb)SrCaCu-oxalates.

<h2 align="center">2. Preparation of the Precursor Powders</h2>

A coprecipitation of oxalates for the BSCCO-system has already been described in the literature (7-9). In all cases an aqueous medium of nitrates was used for the precipitation. For this process a significant dependence on the pH-value has been observed, causing process difficulties especially in large batches. Also the relatively high solubility of the alkaline-earth oxalates in water has to be considered. To minimize this solubility effect and to prevent the dependence on the pH-value we worked with alcoholic solutions of oxalic acid during the precipitation as a precipitating agent. The metal components are handled as nitrates in concentrated aqueous solutions as possible.

Figure 1 shows the accuracy and the precision of this method. In this figure a box plot represents the deviations of dried oxalate precipitates from one Bi(Pb)SrCaCu-target composition of 12 separate runs determined by ICP-AES (Induced Coupled Plasma-Atom Emission Spectroscopy). The median of the values is portrayed by a horizontal line segment within a rectangle and the upper and lower quartiles of the data are portrayed by the top and bottom of the rectangle. The bars indicate the whole range of values. Taking into account that this graph reflects the combined effects of uncertainties in the determination of the raw material concentration, the coprecipitation step itself and the ICP-determination of the composition of the oxalate mixture, this observed accuracy in the range of $\pm 3\%$ for each element is fairly high.

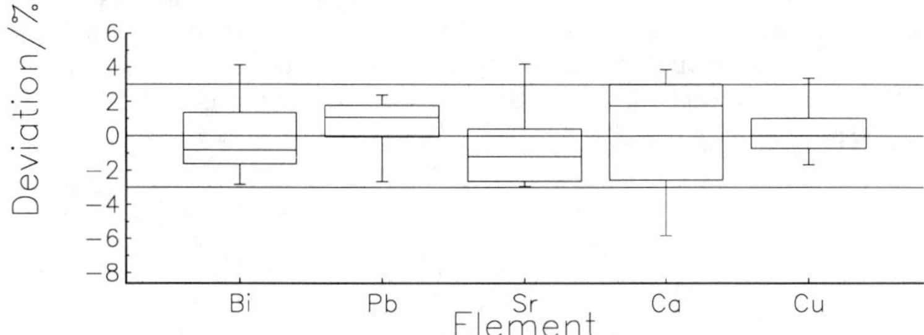

Figure 1   Box plot of the standard deviation of 12 independent samples for each element on a given target composition determined by ICP-AES. An explanation of the markers is given in the text.

3. Properties of the Coprecipitated Powders

To form the oxide precursor, the oxalates are decomposed to carbonates which are further decomposed to the oxides. This process can easily be monitored by a DT/TG analysis. In order to achieve high J_c in the conductor it is important to reduce the carbon content to low levels during the calcination process (10). Figures 2 and 3 show the dependence of the C-content on the calcination temperature and the oxygen partial pressure respectively. Two processes for the decomposition of Metalcarbonates can be taken into account:

1) $MeCO_3 \rightarrow CO_2 + MeO$ Me=Bi, Pb, Sr, Ca, Cu

2) $CO_2 \rightarrow CO + \frac{1}{2}O_2$

Figure 2 shows the dependence of the C-content on the calcination temperature. A higher calcination temperature reduces the C-content. This behaviour can be explained by equation 1). However the temperature must be limited to restrict grain growth and undesired sintering effects, which leads to an inhomogeneous product..

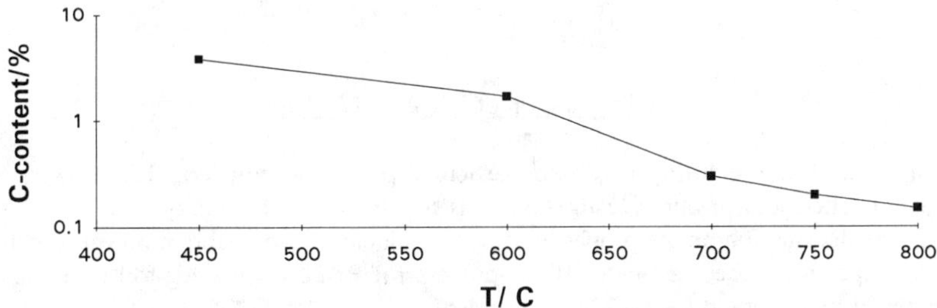

Figure 2   Dependence of the C-content of an oxalate precursor powder on the calcination temperature after 10h heattreatment

Figure 3 shows the impact of the oxygen partial pressure on the residual C-content. The oxygen partial pressure has been adjusted by a mixture of N_2/O_2. A low oxygen partial pressure should shift the equilibrium 2) to the right side. In combination with equilibrium 1) a decreasing oxygen partial pressure causes a decomposition of the particular carbonates. Our measurements showed a decrease of the remaining C-content to a minimum of 500 ppm

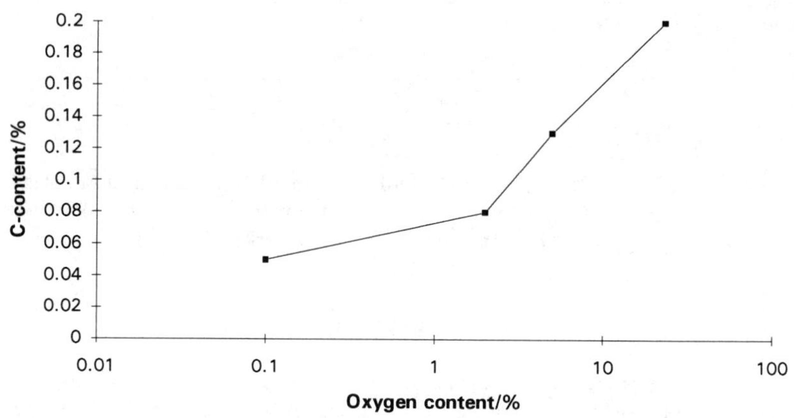

Figure 3   Dependence of the C-content of an oxalate precursor powder on the oxygen partial pressure (750°C/10h heattreatment)

The specific surface area of the oxalate powder measured by N_2-adsorption (BET-measurement) was typically in the range of 28 m^2/g. The primary particle size of this powder is in the subμm-range, observed by electron microscopy. SEM-investigations of powder, calcined at 750°C, show plate like crystals of the Bi-2212 phase. Bi-2212 as the main component was confirmed by XRD analysis. The nucleation and growth of this phase led to a formation of agglomerates, so that the observed d_{50}-values, determined by laser-light scattering, increased to 4-8μm after calcination. The reactivity of this Bi(Pb)SrCaCuO precursor with a composition of $Bi_{1.8}Pb_{0.4}Sr_2Ca_{2.2}Cu_3O_x$ could be demonstrated by a conversion of a pressed pellet of an oxide mixture to 60% Bi-2223 target phase in 10h at 845°C in air.

4. Preparation of B(P)SCCO - Tapes

In order to test the utility of these particular precursor powders, HTSC-tapes were prepared. The focus of our investigations was the impact of the homogeneity and a small grain size distribution on the properties of the conductor. Two different preparation routes for the tapes were used; a Doctor-Blade-process for Bi-2212 on a Ag-foil (1-3) and the Powder-in-tube method for Bi-2223 tapes (4-6).

4.1 Thick-films of Bi-2212 by a Doctor-Blade-Process

Thick films of Bi-2212 were prepared by casting a slurry containing precursor powder with the composition $Bi_{2.18}Sr_2CaCu_2O_x$ and organic additives on a 100μm Ag-foil. We chose a mixture of Ethanol/Toluol as the dispergant (68:32), Genamin CO 20 [®] (Hoechst AG) as a dispersant, Acryloid [®] (Hoechst AG) as a binder and Palatinol C [®] (Rhom + Haas) as a plasticizer. An optimized slurry contained about 60% powder and 40% additives. Because of the fine grain size very thin layers of finally 10-30μm thickness could be cast and dried on a foil with a width of 20cm and a length of more than 1m. After the burn out of the organic additives at 450°C on small samples the coating was partially melted at 890-900 °C for 6-10min followed by a crystallization step by cooling the melt from 870 to 830°C with 1 to 2°C/h. Finally the tape was given an additional treatment in pure N_2 at 670 °C/2h. An XRD pattern of the resulting tape showed single Bi-2212 phase with a nearly perfect c-axis alignment (00l-texture). Fig.4 shows a rocking curve of the (00<u>10</u>) reflection of a typical film sample. The full width half maximum of 4.6° reflects the misalignment of the Bi-2212 crystals.

The midpoint transition temperature of a typical film is $Tc_{50}=90K$ with a transition width (T_c10% to T_c90%) of 0.8K. A backscattered electron image of a polished cross section,

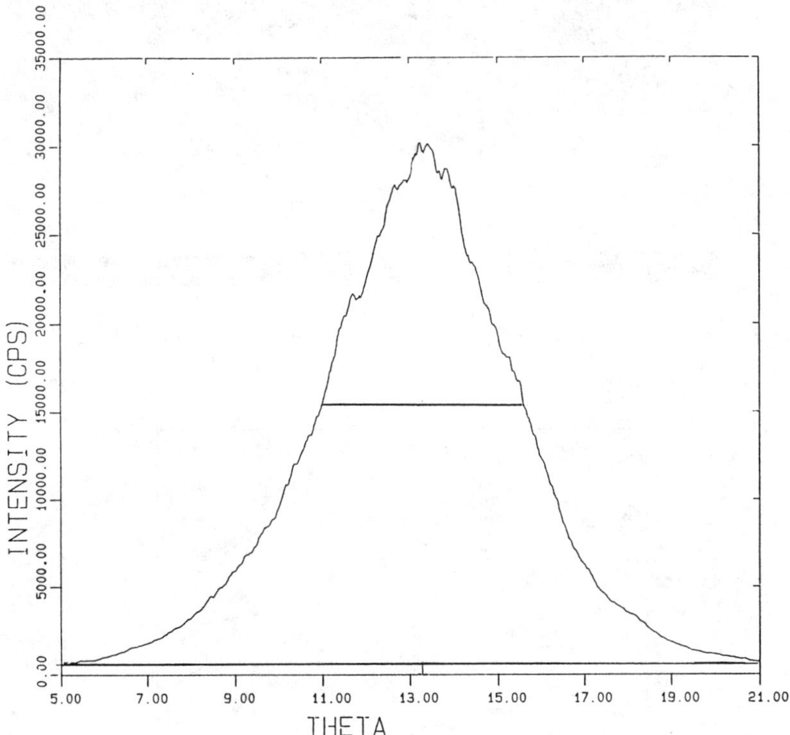

Figure 4   Rocking curve of (00<u>10</u>) reflection of a Bi-2212 tape which shows a full width half maximum of 4.6°

shown in Fig.5, reveals a dense microstructure with only a small amount of secondary phases, mainly Ca_2CuO_3 and some large silver grains.

These promising properties of the films led to J_c-values in the range of 1.0×10^4 A/cm^2 (77k/0T). Figure 6 shows the typical field dependence of a typical film for field applied parallel or perpendicular to the c-axis. Limitations of the critical current density are probably caused by defects as cracks and remaining secondary phases as Ca_2CuO_3.

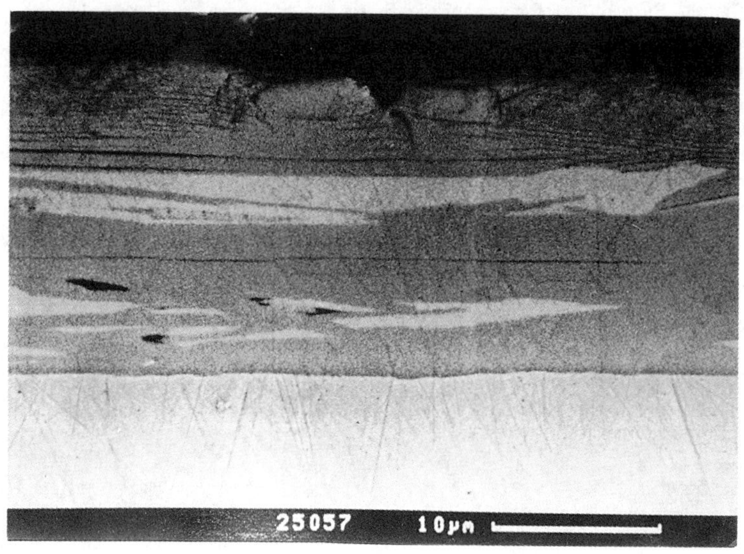

Figure 5   Microstructure of a cross section of a polished Bi-2212 thickfilm

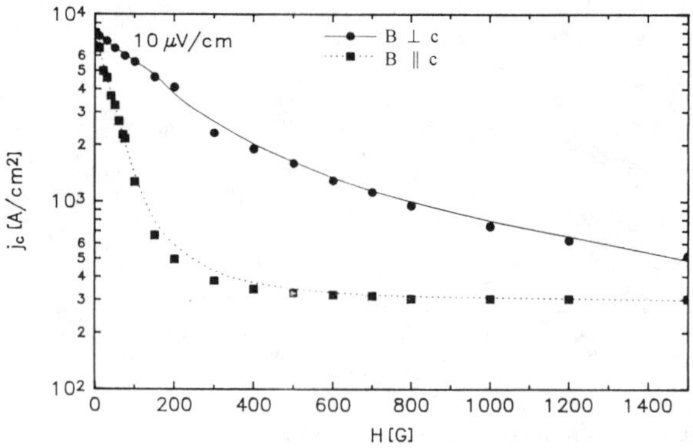

Figure 6   Field dependence of critical current density of a Bi-2212 thick film

4.2 BSCCO-2223 Tapes by the Powder-in-tube Method (PIT)

In order to test the utility of precursor powder derived by the coprecipitation route in the PIT process material of the composition $Bi_{1.8}Pb_{0.3}Sr_{1.9}Ca_{2.0}Cu_{3.1}$ was filled into a Ag-tube and conventionally cold deformed by extrusion, drawing and rolling steps to a tape of 160µm thickness and 2.5mm width (11). Figure 9 shows an optical micrograph of a longitudinal cross section of an unreacted composite tape, showing the fine grain size and the homogeneous distribution of the phases.

Samples, 2.5cm in length, have been processed in a pO_2 of 0.075atm at different temperatures between 805 and 827°C with one intermediate pressing step for an overall time of 96h. J_c values of 15 000 A/cm^2 were typically obtained for samples heat treated at 815°C. Figure 7 b) shows secondary phases like CuO (light particles) and Ca_2CuO_3 and imperfections as residual porosity and some misaligned crystals growing into the silver sheath of this particular sample. It can be expected, that further optimization of the heat treatment and pressing procedure yield higher j_c-values. These investigations have shown the potential of this material for the PIT-process. The high homogeneity and the fine particle sizes of the starting precursor material is especially important for manufacturing multifilamentary wires with reduced diameters of the single filaments.

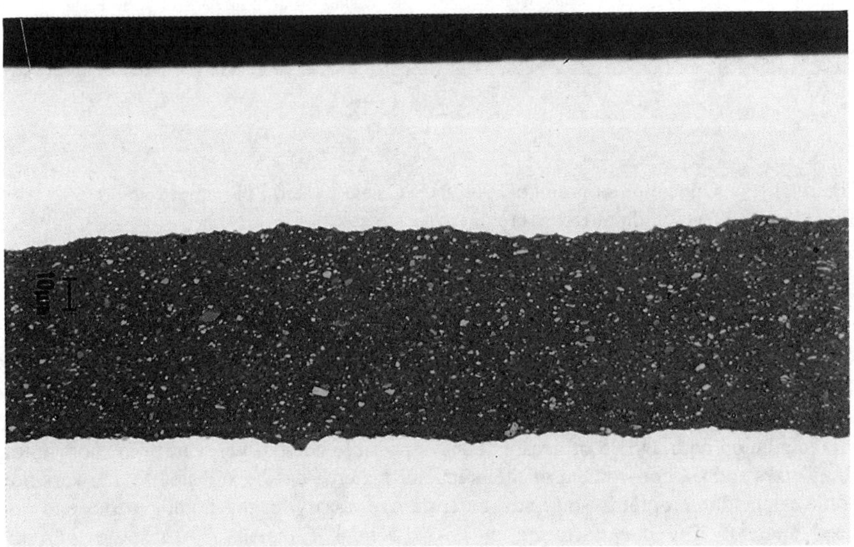

Figure 7a)   Optical micrograph of longitudinal cross sections of as rolled PIT-sample

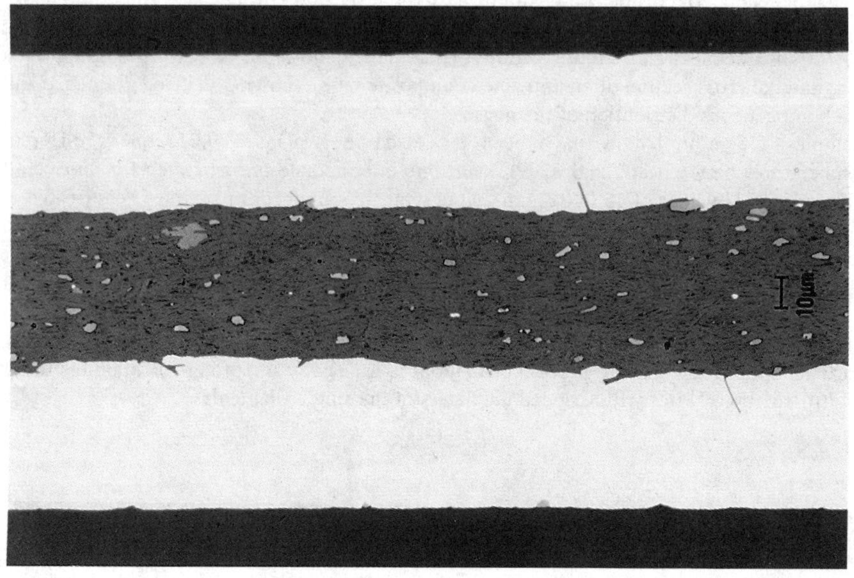

Figure 7b)   Optical micrograph of 96h/815°C heat treated PIT sample
              (longitudinal cross section)

5. Conclusions

We have developed a modified coprecipitation technique for producing Bi(Pb)SrCaCuO for use in the development of conductors for energy applications. This method has several advantages in comparision to the conventional coprecipitation method. The coprecipitation with a yield of nearly 100% is possible without pH-control in short times. Grain sizes and Carbon-content of the precursor material can be adjusted by the variation of the calcination program. No grinding steps are necessary for this homogeneous and fine sized material. The manufacturing of Bi-2212 thick films on Ag-substrate with the Doctor-Blade-process lead finally to 10-15µm coated foils with an optimized microstructure and j_c-values up to 1.0 x 10^4 A/cm^2. The use of Bi-2223 precursor powders in the Powder-in-tube method for short pressed monofilamentary samples shows a high potential for further investigations in multifilamentary wires especially in long length.

References:

1. B. Kress, H. Neumüller, H. Assmann, M. Wohlfart, J. Kase, H. Kumakura and K. Togano, Physica C 185-189(1991)2407
2. J. Shimoyama, T. Morimoto, H. Kitaguchi, H. Kumakura, K. Togano, H. Maeda, H. Nomura and M. Seido, Jpn. Jrnl. Appl. Phys. 31(1992)L163
3. T. Morimoto, J. Shimoyama, J. Kase and E. Yanagisawa Supercond. Sci. Technol. 5(1992)328
4. M. Ueyama, T. Hikata, T. Kato and K. Sato Jpn. Jrnl. Appl. Phys. 30(1991)L1384
5. R. Flükiger, B. Hensel, A. Jeremie, M. Decroux, H. Küpfer, W. Jahn, E. Seibt, W. Goldacker, Y. Yamada and J. Xu, Supercond. Sci. Technol. 5(1992)61
6. M. Wilhelm, H. Neumüller and G. Ries, Physica C 185-189(1991)2399
7. J. Hagberg, A. Kusimäki, J. Levoska and S. Leppävuori, Physica C 160(1989)369
8. G. Gritzner and K. Bernhard, Physica C 181(1991)201
9. Y. Sakka and M. Ohtaguchi, Jrnl. Mat. Sci. Lett. 11(1992)749
10. R. Flukiger, B. Hensel, A. Jeremie and M. Decroux, Supercond. Sci. Technol. 5(1992)61
11. K. Sandhage, G. Riley and W. Carter, JOM, 43(1991)21

Molten Salt Powder Synthesis in the Development of Practical Superconductors

S. S. Bayya, C. Park and R. L. Snyder

Institute of Ceramic Superconductivity
New York State College of Ceramics at Alfred University
Alfred, NY 14802

abstract

This paper presents our results on the growth of thallium containing superconducting phases from molten salts and also some preliminary results on grain orientation by tape casting. $T\ell$-2201, $T\ell$-2212, $T\ell$-1212 and $T\ell$-1223 superconductors were grown successfully from molten salts. The single layer thallium compounds were found to be stable in the molten salt only upon the addition of Pb and Sr. Initial tape casting experiments did not significantly improve the J_c of these superconductors. J_c values of 500 Acm^{-2} for $T\ell$-2212 and 250 Acm^{-2} for $T\ell$-1212 were observed.

Processing of Long Lengths of Superconductors
Edited by U. Balachandran, E.W. Collings and A. Goyal
The Minerals, Metals & Materials Society, 1994

Introduction

Since the discovery of superconductivity in the ceramic oxide systems, extensive research has been carried out by the scientific community to understand the parameters controlling superconductivity in these systems. Various factors that limit the current carrying capacity of these superconductors have been identified and attention has now turned to the use of various processing techniques to design the microstructure of the superconductors in order to produce high critical current densities. High J_c values have been achieved in thin films but the total currents carried are small. Powder-in-tube technology is showing some promise in developing practical wire. However, in most of the bulk processing methods the problems of weak links still need to be solved.

One of the approaches that we have been pursuing to produce high J_c ceramics is a molten salt synthesis[1, 2, 3]. Molten salt synthesis is a flux assisted growth technique where anisotropic crystallites grow from the molten salt acting as a solvent. The crystallite morphology can be controlled by controlling the processing parameters. These crystallites can be used to fabricate grain oriented ceramics by tape casting - lamination - sintering[4].

Our previous experiments[1, 2, 3], which were mainly focused on the double layer thallium compounds, proved the feasibility of using the molten salt process to grow anisotropic crystallites of superconductors in the $T\ell$-Ba-Ca-Cu-O system. Here we present our work on the molten salt synthesis of both the single and the double layer thallium compounds with the initial results of tape casting experiments.

The single layer compounds which have only one $T\ell$-O plane have been getting more and more attention after reports indicating that the stronger coupling between the Cu-O layers in these compounds can lead to improved transport properties in the presence of magnetic fields[5, 6]. Appreciably improved superconducting properties of single-layer compounds have been reported[7, 8, 9].

Experimental Procedure

Growth of $T\ell$-superconductors from salt

Reagent grade $T\ell_2O_3$, PbO, SrO_2, BaO_2, CaO and CuO were used in this study. Reaction batches with cation compositions 2-2-0-1, 2-2-1-2, 2-2-2-3, 1-1-1-2, 1-2-1-2 and 1-2-2-3 were made. 3.0 g of each superconductor precursor batch was mixed with 3.0 g of salt separately. Molten salt reactions were carried out at temperatures given in table-1 and 2. Details of the experiment have been discussed earlier[3]. Briefly, the mixtures are melted in Pt crucibles, soaked at temperature and then, at room temperature, the salt is removed by dissolving it in water. The powders obtained were characterized by X-ray powder diffraction and scanning electron microscopy for the phase assembly and crystallite morphology respectively.

Tape casting and lamination

Tℓ-2212 and Tℓ-1212 powders obtained from the KCl molten salt were used for the tape casting studies. To this powder, commercially available binder[1] with appropriate binder modifier[2] were added along with toluene in order to achieve the necessary rheological properties required in tape casting.

Powder was mixed with the binder, modifier and toluene by ball milling using zirconia balls as the grinding media. Ball milling was done for 12 hours to ensure homogeneity of the slurry. Tape casting was done on a flat glass plate using a hand held doctor-blade. The doctor-blade was used to control the thickness of the tape.

The green tapes were cut into 2.5 cm square pieces. 10-25 such pieces of green tape were laminated in a die at a temperature of 75°C and a pressure of 20 MPa. The 2.5 cm laminated monoliths were cut again with dimensions of 20 x 4 mm approximately. The thickness depends on the number of layers of green tape used.

To remove the binder, the laminated samples were heated up to 500°C/min. in a tube furnace in a flowing oxygen atmosphere. After the binder burnout, samples were again characterized by X-ray diffraction to ensure the stability of the superconducting phase during the binder burnout cycle.

The laminated bars were sintered from 800°C to 900°C for 1-3 hrs. Sintering was carried out in a closed crucible with loose powder of the same composition around it. These sintered samples were characterized by X-ray diffraction, SEM, resistivity measurements and critical current density measurements.

Results & Discussion

Growth of Tℓ-superconductors from salt

Table-1 illustrates the growth of superconducting phases from the precursor-salt system. The "*" in table-1 represents the presence of some unidentified peaks in the XRD patterns. As can be seen from table-1 the Tℓ-2201, and Tℓ-2212 superconductors are very stable in the salt systems in the absence of Pb & Sr and they can be formed from their precursor batches by molten salt synthesis. The Tℓ-2212 phase was found to be very stable over a wide compositional range as it was the dominant phase upon starting with the initial stoichiometries of 2-2-1-2, 1-1-1-2, 1-2-2-3 and 2-2-2-3. The 2223 phase is not particularly stable in the molten salts.

---

[1] Metoramic Sciences, Inc., Cladan Product No. B-73305

[2] Metoramic Sciences, Inc., Cladan Product No. M-1117

Table 1: Growth of Tℓ-superconducting phases from (Tℓ-Ba-Ca-Cu-O) precursor-salt systems

Precursor (Tℓ-Ba-Ca-Cu)	Salt System	Reaction Temperature	Reaction Time	XRD Results
2201	KCl-NaCl	900°C	1 hr.	Tℓ-2201, *
2212	KCl	900°C	1 hr.	Tℓ-2212
2212	KCl-NaCl	900°C	1 hr.	Tℓ-2212
2223	NaCl-KCl	850°C	1 hr.	Tℓ-2212, Tℓ-2223, CuO, *
2223	NaCl-KCl	900°C	30 min.	Tℓ-2212, CuO, Tℓ_2O$_3$
1112	NaCl-KCl	850°C	1 hr.	Tℓ-2212, CuO, *
1112	NaCl-KCl	900°C	1 hr.	Tℓ-2212, CuO, *
1223	NaCl-KCl	850°C	1 hr.	Tℓ-2212, Tℓ-2201, Tℓ_2O$_3$, *

Table 2: Growth of Tℓ-superconducting phases from (Tℓ-Pb-Ba-Sr-Ca-Cu-O) precursor-salt system (reaction time:1 hour). All the samples have a small amount of $(Ca_xSr_{1-x})_2CuO_3$. * indicates small amount.

Starting composition						Temp.	Phases obtained
Tℓ	Pb	Sr	Ba	Ca	Cu	(°C)	
0.5	0.5	2	0	1	2	800	1212
0.5	0.5	2	0	1	2	850	1212
0.5	0.5	2	0	1	2	900	1212
0.5	0.5	1.6	0.4	1	2	900	1212 + BaPbO$_3$
0.5	0.5	1.8	0.2	1	2	900	1212 + BaPbO$_3$
1	0	1.6	0.4	1	2	900	1212
0.5	0.5	2	0	2	3	900	1212 + 1223 + Ca$_2$PbO$_4$
0.5	0.5	1.6	0.4	2	3	900	1223 + BaPbO$_3$ + Ca$_2$PbO$_4$
0.5	0.5	1.6	0.4	2	3	850	1223 + BaPbO$_3$ + Ca$_2$PbO$_4$
0.5	0.5	1.0	1.0	2	3	900	1223 + BaPbO$_3$ + 1212*
0.5	0.5	1.3	0.7	2	3	900	1223 + BaPbO$_3$ + 1212*
0.5	0.5	1.8	0.2	2	3	900	1223 + 1212 + Ca$_2$PbO$_4$ + BaPbO$_3$*

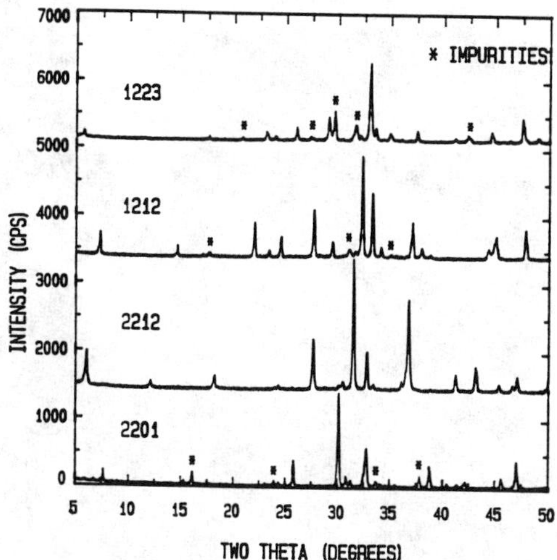

Figure 1: XRD patterns showing the growth of thallium superconductors from the salt melts

The single layer thallium compounds could only be stabilized by the addition of Pb and Sr to the system (table-2). This is consistent with other studies which suggest stabilization via partial substitution of Pb for Tℓ and partial or complete substitution of Sr for the Ba[7, 8, 10, 11, 12, 13, 14]. In comparison to the double layer Tℓ-2212 superconductor, the single layer Tℓ-1212 phase was also found to be quite stable but only in the presence of Pb and Sr. Tℓ-1223 phase was also observed from its stoichiometric ratio but it was not obtained as a single phase. In the presence of Pb & Sr in the starting composition all samples, after molten salt synthesis, contained a small amount of $(Sr,Ca)_2CuO_3$. It was also observed that in the presence of both Pb and Ba together, $BaPbO_3$ resulted as one of the impurity phases. Other phases obtained upon Pb & Sr additions are listed in table-2.

Figure-1 shows the X-ray diffraction pattern of powders obtained by molten salt synthesis. The reflections from the 00ℓ planes are enhanced in the diffraction patterns indicating a high degree of crystallite picture of the powder (figure-2). The picture shows a Tℓ-2212 powder grown from molten salt at 900°C where the Tℓ-2212 crystals are about 1-2μm thick and 10-15 μm in length and width. The crystallite shape and size can be controlled by varying the processing parameters and has been discussed elsewhere in greater detail[2, 3].

Tape casting and lamination   The average thickness of green tapes was about 100μm and were cut in 2.5 sq. cm size for lamination. Figure-3 shows the X-ray diffraction pattern of both surfaces of a green Tℓ-2212 tape. It was observed that the

Figure 2: SEM picture of a typical powder obtained from molten salt

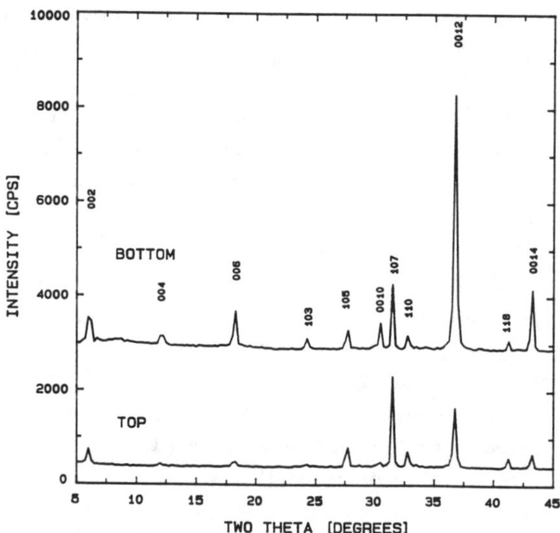

Figure 3: XRD patterns showing different degree of orientation on the two surfaces of a green tape

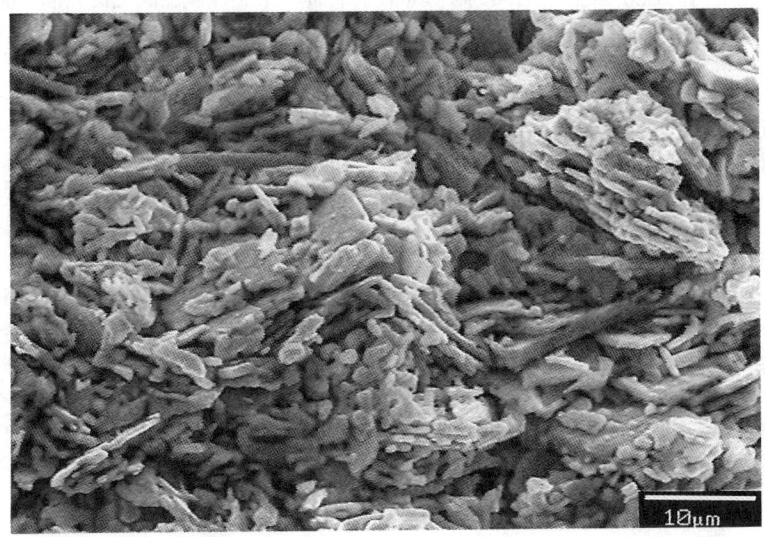

Figure 4: SEM picture of the cross-section of a laminated-sintered Tℓ-1212 sample

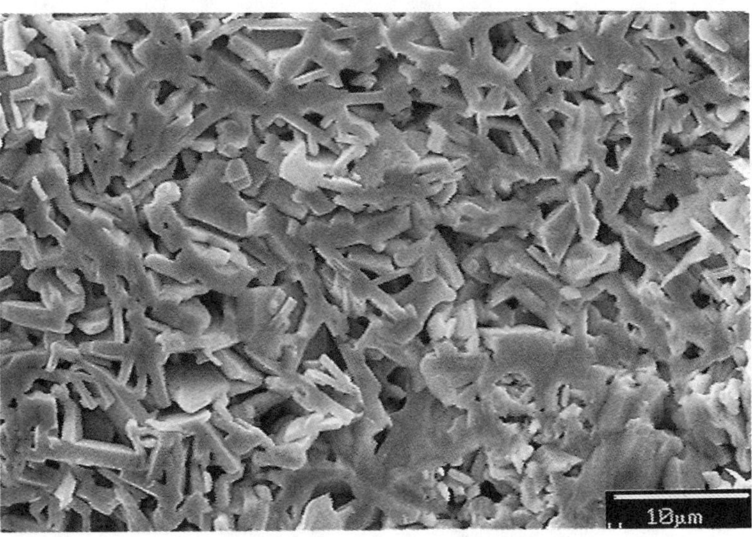

Figure 5: SEM picture of the cross-section of a pressed-sintered Tℓ-1212 sample

degree of orientation was different for different surfaces of the tape; the upper face (facing air) had a smaller degree of orientation than the lower face (facing the glass plate). The crystallite alignment in the lower face of the green tape was caused by particle sedimentation. To quantify the degree of preferred orientation, the Lotgering factor[15] (F) was calculated and was found to be 0.25 for the top surface while that at the bottom surface it was 0.76.

The samples after lamination were slowly heat treated in flowing oxygen for binder removal and were characterized by X-ray diffraction for phase stability. The samples were found to be stable after the binder burnout and showed the presence of the starting superconducting phase. The samples were again characterized by X-ray diffraction after sintering and were found to preserve the superconducting phase. Figure-4 shows an SEM picture of the cross-section of a laminated sample after sintering. For comparison, another sample was pressed (instead of tape casting and lamination) and sintered using the molten salt synthesized powder. Figure-5 shows the SEM picture of the cross-section of this sample. By comparison, some improvement in the grain orientation is observed in the tape-cast sample. However, the grains are still not oriented enough to improve the J_c values significantly. This is due to the fact that the thickness of the green tapes was about $100\mu m$ where as the thickness of the individual grain was about 1-2 μm. In general the microstructure can be improved substantially by reducing the thickness of the individual tapes to about 10 μm and also by varying the tape casting conditions to improve the orientation in the individual tapes; these studies are in progress. The grain orientation in the individual tapes was also observed to be different on either side of the tape. These can also be improved by changing the binder composition, viscosity of the slurry and powder loading.

T_c and J_c measurements on these sintered samples were carried out and are given in table-3. As can be expected from the microstructure of these samples the J_c values are still quite low. Further improvement should be possible by varying the tape casting parameters discussed above and this is the subject of current study.

Table 3: Electrical characterization of sintered samples

Composition	Sample	T_c (K)	J_c (Acm^{-2})
Tℓ-2212	Laminated bar	108	500 at 77 K
Tℓ-1212	Laminated bar	78	250 at 40 K
Tℓ-1212	Sintered bar	76	170 at 40 K

Conclusions

Tℓ-2201, Tℓ-2212, Tℓ-1212 and Tℓ-1223 superconductors were grown successfully from molten salts. The single layer thallium compounds were found to be stable in the molten salt only upon the addition of Pb and Sr. Tℓ-2212 and Tℓ-1212 phases showed greater stability compared to the other superconducting phases.

Initial tape casting experiments did not significantly improve the J_c of these superconductors. J_c values of 500 Acm^{-2} for Tℓ-2212 and 250 Acm^{-2} for Tℓ-1212 were observed. Further improvement in J_c should be possible by varying the processing parameters during tape-casting and by improving the final microstructure obtained.

Acknowledgments

We would like to thank the New York State Science and Technology Foundation and their Center for Advanced Ceramic Technology along with the New York State College of Ceramics for their sponsorship of this work.

References

1. S.S. Bayya, G.C. Stangle and R.L. Snyder, "Synthesis of Superconducting Phases in Tℓ-Ba-Ca-Cu-O System," Superconductivity and its Applications, Vol. 4, Y.H. Kao, H.S. Kwok, and A.E. Kaloyeros editors, (Am. Phys. Soc. 1992), 261.

2. S.S. Bayya and R.L. Snyder, "Growth of Anisotropic Tℓ-2212 crystallites by Molten Salt Synthesis", AIP Conference Proceedings 273, Superconductivity and its Applications, Hoi S. Kwok, David T. Shaw and Michael J. Naughton editors, (Am. Phys. Soc. 1993), 575.

3. S.S. Bayya and R.L. Snyder, "Growth of Anisotropic shaped Superconducting Particles in the Tℓ-Ba-Ca-Cu-O System," Physica C, **208**, (1993), 69.

4. M. Homes, R.E. Newnham and L.E. Cross, "Grain Oriented Ferroelectric Ceramics", Am. Ceram. Soc. Bull., **58**, [9] (1979), 872.

5. K.E. Gray and D.H. Kim, Physica C, **180** (1991), 139.

6. T.Nabatame, J.Sato, Y.Saito, K.Aihara, T.Kamo, and S.P.Matsuda, Physica C, **193** (1992), 390.

7. T.Kamo, T.Doi, A.Soeta, T.Yuasa, N.Inoue, K.Aihara, and S.P.Matsuda, Appl. Phy. Lett., **59** (1991), 3186.

8. R.S.Liu, D.N.Zheng, J.W.Loram, K.A.Mirza, A.M.Campbell, and P.P.Edwards, Appl. Phy. Lett., **60** (1992), 1019.

9. J.A.Deluca, P.L.Karas, J.E.Tkaczyk, C.L.Briant, M.F.Garbauskas, and P.J.Bednarczyk, Physica C, **205** (1993), 21.

10. M.A.Subramanian et al., Science, **242** (1988), 249.

11. J.C. Barry, Z. Iqbal, B.L. Ramakrishna, R. Sharma, H. Eckhardt, & F. Reidinger, J. Appl. Phy., **65** (1989), 5207.

12. C. Martin, J. Provost, D. Bourgault, B. Domenges, C. Michel, M. Hervieu, & B. Raveau, Physica C, **157** (1989), 460.

13. S.P.Matsuda, T.Doi, A.Soeta, T.Yuasa, N.Inoue, K.Aihara, and T.Kamo, Physica C, **185-189** (1991), 2281.

14. T.Doi, M. Okada, A. Soeta, T.Yuasa, K.Aihira, T.Kamo, and S.P.Matsuda, Physica C, **183** (1991), 67.

15. F.K. Lotgering, "Topotactical Reactions with Ferrimagnetic Oxides having Hexagonal Crystal Structures-I", J. Inorg. Nucl. Chem., **9** (1959), 113.

LIQUID AND SOLID SOLUBILITY OF Ag IN $Bi_2Sr_2CaCu_2O_8$

R. W. McCallum, K. W. Dennis, L. Margulies, and M. J. Kramer

Ames Laboratory,
Iowa State University
Ames, IA 50011

Abstract

Currently, the leading candidates for usable high temperature superconducting wires are Bi-Sr-Ca-Cu-O tapes produced by modifying the powder in tube process. In order to obtain crystallographic texture, these tapes are processed by partial melting then slow cooling through the peritectic. While it is well known that Ag suppresses the lowest liquidus temperature of the Bi based materials, no serious investigation of the Bi-Sr-Ca-Cu-O-Ag phase diagram has been undertaken. Solubilities of Ag in Bi2212 melt was determined by ICP-AES and TEM/EDS on samples of Bi2212 and 20 wt.% Ag which were melt-spun in air at temperatures of 900 to 1200°C. The most prominent feature of the phase diagram was found to be a liquid immiscibility at all temperatures. There exists a eutectic on each side of this gap. For the Ag rich side the eutectic is at 98 at.% Ag and depresses the melting by 15°C over that for pure silver for the same oxygen partial pressure. The oxide rich eutectic contains no more than 4 at.% Ag and depresses the solidus by 30°C. The solid solubility of Ag in Bi-Sr-Ca-Cu-O glass (oxide glass) is found to be less than the detection limits of TEM/EDS and Ag is systematically excluded from the oxide glass.

Processing of Long Lengths of Superconductors
Edited by U. Balachandran, E.W. Collings and A. Goyal
The Minerals, Metals & Materials Society, 1994

<u>Introduction</u>

Currently, the leading candidates for usable high temperature superconducting wires are Bi-Sr-Ca-Cu-O tapes produced by modifying the powder in tube process. In order to obtain crystallographic texture, these tapes are processed by partial melting then slow cooling through the peritectic. There exist in the literature numerous, sometime conflicting, reports of the effect of Ag additions to the $Bi_2Sr_2CaCu_2O_8$ (Bi2212) system. From the standpoint of superconductivity, it has been reported that Ag_2O doped Bi2212 prepared via melt-quenching and heat treated in a vacuum shows a decrease in T_c and J_c (1) while others have reported little to no effect on superconducting properties when Ag is added to sintered Bi2212 (2,3). Numerous attempts to introduce Ag inclusions into the superconducting core of these wires have been made to try and improve flux pinning and mechanical properties. If small Ag precipitates can be formed within Bi2212 grains, flux pinning may be enhanced. Alternatively, if Ag can be concentrated at the grain boundaries, it may increase both the mechanical behavior and the weak linked nature of the ceramic. Obviously, the effect of Ag on both the mechanical and superconducting properties of Bi2212 will depend on how the Ag is distributed in the superconducting matrix. Tracer diffusion studies of [110]Ag in Bi2212 shows high diffusivity rates in the a-b plane (4). Therefore, the amount of Ag that goes into the Bi2212 structure will be limited solely by its solubility. While it is well known that Ag suppresses the lowest liquidus temperature of the Bi based materials, no serious investigation of the Bi-Sr-Ca-Cu-O-Ag phase diagram has been undertaken. A detailed knowledge of the Ag-Bi2212 phase diagram is important in recognizing the inherent limitations on Ag-Bi2212 compounds and how they should be processed. When considering the Bi-Sr-Ca-Cu-O-Ag phase diagram, the oxygen partial pressure dependence of the Bi2212 and Bi2212+Ag thermal events must be considered (5).

<u>Experimental Procedures</u>

In order to determine the high temperature solubility of Ag in a Bi2212 melt, samples were quenched from the temperature of interest by chill block melt spinning. Melt-spinning charges were prepared by blending Kalichemie $Bi_2Sr_2CaCu_2O_x$($< 30\mu$m) with 20 wt% Ag powder(<5 μm) then pressing into powder pellets for ease of handling and greatest contact of components. SiO_2 was used as crucible material at 1100°C and below and Al_2O_3 was used at 1100°C and above. Orifice diameters ranged from 1.0 mm at lower temperatures to 0.85mm at higher temperatures to control self-purging of the charge due to viscosity changes. All melt-spinning was performed in air. Direct radio-frequency (RF), at 450kHz, induction heating of the pellets was unsuccessful, as the silver would melt, segregate and flow without conductively heating the ceramic oxide sufficiently enough to melt. Charges were, therefore, heated radiantly using an Inconel susceptor heated by RF. This susceptor also acted as a shielding antenna, eliminating RF noise from the thermocouple signal and excluding direct RF heating of the charge. Temperature was monitored simultaneously by a K-type thermocouple in a SiO_2 sheath embedded in the melt and by a two-color infared pyrometer aimed at the susceptor. On reaching the desired temperature, the molten stream was quenched onto a copper chill block, at ambient temperature, rotating at 20 meters per second. Ejection pressures were 3 - 4 psi.

TEM was performed using a Philips CM30 at 300kV equipped with a Be window energy dispersive detector (EDS). The samples were prepared by floating lightly ground ribbon pieces in ethanol onto a gold grid. Thermal analyses were done on lightly ground ribbon pieces using a Perkin-Elmer DTA1700 in flowing (50 cc/min.) dry CO_2-free air Chemical analyses were performed using ICP-AES on ribbon pieces which had been optically examined at 50x to assure separation of product types. However, this was not always possible.

Results

Melt-Spinning

In initial experiments, it was observed that in cases where after reaching temperature, the Ag-Bi2212 melt failed to shoot from the crucible, the solidified melt consisted of a Ag˙ nodule at the bottom of the crucible surrounded by oxide material. This suggests an extremely low solid solubility of Ag in Bi2212 melt. Following optimization of the furnace parameters, the melt spun material that resulted from a run could be separated into two distinct types. Long lengths of Ag ribbon (10 - 50 cm) were thrown into a collection tube. Of these ribbons, some showed a double-sided morphology. The remainder of the ribbons were nearly purely metallic with small streaks of oxide glass near the ribbon edges. It seems clear that the double sided ribbons resulted from a phase separated liquid stream. It is not clear, though, if the oxide glass seen on the metal ribbons was due to separation in the liquid or subsequent separation upon solidification. The second distinct type of melt-spun material was an oxide glass flake, 5 - 10 mm in length. The oxide glass flakes were thrown off the wheel at a lower trajectory than the metal ribbons, and as a result did not land in the collection tube. Instead the oxide glass flakes collected in the wheel chamber. Often, a large clump of tangled oxide glass ribbon would collect just below the opening to the collection tube. XRD analysis of this material showed significant crystallinity.

ICP-AES

Chemical analysis of the oxide and metal ribbons are presented in Table 1. The oxide ribbons show an increase in Ag content from 6.1 at. % at 900°C to 9.8% at 1200°C. The two samples processed at 1100°C showed anomalously low Ag contents of 6.3% and 4.9%. In both these cases the ribbons agglomerated, which may have resulted in a slow quench and possible phase separation, explaining the lower Ag contents. The Bi content of the oxide ribbons steadily decreased with increasing temperature from 30% at 900°C to 24% at 1200°C. Similarly, the Cu content decreased from 28% at 900°C to 25.2% at 1200°C. The Sr and Ca content of the oxide ribbons showed little systematic variation with temperature. The Si and Al contamination was minimal for samples processed in quartz, while samples processed in Al_2O_3 showed 4.5% Al and 0.3-2.1% Si in the oxide ribbons and negligible Si and Al content in the metal ribbons. The metal ribbons showed a decrease in Ag content with increasing temperature from 99% at 950°C to 96.8% at 1200°C. Again, the samples processed at 1100°C gave slightly anomalous results. Finally, the amount of Cu in the metal ribbons increased with temperature from 0.8% at 950°C to 2.2% at 1200°C. This may reflect an increasing tendency to form metallic Cu with increasing temperature, which in turn is soluble in Ag.

TABLE 1. ICP analysis, in mole % and cation ratio of Bi:Sr:Ca:Cu with Cu fixed at 2, of the precursor Bi2212, oxide ribbons, and metal ribbons. Mole percentages are accurate to 1 to 3% relative.

MOLE %

	Bi	Sr	Ca	Cu	Ag	Si	Al	Bi/Sr/Ca/Cu
PRECURSOR								
	25.5	27.2	13.5	27.2	-	-	-	1.88/2.00/1.00/2.00
OXIDE RIBBONS								
Temp(C)								
900	30.0	24.5	11.2	28.0	6.1	0.1	0.0	2.14/1.75/0.80/2.00
950	28.4	24.8	12.5	26.4	7.6	0.2	0.1	2.15/1.87/0.95/2.00
1000	28.4	24.0	12.4	27.6	7.5	0.1	0.0	2.05/1.74/0.90/2.00
1100#	28.0	25.2	12.7	27.5	6.3	0.3	0.0	2.04/1.84/0.93/2.00
1100*#	24.7	25.2	12.7	25.9	4.9	2.1	4.5	1.91/1.95/0.98/2.00
1200*	24.0	23.5	11.7	25.2	9.8	0.3	4.6	1.99/1.86/0.93/2.00
METAL RIBBONS								
950	0.0	0.0	0.0	0.8	99.0	0.0	0.0	0.00/0.00/0.00/2.00
1000	0.3	0.2	0.1	1.2	98.2	0.0	0.0	0.50/0.33/1.68/2.00
1100#	0.1	0.0	0.0	1.2	98.6	0.0	0.0	0.17/0.00/0.00/2.00
1100*#	0.4	0.2	0.1	4.4	94.9	0.0	0.0	0.18/0.09/0.05/2.00
1200*	0.3	0.3	0.2	2.2	96.8	0.1	0.1	0.27/0.27/0.18/2.00

* in Al_2O_3 crucible with thermocouple
# Ribbons aglomerated, resulting in a slow quench and possible phase separation.

<u>DTA</u>

Differential thermal analyses (DTA) were done on product ribbons to determine melting events and, therefore, invariant points on the pseudo-binary phase diagram. The DTA curve of the Ag-rich flake from the melt-spin run at 950°C showed an onset of the liquidus at 933°C in air (Fig. 1). This compares to the onset of the liquidus of pure silver in air at 947°C. The suppression of onset temperature is indicative of a eutectic between Ag and the Bi2212 component. The DTA curve of the Bi2212-rich oxide glass flake from the melt-spin run at 950°C showed much more complex melting events (Fig. 2). The endothermic event defining the liquidus deviates from the baseline at 810°C with an initial onset of 830°C. This is followed by a compound melting event indicative of two, and probably more, phase boundaries being crossed. Normally, a melting event may occur over a 10 to 15°C range, but this one occurs over at least a 75°C range, corroborating the assumption that this complex peak represents a number of phase fields being crossed. An endothermic event also occurs at an onset of 935°C, which corresponds quite well with the measured onset temperature of the Ag-rich flakes. The 935° endotherm further demonstrates the low solubility of Ag into Bi2212.

<u>TEM</u>

The oxide rich ribbons from the 1000°C run were examined using TEM. The material is a two-phase mixture of an oxide glass matrix with inclusions of Ag (Fig. 3). The inclusions of crystalline Ag varied from 5 to 50 nm and tended to be spherical. EDS of the matrix region between the Ag particles showed no Ag, while amounts of the ceramic oxide elements, in stoichiometric proportions, were variable when the probe was placed on a Ag particle. This variation scaled inversely with the Ag particle size. The lack of Ag in the spectra for the glass matrix puts the solubility limit of Ag into the oxide glass to < 1 %. The variation in Ag particle size most likely reflects variation in quench rates across the ribbon. These observations strongly suggest that the affinity of Ag for the oxide ceramic is very low.

Discussion

To deal correctly with the Bi-Sr-Ca-Cu-O-Ag phase diagram it is necessary to consider the consequences of a six component system. For such a system, the number of phases plus the number of degrees of freedom is eight. Restricting ourselves to atmospheric pressure, one degree of freedom is eliminated. Thus, six phases can be in equilibrium at a fixed temperature if the compositions of all six phases are fixed. As it is beyond the scope of this paper to deal with a six dimensional phase diagram, we will consider a schematic pseudo-binary isopleth through the Bi2212 - Ag system (Fig. 4). In this isopleth we must make the approximation that both end points are components, that is that they are either elements or congruently melting compounds. As we know that this is not the case for Bi2212, this representation will not follow the normal rules for binary phase diagrams. In particular, the compositions of all phases presented in the diagram do not lie in the plane of the isopleth and the lever law is not valid for the individual oxide phases. Starting with the Bi2212 compound, we know from the high temperature x-ray work of Ming Xu et. al. (6) that Bi2212 decomposes to $(Sr_{1-x}Ca_x)CuO_2$ (11)+ liquid to $(Sr_{1-x}Ca_x)_2CuO_3$ (21) + liquid

199

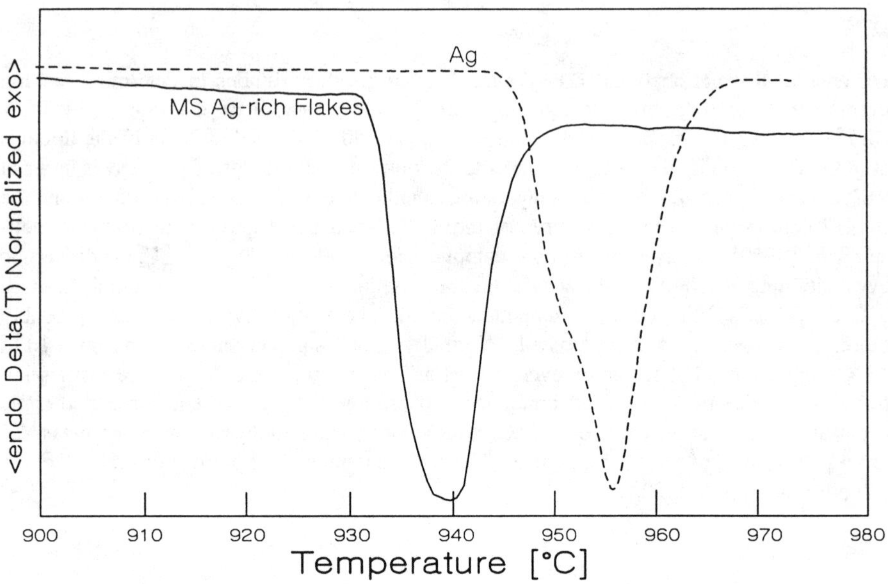

Figure 1 - DTA of Ag-rich ribbon pieces from the 950°C melt-spin run.

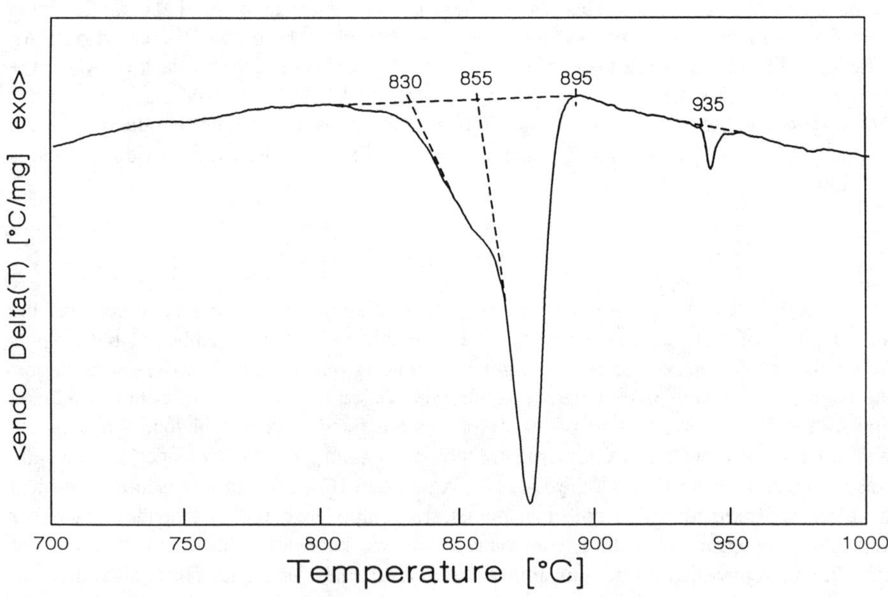

Figure 2 - DTA of oxide-rich glass ribbon pieces from the 950°C melt-spin run.

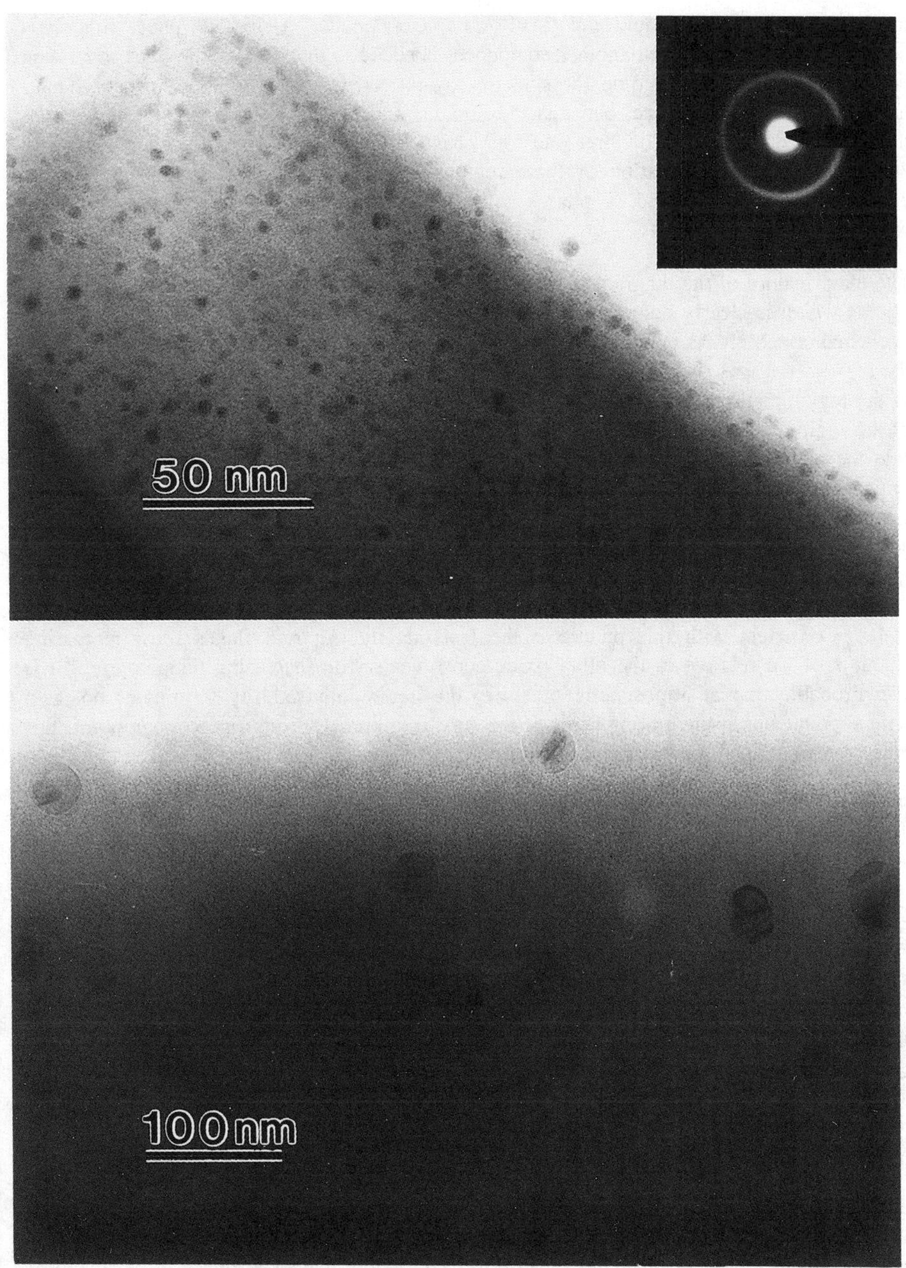

Figure 3 -   TEM micrograph of the 1000°C oxide flakes showing the variability in distribution of the Ag precipitates within the glassy matrix. The broad ring in selected area diffraction pattern indicates the oxide is fully amorphous but the discrete spots index to Ag, indicating the Ag is crystalline.

to $(Sr_{1-x}Ca_x)O$ (10) + liquid and finally completely melts. Using the phase rule, it is possible to derive that when another component is added to the system, the lines separating the two phase regions become three phase regions as we move into our pseudo-binary isopleth. As a consequence, between the liquidus and the Bi2212 + Ag solidus we pass through seven different two, three and four phase regions within a space of approximately 80°C. As the differentiation of these regions is beyond the resolution of the current experiment, the 11, 21 and 10 phases will all be lumped together as OX meaning solid oxide phases (Fig. 4).

The main feature of the diagram is the large liquid immiscibility between the oxide and Ag liquids which is clearly demonstrated by the segregation of Ag and Bi2212 melt in the air quenched crucible. As demonstrated by the two types of flakes obtained on melt spinning, the segregation was such that the composition at the limits of immiscibility for both sides of the region could be obtained from a single run at fixed temperature. On the Bi2212 side of the region, the slope of the boundary is approximately 1 at.%/100°C, while on the Ag side the slope is somewhat steeper. From this data it is not possible to extrapolate the critical point for the immiscibility region which is well outside the range of the experiments. From drop tube melt spinning in $REBa_2Cu_3O_{7-\delta}$ systems (7), we have determined that at temperatures on the order of 1600 - 1700 °C, CuO begins to reduce to form Cu metal and the melt phase segregates into metallic and oxide melts. Thus, the critical point for the immiscibility will most likely occur in what is essentially a metallic melt. Consistent with this picture is the fact that the Ag rich flakes show increasing amounts of Cu relative to the other oxide constituents with increasing temperature. It is also probable that at atmospheric pressure, the liquid immiscibility terminates not at a critical point, but at the boiling point of the Ag-Cu mixture. From a processing point, this limits the amount of Ag precipitates within the oxide glass to less than ten percent.

From the DTA measurements on both the oxide rich and Ag rich flakes, it is clear that a psuedo-eutectic exists on each side of the diagram. For the oxide rich side the existence of this eutectic has been kn

ow for quite some time but the composition at which it occurs has not been extensively investigated. The exact depth of the eutectic is not an easily determined number since it involves a cross section diagram containing an eutectic trough. In order to accurately determine the relative eutectic depth, samples with equal oxygen content must be compared as the effects of oxygen content are at least a factor of five larger than those of Ag (5). Comparing samples with as similar of a thermal history as possible, yields a value of 30°C for this eutectic trough. Extrapolating along the liquid immiscibility line gives a Ag composition of 4 at.% for the eutectic. It should be noted that this is an upper limit to the Ag composition, but the lower limit is not well defined at this point.

Extrapolating the liquid immiscibility line on the Ag rich side gives an eutectic composition of 98 at.% Ag. As no additional phase boundaries are crossed, this extrapolation should be valid. The temperature of the eutectic is 15°C below that of pure Ag under identical oxygen conditions.

From a detailed examination of the melt spun oxide rich flakes, it is possible to determine the limits of solid solubility of Ag in the oxide glass. On the micron scale these materials are homogeneous. Therefore chemical analysis or EDS using SEM gives a value determined by the temperature from which the material was quenched. TEM observations reveal that these analyses are misleading. On this scale, the Ag segregates into spherical precipitates, and the amorphous matrix contains no Ag to the limits of detectability. In situ crystallization of the glass in the TEM (7) indicates that the Ag coarsens at temperatures below the crystallization temperature and given the opportunity, migrates out of the sample. In no case were Ag inclusions found within Bi2212 grains. Determination of the solid solubility of the oxides in Ag has not been done at this time due to the difficulty of preparing TEM samples of the ductile material.

In using the temperatures presented in this phase diagram, one should be cognizant of the fact that both the oxide phases and Ag have reasonably large dependences of their decomposition or melting temperatures on oxygen content (5). While oxygen diffusion in Ag is high near the melting temperature, this is not the case for the oxide materials. The effects of oxygen non-equilibrium are readily observable in processing these materials. Runde et. al. have performed tracer diffusion measurements for oxygen in Bi2212 up to 650°C (8). If one takes the rather risky approach of evaluating the fit they obtained for temperatures near the melting, the time required for oxygen to diffuse 10 microns is of the order of 10 minutes. Clearly the melting temperatures obtained for bulk samples are going to be highly dependent on the exact heating sequence and oxygen partial pressure.

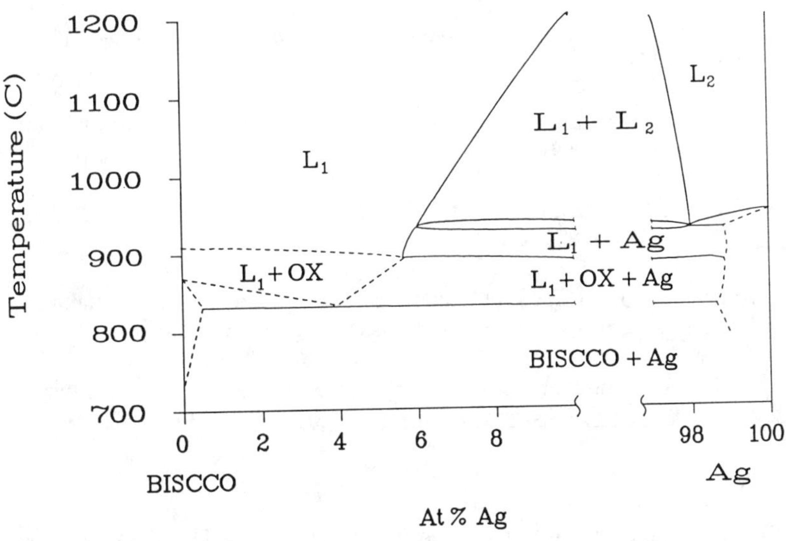

Figure 4 - Simplified pseudo-binary isopleth in the Bi-Sr-Ca-Cu-O-Ag system treating both endpoints as components.

Conclusions

We have demonstrated the existence of a liquid immiscibility gap in the Bi2212+Ag system which covers the region from approximately 8 to 98 at.% Ag. There exists an eutectic on each side of this gap. For the Ag rich side the eutectic is at 98 at.% Ag and depresses the melting by 15°C over that for pure silver for the same oxygen partial pressure. The oxide rich eutectic contains no more than 4 at.% Ag and depresses the solidus by 30°C. The solid solubility of Ag in oxide glass is found to be less than the detection limits of TEM/EDS and Ag is systematically excluded from the Bi2212 during crystallization. All phase transition temperatures in the systems are sensitive to oxygen content and times to reach oxygen equilibrium are long.

Acknowledgments

The work was performed at Ames Laboratory, Iowa State University and was supported by the Director of Energy Research, Office of Basic Sciences, U.S. Department of Energy under Contract No. W-7405-ENG-82

References

1. W. H. Lee, Yoshihiro Abe, and Eikichi Inukai, "Ag$_2$O-Doped Bi$_2$Sr$_2$Ca$_1$Cu$_2$O$_x$ Superconductors Prepared via Melt-Quenching," _Journal of the American Ceramic Society_, 76 (4) (1993), 849-856.

2. T. E. Jones et al., "Effect of silver additions on sintered Bi$_2$Sr$_2$CaCu$_2$O$_8$," _Physica C_, 201 (1992), 279-288.

3. S. X. Dou et al., "The Interaction of Ag with Bi-Pb-Sr-Ca-Cu-O Superconductor," _Physica C_, 160 (5-6) (1989), 533-540.

4. Y. Fang et al., "Tracer diffusion of [110]Ag in Bi$_2$Sr$_2$CaCu$_2$O$_x$," _Applied Physics Letters_, 60 (18) (1992), 2291-2293.

5. M.J. Kramer et al. "Effects of oxygen partial pressure of the crystallization pathways of Bi$_2$Sr$_2$CaCu$_2$O$_x$ and Bi$_2$Sr$_2$CaCu$_2$O$_x$+Ag" (in preparation).

6. M. Xu et al., "Investigations of Crystalline Phases in the Melting of Bi$_2$Sr$_2$CaCu$_2$O$_x$," _Applied Superconductivity_, 1 (1/2), (1993), 53-60.

7. Work in progress.

8. M. Runde et al. "Tracer diffusion of oxygen in Bi$_2$Sr$_2$CaCu$_2$O$_x$" _Physical Review B_, 45 (13) (1992), 7375-7382.

PINNING IN BSCCO SUPERCONDUCTORS

Peter Majewski, Steffen Elschner*, Bernhard Hettich*,Stefanie Kaesche,
Christoph Lang *, and Fritz Aldinger

Max – Planck – Institut für Metallforschung, Institut für Werkstoffwissenschaft,
Heisenbergstr. 5, 70569 Stuttgart, and
*Hoechst AG, Corporate Research, D – 65926 Frankfurt am Main

Abstract

Considering the phase equilibrium diagram of the system $(Bi_2O_3, PbO) - SrO - CaO - CuO$, single phase ceramics of the 2212 and 2223 phase with the composition $Bi_{2.18}Sr_{1.75}Ca_{1.25}Cu_2O_{8+d}$, $Bi_{2.3}Sr_2CaCu_2O_{8+d}$ and $Bi_{1.8}Pb_{0.4}Sr_2Ca_{2.2}Cu_3O_{10+d}$ have been transformed by a simple annealing procedure into multi phase samples. The transformation results in the formation of secondary phases and in an increase of the critical current density at 1 T of five to ten times (2212) and at 2 T of three times (2223). These increases are believed to express improved pinning properties of these ceramics. The nature of the produced pinning centres is not clarified yet.

Processing of Long Lengths of Superconductors
Edited by U. Balachandran, E.W. Collings and A. Goyal
The Minerals, Metals & Materials Society, 1994

Introduction

"$Bi_2Sr_2CaCu_2O_8$" (2212 phase, $T_c \leq 94$ K) and "$(Bi,Pb)_2Sr_2Ca_2Cu_3O_{10}$" (2223 phase, $T_c = 110$ K) exhibit weak internal pinning at temperatures above 20 K resulting in a decrease of the critical current density of about two orders of magnitude in an applied magnetic field of up to 2 T /1,2,3/. Therefore, an application of these materials in devices under magnetic fields is still not possible yet. Considering the observation of Murakami et al. /4/, who increased the pinning of Y–Ba–Cu–O ceramics by the introduction of the secondary phase Y_2BaCuO_5, Ca_2CuO_3 has been claimed to increase the pinning of 2212 and 2223 phase ceramics /5,6/. However, the grains of powdermetallurgically prepared ceramics are much to large (1 – 100 μm) to act as effectiv pinning centres which are assumed to may not exceed about 10 nm /7,8/. In this article a processing route is presented resulting in superconducting bulk ceramics with very fine ($\ll$ 1 μm) precipitates of secondary phases and enhanced pinning properties using temperature dependend solubility lines.

Experimental

Samples with the composition $Bi_{2.18}Sr_{1.75}Ca_{1.25}Cu_2O_{8+d}$ (batch 1), $Bi_{2.3}Sr_2CaCu_2O_{8+d}$ (batch 2) and $Bi_{1.8}Pb_{0.4}Sr_2Ca_{2.2}Cu_3O_{10+d}$ (batch 3), using Bi_2O_3, PbO, $SrCO_3$, $CaCO_3$ and CuO as starting material (purity 99 %) have been prepared. The mixed and grinded powders were calcined at 750 °C and 800 °C for 24 hrs and pressed into cylindric pellets (15 mm long and 4 mm in diameter, 625 MPa). The pellets of batch 1 and 2 were sintered at 820 °C in air for 48 hrs, furnace cooled and subsequently annealed at 885 °C for 10, 15, 22.5, 25, 30, 37.5, 45 and 60 minutes (batch 1) and 15, 30, 45 and 60 (batch 2) min in air and finally quenched on a cold copper plate in air (Fig 1 and 2).

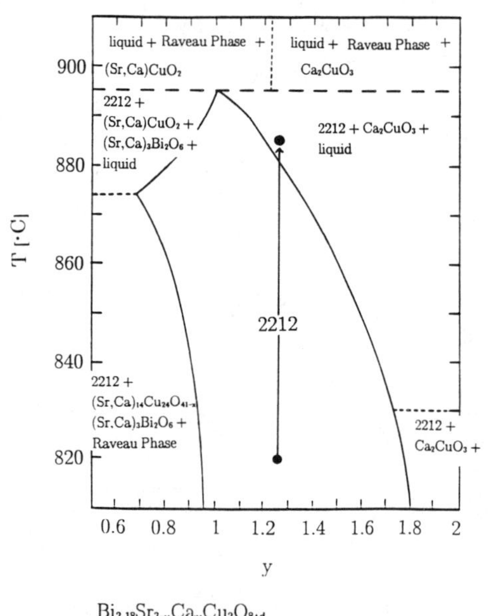

Fig. 1   Temperature vs. Ca content [9,10] including the performed annealing step to precipitate Ca_2CuO_3 + liquid.

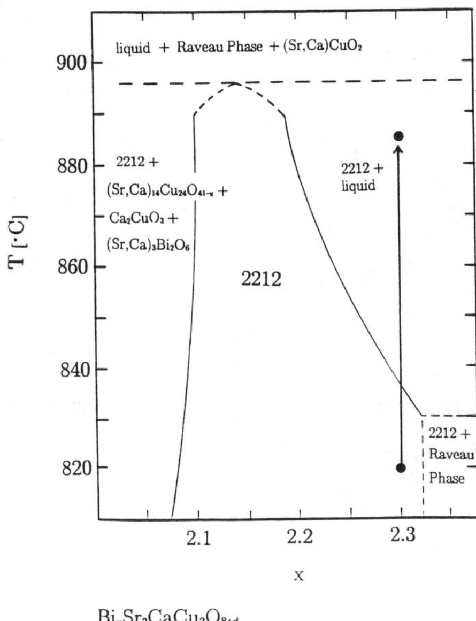

Fig. 2:   Temperature vs. Bi content [10] including the performed annealing steps to precipitate liquid.

The samples of batch 3 were sintered at 840 °C and subsequently annealed in air at 870 °C for 15, 22.5, 30, 37.5 45 and 60 minutes and quenched (Fig. 3). The critical current densities of the pellets have been obtained by measuring of the magnetic susceptibility at 30 K up to 2 T and calculating J_c from these data using Bean's model /11/. The magnetic susceptibility has been measured at 30 K, as at this temperature an increase of the pinning properties is expected to become most obviousely. The secondary phase content of the samples have been determined by optical and electron microscopy. The critical temperatures (onset) of the pellets have been determined by AC susceptibility measurements. Phase identification has been performed using electron microscopy with energy dispersive x–ray analysis (EDX) and optical microscopy using polarized light.

<u>Results</u>

The as sintered 2212 samples are single phase (> 99 vol. – %) refering x–ray analysis, as well as, electron and optical microscopy. According to the phase equilibria, the annealing of the samples of batch 1 results in the formation of Ca_2CuO_3 and liquid (Fig. 1). The samples of batch 2 contain in the annealed state only a liquid phase besides the 2212 phase but no cuprate (Fig. 2). The formation of these secondary phases is combined with a reduction of the Ca– (batch 1) and Bi content (batch 2) of the 2212 phase. The annealing of the sintered starting sample of batch 3 results in the formation of 2212 + Ca_2CuO_3 + (Sr,Ca)CuO₂ + liquid combined with a decrease of the Pb content of the 2223 phase (Fig. 3), as it is described in /12/.

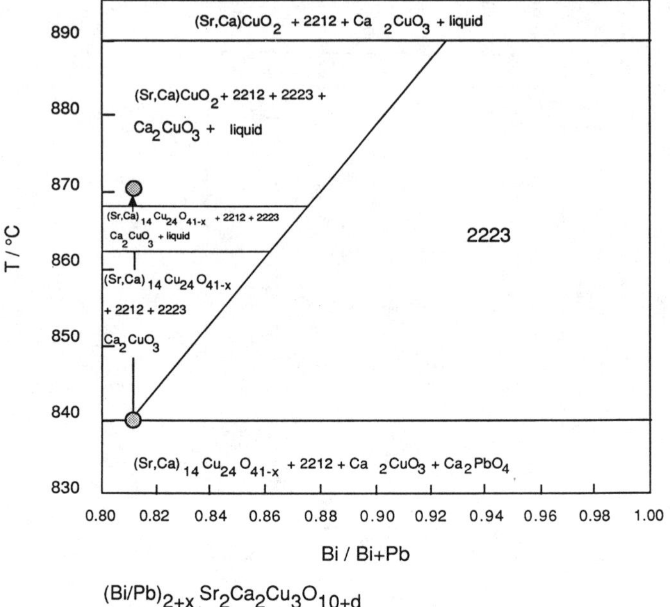

Fig. 3: Temperature vs. the Bi/(Bi+Pb) ratio including the performed annealing step to precipitate 2212 + Ca$_2$CuO$_3$ + (Sr,Ca)CuO$_2$ + liquid.

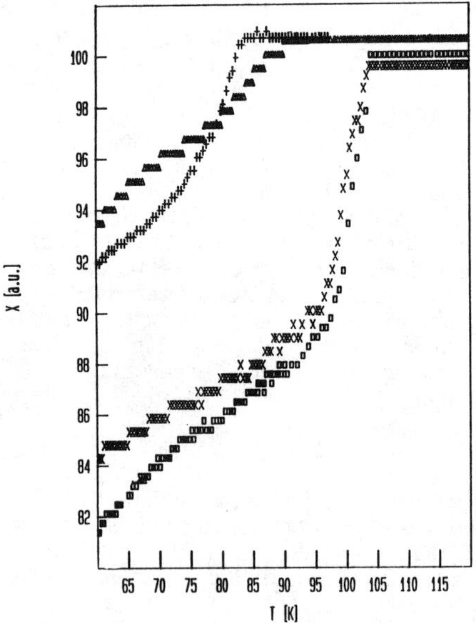

Fig. 4: Susceptibility vs. temperature plot of the starting samples (squares: batch 3, crosses: batch 1) and the 30 min annealed samples (×: batch 3, triangle: batch 1).

Fig. 4 shows the susceptibility vs. temperature plots of a starting sample and the 30 min post annealed sample of batch 1 and 3. It is seen that the critical temperatures T_c of the samples and the trend of the susceptibility lines have not be changed significantly by the post annealing ($\Delta T_c \simeq 5$ K).

The graphic analysis (Fig. 5) clearly shows that the $J_{c(B)}/J_{c(0\ T)}$ values, which represent the pinning force, exhibit a distinct maximum at an annealing time of about 20 min for batch 1 and at 30 min for batch 2 and 3. With increasing and decreasing annealing time the values decrease.

The J_c vs. B plot of the starting samples and the 15 min annealed sample of batch 1, the 30 min annealed sample of batch 2 and the 37.5 min annealed sample of batch 3 are depicted in Fig. 6. The samples of batch 1 and 2 have a J_c of 4000 A/cm^2 and 1100 A/cm^2 at zero magnetic field, respectively. The samples of batch 3 have a J_c of 10500 A/cm^2 at zero magnetic field. The relativ difference between the J_c of the starting samples and the annealed samples increases with increasing magnetic field (Fig. 7)

Discussion and Conclusions

The increase of the J_c at magnetic fields above 0.2 T with annealing showes a strong evidence for the improvement of the pinning properties of such samples by the precipitation of secondary phases. The fact, that the annealed samples of batch 2 exhibit an increased pinning indicates that not only crystaline precipitates may cause this effect but also paricles of the quenched liquid phase. The decrease of the pinning effect with increasing annealing time might be due to coarsening of the secondary phases.

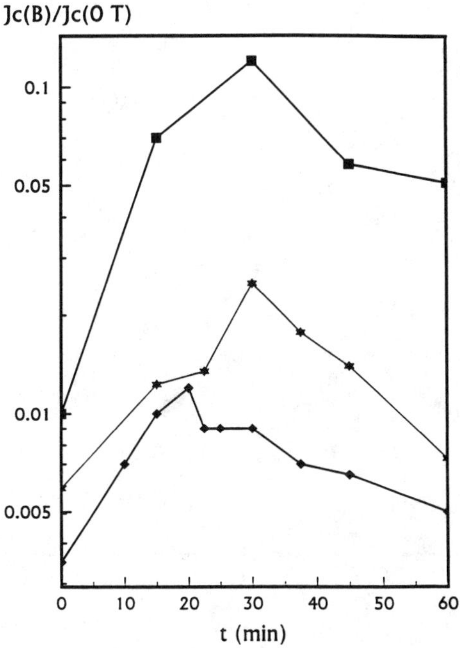

Fig. 5: $J_{c(B)}/J_{c(0\ T)}$ at 30 K vs. the post annealing time of the samples of batch 1 (rhombs, B = 1 T), batch 2 (squares, B = 1 T) and batch 3 (stars, B = 2 T).

However, it has to be taken into account that besides the effect of such secondary phases the increased pinning can also be caused by defects in the 2212 phase itself resulting from the phase transformation due to the reactions:

2212 (Ca rich)->2212 (lower Ca content)+Ca_2CuO_3+liquid (batch 1)
2212 (Bi rich)->2212 (lower Bi content)+liquid (batch 2)
2223 (Pb-rich)->2223 (lower Pb content)+2212++Ca_2CuO_3+(Sr,Ca)CuO_2+liquid (batch 3)

Such defects could be e.g. lattice distortions which may act as pinning centres. In this case, the observed decrease of J_c with increasing annealing time could be explained by the defect healing.

However, none of the potential pinning centres have been proofen directly and further detailed investigations have to be performed to clarify the nature of the produced pinning centres. Nevertheless, the results show that it is possible on principle to introduce artifical pinning in a bulk ceramics of the 2212- or 2223 phase.

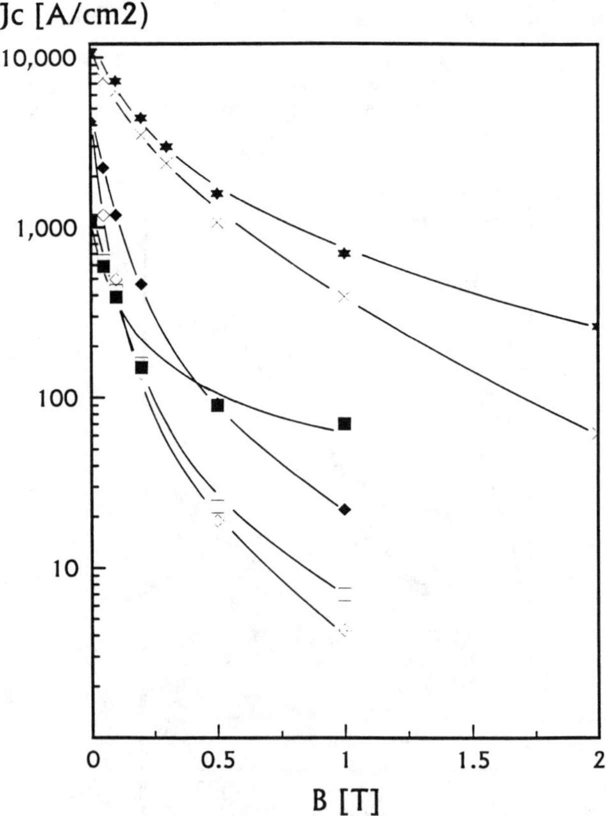

Fig. 6: J_c vs. the applied magnetic field of the starting sample (rhombs: batch 1; squares: batch 2; crosses: batch 3) and the 15 min annealed sample (filled rhombs) of batch 1, the 30 min annealed sample (filled squares) of batch 2 and the 37.5 min annealed sample (stars) of batch 3. T = 30 K.

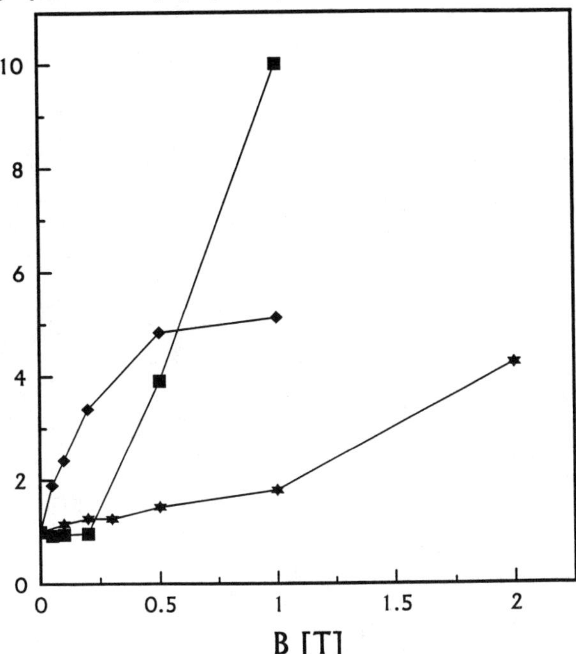

Fig. 7: The quotient of J_c of the annealed sample ($J_{c\ annealed}$) over J_c of the starting sample ($J_{c\ start}$) vs. the magnetic field B. rhombs: batch 1; squares: batch 2; stars: batch 3.

Reference

1. H. Krauth, K. Heine, J. Tenbrink, High-Temperature Superconductors - Materials Aspects, Ed.: H.C. Freyhardt, R. Flükiger, M. Peuckert, DGM, Oberursel, FRG, 1991, 29.
2. K. Togano, H. Kumakura, H. Maeda, J. Kase, Chemistry of High Temperature Superconductors, Ed.: C.N.R. Rao, World Scientific, London, 1991, 399.
3. J. Bock, S. Elschner, E. Preisler, Adv. in Superconductivity III, Editors: K. Kajimura, H. Hayakawa, Proc. ISS 1990, Sendai, Springer Verlag, Tokyo, 1991, 797.
4. M. Murakami, H. Fujimoto, T. Oyama, S. Gotho, Y. Shiohara, N. Koshizuka, S. Tanaka, High-Temperature Superconductors - Materials Aspects, Ed.: H.C. Freyhardt, R. Flükiger, M. Peuckert, DGM, Oberursel, FRG, 1991, 13.
5. D. Shi, M.S. Boley, U. Welp, J.G. Chen, Y. Liao, Phys. Rev, B 40 (1989), 5255.
6. T. Umemura, K. Egawa, S. Kinouchi, M. Wakata, S. Utsunomiya, Physica. C, 185-189 (1991), 2219.
7. J. Boiko, P. Majewski, F. Aldinger, Zeitschrift f. Metallkde, 1993, in press.
8. E.H. Brandt, Physica C, 195 (1992), 1.
9. P. Majewski, B. Hettich, MRS Proc., 275 (1992), 627.
10. P. Majewski, B. Hettich, N. Rüffer, F. Aldinger, J. Elec. Mat, 1994, in press.
11. C.P. Bean, Phys. Rev. Lett., 8 (1962), 250.
12. P. Majewski, S. Kaesche, H.-L. Su, F. Aldinger, to be published.

Processing and Properties of Bi-2212 Thick Films

B. Heeb, D. Buhl, Th. Lang, L.J. Gauckler

ETH Zürich, Nichtmetallische Werkstoffe, CH-8092 Zürich

Abstract

Bi-2212 thick films were produced by tape casting and partial melting on Ag-substrates. Highly aligned microstructures throughout the oxide thickness of 20 µm were achieved. A reducing heat treatment optimizing the oxygen stoichiometry of the Bi-2212 phase led to a T_c of 90 K and to a j_c of $1.75 \cdot 10^4$ A/cm^2 at 77 K/0 T using the 1 µV/cm criterion. The current-voltage characteristic of the thick films follows the power law $E \propto j^\alpha$ at 77 K in zero field. The exponent $\alpha=5$ for the films is considerably smaller than those found for melt processed bulk material.

Processing of Long Lengths of Superconductors
Edited by U. Balachandran, E.W. Collings and A. Goyal
The Minerals, Metals & Materials Society, 1994

1 Introduction

Great effort has been concentrated on fabrication techniques of Bi-based high-T_c superconductors in form of tapes, which are very promising for technological applications. For these applications, the most significant parameter is the critical current density j_c. High current densities require intergranular connectivity avoiding weak link behavior as well as a high degree of grain orientation due to the anisotropic properties of these materials.

Melt processed $Bi_2Sr_2CaCu_2O_x$ (Bi-2212) is well known as a material without weak links, with good ability to form textured microstructures and thus favorable for current carrying applications. Different techniques such as tape casting [1], dip coating [2], screen printing [3] and drying alcohol suspensions [4] are used to produce Bi-2212 thick films on Ag-substrates with a final oxide layer thickness of 10 to 100 µm. The heat treatment programs of these tapes include partial melting, slow cooling to the annealing temperature followed by short annealing. These procedures result in highly textured microstructures of Bi-2212 with small amounts of secondary phases.

Tape casting and dip coating are very promising for practical use because of the possibility of producing long devices of superconducting material. At low temperature (4.2 K) in short samples high critical current densities up to $5.9 \cdot 10^5$ A/cm^2 in zero field and $2.6 \cdot 10^5$ A/cm^2 in high magnetic fields (12 T) have been achieved, wheras in long tapes (1 m) $1 \cdot 10^5$ A/cm^2 in zero field has been reported [2]. In contrast to these results, it seems to be difficult to reach high critical current densies with good reproducibility at 77 K. So far, the best values are published by Shimoyama et al. ($3.2 \cdot 10^4$ A/cm^2) [2]. In addition to the annealing at high oxygen partial pressures, a reducing heat treatment is required to adjust the optimal oxygen content in order to raise T_c above 77 K [5].

In this paper, we report processing procedures to fabricate textured Bi-2212 thick films on Ag-substrates by tape casting via the partial melting route. Emphasis is laid on the parameters enhancing the critical current densities and their reproducibility at 77 K, as well as the correlation between these properties and the microstructures of the tapes.

2 Experimental

Fabrication of tapes via tape casting

The Bi-2212 powder was prepared by the standard calcination process. Appropriate amounts of Bi_2O_3, $SrCO_3$, $CaCO_3$ and CuO were mixed in the ratio of Bi:Sr:Ca:Cu = 2.2:2:1:2 and calcined at 780 °C, 800 °C, and 820 °C in air for 12, 12, and 70 hours, respectively, with intermediate grindings. The Bi-2212 powder was mixed with the organic formulation consisting of solvent (ethanol), dispersant (triolein), plasticizer (polyethylene glycol, phtalic acid ester) and binder (polyvinyl butyral) and milled for 4 hours. Special attention was paid to use only non-toxic organic additives. The resulting slurry was cast into 1.7 m long, 150 mm wide and 50 μm thick tapes by doctor blade tape casting. Samples were then cut from the green tape, put on a 50 μm thick silver foil and subjected to the heat treatment.

The heat treatment consisted of removal of the organic additives, partial melting, slow cooling to achieve aligned microstructures, and annealing to increase the 2212 volume fraction and to optimize the oxygen stoichiometry.

The organic formulation was removed at 500 °C. The tapes were then partially molten at 875 to 880 °C for 5 to 10 minutes and cooled to 820 °C at the rate of 5 °C/h to achieve aligned microstructures. The maximum heat treatment temperature was chosen a few degrees above the onset of melting of the tapes on the Ag substrates at 873 °C in order to avoid large oxygen loss, Bi-evaporation and phase separation. It was assumed that at this temperatures a sufficient amount of liquid is present to guarantee complete densification and good grain alignement. At 820 °C the tapes were annealed for 1 to 50 h to increase the 2212 phase content and subsequently cooled to 300 °C with 500 °C/h and then air cooled. The slow cooling rate was chosen to avoid bending and cracking of the samples. The entire heat treatment was carried out in oxygen at atmospheric pressure. The oxygen stoichiometry of thick films produced in this way, containing mainly the 2212 phase, was adjusted by a subsequent annealing at 500 °C for 8 to 10 h in nitrogen in order to maximize T_c.

Sample characterization

The microstructures of the samples were investigated by light microscopy, X-ray diffraction analysis with CuK_α radiation and Si as internal standard, and scanning electron microscopy. Chemical compositions of phases were determined by energy dispersive X-ray diffraction at an acceleration voltage of 25 kV using the BiM_α, SrL_α, CaK_α and CuK_α lines. The superconducting properties T_c, T_{irr}, $j_c(T)$ and the E(j)-characteristics were determined by resistive (criterion $\geq$ 1 μVcm^{-1}) and magnetic measurements (criterion < 1 μVcm^{-1}). The transport current density was measured by the conventional four probe technique (sample size 50 x 15 mm^2). The current densities were also determined magnetically with a vibrating sample magnetometer (sample size 3 x 3 mm^2) and were derived from the width of the measured M(H) loops by integration of a critical state like current distribution [6]. As we never found any granularity in melt-processed Bi-2212, ΔM is purely due to macroscopic intergranular currents.

3 Results and discussion

Microstructure

The thickness of the final Bi-2212 layer is approximately 1/3 of the original cast green tape thickness. Fig.1 shows the fractured cross section of a melt processed tape with the oxide thickness of 20 µm. The highly aligned microstructure is attributed to the slow cooling from the temperature of partial melting, the Ag substrate and the low oxide layer thickness < 30 µm [1]. The Bi-2212 layer is composed of stacks of plate like grains with the aspect ratio exceeding 50.

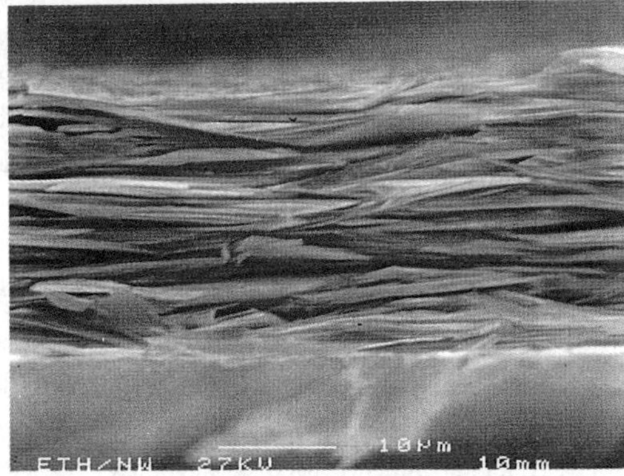

Fig.1: SEM micrograph of the fractured cross section of the Bi-2212/Ag tape.

The high degree of orientation of the (ab)-planes parallel to the Ag substrate was confirmed by XRD measurements (fig.2).

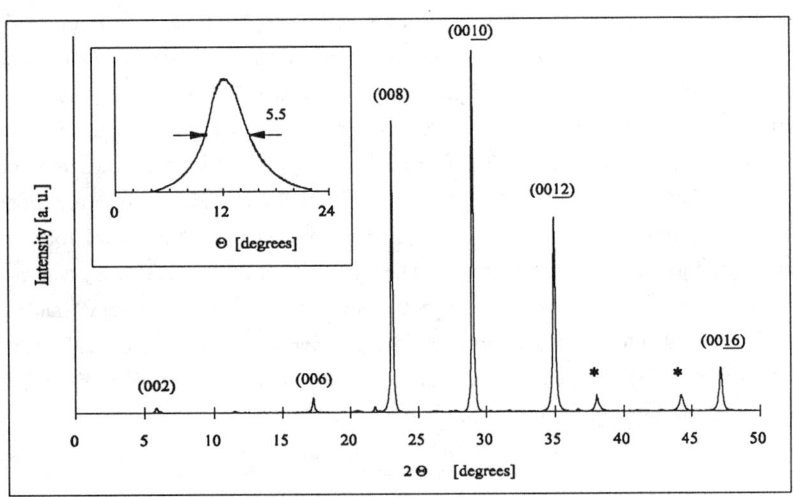

Fig.2: XRD pattern of the melt processed Bi-2212 thick film (Bi-2212 reflections are indexed, * = Ag). The inset shows the rocking curve of the (008)-reflection.

Only (00l) reflections of the Bi-2212 phase are observed as shown in Fig.2. Additionally, the inset in Fig.2 shows the rocking curve of the (008) reflection. Its FWHM may serve to estimate the misalignment of the 2212 crystals. The value of 5.5° is in good agreement with the directly measured misorientation of the 2212 grains which was found to be in the range of $\pm 4 < \varphi < \pm 6$ from Fig.1.

The tapes are nearly single phase Bi-2212. Only small amounts of the Bi-free primary phase $(Sr,Ca)_{14}Cu_{24}O_y$ can be found, which is explained by the Bi-rich powder composition [7] and the low maximum heat treatment temperature minimizing the Bi-evaporation. The Bi-2212 phase has the chemical composition of $Bi_{2.3}Sr_{2.2}Ca_{1.0}Cu_2O_x$. The Bi- and Sr-rich composition is known as a typical feature of the melt processing of Bi-2212 and is due to the $2201 \rightarrow 2212$ transformation during cooling and annealing [8]. The length of the c-axis of the 2212 phase is 30.90 Å.

Critical temperature T_c and irreversibility temperature T_{irr}

Due to the slow cooling at high oxygen partial pressure, the critical temperature of the thick films after the heat treatment in oxygen was as low as 77 K. The high surface/volume ratio of the samples promoted the oxygen diffusion leading to an oxygen excess of the 2212 phase and a low T_c [5]. The reducing heat treatment in nitrogen led to a considerably enhanced T_c of 90 K. However, this value is still several Kelvin below the maximum possible T_c of 95 K for Bi-2212. Therefore, prolonged heat treatments at 500 °C or higher annealing temperatures are required to optimize the oxygen stoichiometry to improve T_c of these thick films. On the other hand, the enhancement of T_c led simultaneously to the enhancement of the irreversibility temperature, i.e. the temperature at which the flux pinning and the critical current density vanished. After the reducing heat treatment T_{irr} was found to be at 85.5 to 86.5 K. The temperature difference between T_c and T_{irr} remained constant at about 3 K.

Critical current density j_c

Fig.3 shows the critical current density of the Bi-2212 thick films before and after the reducing heat treatment in nitrogen as a function of temperature. At temperatures below 50 K j_c of both tapes are identical. At 10 K we achieved $j_c = 8 \cdot 10^4$ A/cm^2 using the 1 nV/cm criterion. At higher temperatures, j_c is markedly improved due to the enhancement of T_c and T_{irr}. In contrast to the oxygenated tape, in which j_c vanished at $T_{irr} = 74$ K, the reduced tapes were irreversible up to 86 K. At 77 K $j_c = 4.4 \cdot 10^3$ A/cm^2 (1 nV/cm criterion) was obtained corresponding to $j_c = 1.75 \cdot 10^4$ A/cm^2 with the 1 µV/cm criterion. This value was derived using the relation $E = j^{\alpha}$ with $\alpha = 5$ (see next chapter).

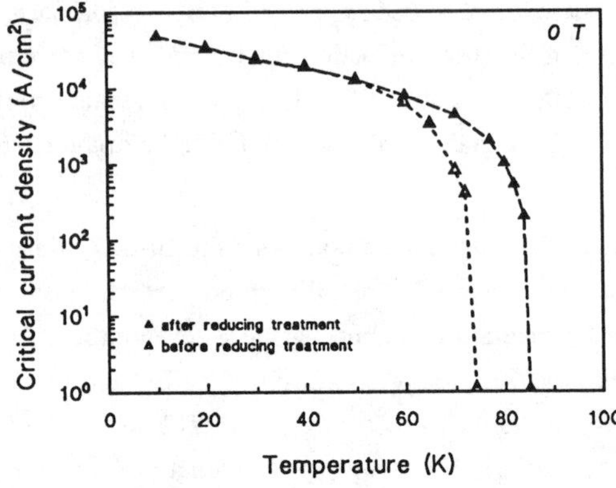

Fig.3: j_c of Bi-2212 thick films (20 μm) as a function of temperature (1 nV/cm criterion)

In addition, we found that the critical current density was strongly dependent on the maximum heat treatment temperature during partial melting and the time in the partially molten state as reported previously [9]. If the heat treatment temperature was too low, incomplete melting led to $T_{irr} < 77$ K and j_c (77 K/0 T) = 0, at too high temperatures and/or after too long melting time, increasing amounts of secondary phases ($(Sr,Ca)_{14}Cu_{24}O_x$ and $Bi_3Sr_4Ca_3O_y$) and low j_c were obtained.

Current voltage characteristic E(j)

The current voltage characteristics in melt processed *bulk* Bi-2212 obey the power law

$$E = E_0 \cdot \left(\frac{j}{j_0(H,T)}\right)^{\alpha(H,T)}$$

over a wide range of T, H and E [6,10]. The E(j) characteristic of the *thick film* was determined at T = 77 K for E = 10^{-9} to $5\cdot10^{-5}$ V/cm. The straight line in a log-log plot (fig.4) shows that the above power law is also valid for this material. The exponent α derived from the slope of the straight line is 5. This value is considerably smaller than $\alpha=10$ found for bulk material previously [6,10]. Because of the strong temperature dependence of α near T_c [10], this may be due to the lower T_c and T_{irr} of the tapes compared to the bulk material (T_c = 95 K, T_{irr} = 93 K). However, it cannot be excluded, that the improved grain alignment influences the E(j) characteristic also. Further studies will clarify this.

Conclusions

Bi-2212 thick films on Ag-substrates were produced by tape casting using a non-toxic organic formulation. Partial melting carried out in oxygen and a subsequent annealing in nitrogen to adjust the oxygen stoichiometry led to highly oriented microstructure of almost pure Bi-2212 with T_c of 90 K.

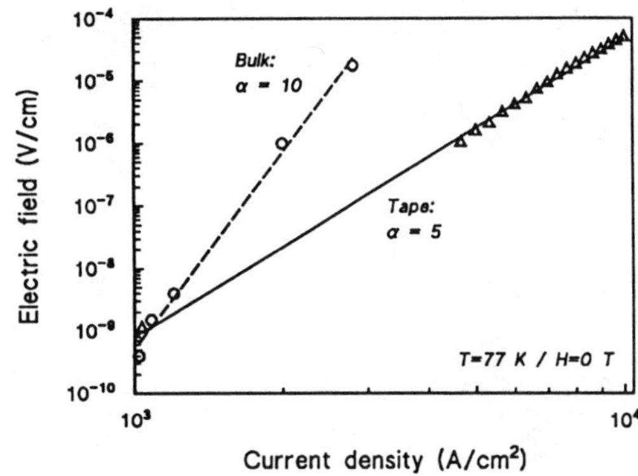

Fig.4: E(j) characteristic of Bi-2212 thick film and bulk material [6] at T=77 K for H=0.

Controlling the maximum heat treatment temperature within the range of 2 to 7 °C above the onset of melting is crucial to achieve nearly single phase Bi-2212 oxide layers and high critical current densities. J_c as high as $1.75 \cdot 10^4$ A/cm^2 was obtained at 77 K in zero field. The current-voltage characteristic of the Bi-2212/Ag tapes obeys the power law E$\propto$j$^\alpha$ with the exponent $\alpha = 5$.

Acknowledgement

The authors gratefully acknowledge the financial support of the Schweizer Nationalfonds (NFP 30) and the Schwerpunktprogramm Werkstoffe (Modul 3B).

References

[1] J. Kase, T. Morimoto, K. Togano, H. Kumakura, D.R. Dietderich, H. Maeda, *IEEE Trans. Mag.* **27** (1991) 1254-1257.

[2] J. Shimoyama, N. Tomita, T. Morimoto, H. Kitaguchi, H. Kumakura, K. Togano, H. Maeda, K. Nomura, M. Seido, *Jpn. J. Appl. Phys.* **31** (1992) L1328-L1331.

[3] N. Noji, W. Zhou, B.A. Glowacki, A. Oota, *Physica C* **205** (1993) 397-405.

[4] F.A. List, H. Hsu, O.B. Cavin, W.D. Porter, C.R. Hubbard, D.M. Kroeger, *Physica C* **202** (1992) 134-140.

[5] G. Triscone, J.-Y. Genoud, T. Graf, A. Junod, J. Muller, *Physica C* **176** (1991) 247-256.

[6] W. Paul, B. Heeb, Th. Baumann, M. Guidolin, L.J. Gauckler, in *Layered Superconductors: Fabrication, Properties and Applications* (MRS Symp. Proc. **275**, Pittsburgh, 1992), pp. 383-387.

[7] S. Wu, J. Schwartz, G.W. Raban Jr., *Physica C* **213** (1993) 483-489.

[8] B. Heeb, L.J. Gauckler, H. Heinrich, G. Kostorz, *J. Mater. Res.* **8** (1993) 2170-2176.

[9] J. Shimoyama, K. Kadowaki, H. Kitaguchi, H. Kumakura, K. Togano, H. Maeda, K. Nomura, *Appl. Supercond.* **1** (1993) 43-51.

[10] W. Paul and J.P. Meier, *Physica C* **205** (1993) 240-246.

CRITICAL CURRENT DENSITY AND MICROSTRUCTURE OF SCREEN-PRINTED Ag-$(Bi,Pb)_2Sr_2Ca_2Cu_3O_x$ TAPES: A COMPARISON WITH THE Ag-SHEATHED TAPES

A.Oota, H.Matsui, M.Funakura, J.Iwaya and J.Maeda

Toyohashi University of Technology, Tempaku-cho, Toyohashi, Aichi 441, Japan

Abstract

The critical current density J_C and microstructure have been investigated as a function of overall tape thickness d between 0.06 mm and 0.22 mm on screen-printed Ag-$(Bi,Pb)_2Sr_2Ca_2Cu_3O_x$ tapes with a mono-layer ceramics core, prepared by three kinds of fabrication process which consists of a combination of cold working (rolling and/or pressing) and sintering. The thickness dependence of J_C at 77K is characterized by two distinctive stages. As d decreases, J_C increases up to 0.1 mm by the densification of ceramics core, while decreasing gradually for d $\leq$0.1 mm by the distortion of Ag/ceramics interface. The factors which dominate J_C in screen-printed "2223" tapes are discussed in comparison with the Ag-sheathed "2223" tapes.

Processing of Long Lengths of Superconductors
Edited by U. Balachandran, E.W. Collings and A. Goyal
The Minerals, Metals & Materials Society, 1994

1. INTRODUCTION

Screen printing is a simple, convenient and economical technique to fabricate tapes and/or thick films using high-T_c superconductors towards superconductivity applications such as circuits, devices and magnetic shields [1-3]. The enhancement of transport critical current density J_c is one of the most significant issues for practical use. Recently, we have succeeded in obtaining c-axis-oriented microstructures (average misalignment angle of 4-6°) with high J_c values of 1.5×10^4 A/cm^2 (77K, 0T) and 9×10^4 A/cm^2 (23K, 0T) in screen-printed $(Bi,Pb)_2Sr_2Ca_2Cu_3O_x$ "2223" tapes sandwiched between Ag substrates [1,3], by applying a combination process of cold working (rolling and/or pressing) and sintering that has shown a great success for Ag-sheathed "2223" tapes [4].

In this paper, we prepare a number of screen-printed "2223" tapes with the overall tape thickness d (0.06 mm$\leq$ d$\leq$ 0.22 mm), by three kinds of fabrication process which consists of a combination of cold working and sintering. We present a systematic study on the J_c value, morphology and microstructure as a function of d. The factors which dominate J_c in screen-printed "2223" tapes are discussed in comparison with the Ag-sheathed "2223" tapes.

2. EXPERIMENTAL

2.1. Sample preparation and measurements

The screen-printed "2223" tapes with a mono-layer ceramics core were prepared. The superconducting paste, which consists of a calcined powder (average particle size of $3\,\mu$m) with cation ratios Bi: Pb: Sr: Ca: Cu = 1.85: 0.35: 1.90: 2.05: 3.05 and organic binders (acrylic resin, telpineol, etc.), was purchased from Dowa Mining Co., Ltd.. The paste was screen-printed 10 μm thick and 6 mm wide on one side of the Ag substrate (0.1 mm thick and 10 mm wide), and heat-treated at 500°C for 1h to evaporate organic binders. After that, two printed-tapes were sandwiched with Ag substrates so as to make contact between the ceramics parts. The result was pre-rolled into tapes with overall tape thickness between 0.08 mm and 0.22 mm through several rolling runs, keeping the reduction rate nearly constant (10-15%). When the tape thickness reached an appropriate value (20-30% greater than the target tape thickness), it was sintered at the temperature T_S=830-860°C for 100 h in air (pre-sintering). Subsequently, it was subjected to three different processes shown in Fig. 1. In process 1, it was rolled again so as to be the target thickness and sintered at the same temperature T_S as the pre-sintering for a period of time t_S=25-100 h. In process 2, after the pre-sintering, it was uniaxially pressed under P=2 GPa and sintered at T_S for t_S=25-100 h. In process 3, after the pre-sintering, it was uniaxially pressed under 2 GPa and sintered at T_S for 25 h. Such a combination process of pressing and sintering was repeated six times maximum. A sintering period of time t_S for this process, which will be used later, is defined as 25h multiplied by a repetition number n_p for pressing and sintering.

To establish a basis for comparison, the Ag-sheathed "2223" tapes with the overall tape thickness d (0.05 mm$\leq$ d$\leq$1.0 mm) were prepared by a powder-in-tube method [5]. They were fabricated by a process of combined drawing and rolling with intermediate sintering steps [5], where no uniaxial pressing technique was applied. The co-precipitation powder with cation ratios Bi: Pb: Sr: Ca: Cu = 1.8: 0.4: 2.0: 2.2: 3.0, supplied from Dowa Mining Co., Ltd., was calcined twice at 800°C for 12h in air. The calcined powder was packed into Ag tubes (o.d. 6 mm and i.d. 4 mm). The composites were drawn into round wire of 2.0 mm in diameter and rolled into tapes with the overall tape thickness between 0.07 mm and 1.2 mm through a number of cold working runs, until reaching appropriate thickness (20-30% greater than the target thickness). After that, the tape was sintered at the temperature T_S=800-860°C for 100h in air (pre-sintering). Subsequently, it was rolled into the target thickness d (0.05 mm$\leq$ d$\leq$1.0 mm) and finally sintered at the same temperature T_S as the pre-sintering for a period of time t_S=150 h in air.

The morphology and microstructure of ceramics core were examined by x-ray diffraction (XRD), scanning electron microscope (SEM) and an image analyzer. Critical current I_C was measured by the current vs voltage (I-V) characteristics with an electric field criterion 1 μV/cm. The J_C value was determined from the I_C value using the cross-sectional area of ceramics core. The J_C anisotropy at 77K, as a function of field direction in a fixed B, was measured using a computer-controlled stepping motor. The temperature dependence of J_C in B=0T was measured by a cryogenic refrigerator (AISIN SEIKI; TAS 2001).

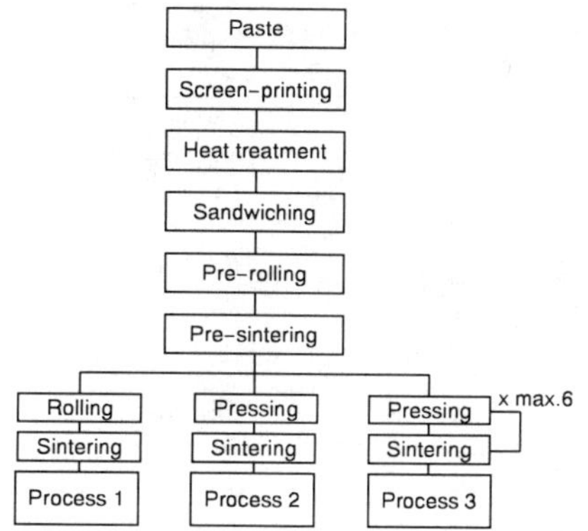

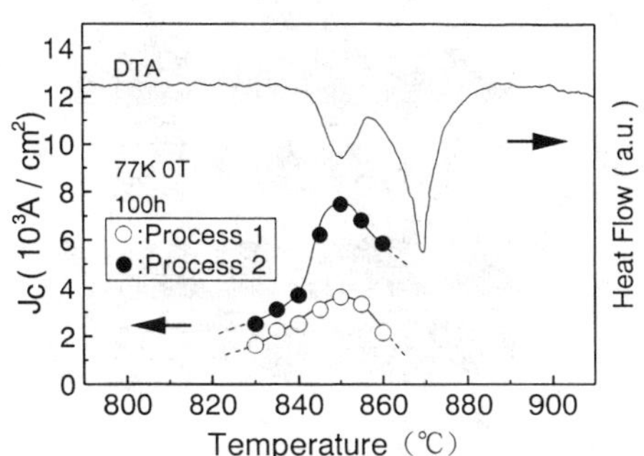

Fig.1.Fabrication process of screen-printed tape.

Fig.2.Influence of sintering temperature on the J_C value at 77K and 0T for processes 1 and 2 on screen-printed tape with d=0.15 mm. Also shown is the DTA curve.

2.2. Characterization of screen-printed "2223" tapes

In conjunction with the differential-thermal-analysis (DTA), the influence of sintering temperature T_S on the J_{C0} value (defined as J_C at 77K and 0T) is shown in Fig. 2 for screen-printed "2223" tapes with d=0.15 mm, fabricated by processes 1 and 2. Note that sintering time t_S was optimized and fixed to 100 h. As can be seen, the DTA curve, upon heating at a rate of 10°C/min, shows a faint endothermic peak due to partial melting [6,7], before exhibiting a pronounced melt peak. The maximum values of J_{C0} at processes 1 and 2 are achieved in the partial melting state, by sintering at T_S=850°C. Even when sintering was made in the partial melting state (840°C $\leq T_S \leq$ 860°C), a deviation from T_S=850°C lowers the J_{C0} value considerably.

According to our recent study [5], the same story holds good for Ag-sheathed "2223" tapes, although there is the difference in the optimum value of T_S between both cases (i.e., 850°C for screen-printed tape and 835°C for Ag-sheathed tape), which is possibly due to the difference in the starting materials (form and composition), fabrication process, etc..

The screen-printed tapes subjected to processes 1 and 2 were prepared under the optimum sintering temperature (T_S=850°C) by changing d between 0.06 mm and 0.22 mm, where t_S was fixed to 100h. The transverse cross-section for a polished surface, perpendicular to a long direction, is shown in Fig. 3 and the result is investigated using an image analyzer. As d decreases from 0.22 to 0.06 mm, the thickness of ceramics core decreases from ~ 20 to far below 10 μm and the cross-sectional area from ~1.2 $\times 10^{-3}$ to ~0.3 $\times 10^{-3}$ cm². As can be seen from Fig. 3, the ceramics core is embedded by silver without severe deformation on the Ag/ceramics interface in thicker tapes with d $\geq$ 0.1 mm, while showing a considerable deformation on the interface in thinner tapes with d $\leq$ 0.1 mm.

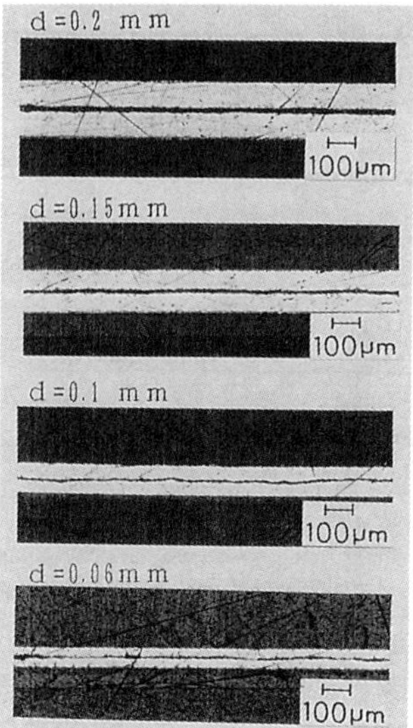

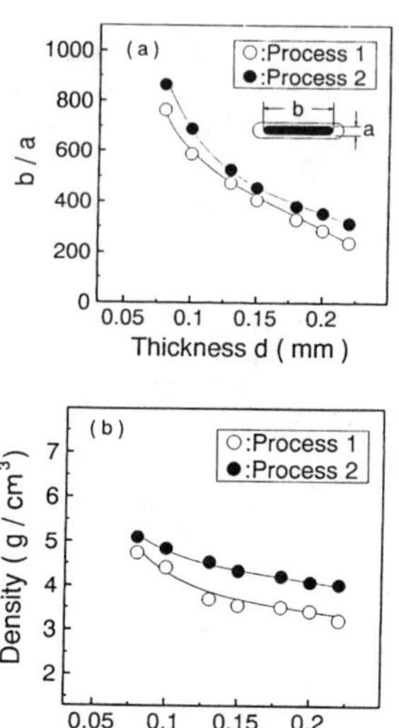

Fig.3.Transverse cross-section for a polished surface of screen-printed tape.

Fig.4.Thickness dependence of (a) aspect ratio b/a and (b) density ρ for screen-printed tape.

The aspect ratio b/a, defined as the ratio of width to thickness for ceramic core in the tape, is shown in Fig. 4(a). As d decreases, b/a shows a monotonic increase. The area ratio S_A/S_C, defined as the ratio of cross-sectional area of silver to ceramic core, shows a similar variation to the ratio b/a shown in Fig. 4(a). This suggests the progress of densification of ceramics core as d decreases. Incidentally, the density ρ for ceramic core, determined by an Archimedes method, shows a monotonic increase with a decrease of d as shown in Fig. 4(b). It is notable that pressing rather than rolling progresses the densification of ceramics core.

The ceramics core of the tape was characterized by XRD patterns with Cu Kα radiation and the SEM observations. The XRD patterns for a wide surface of ceramics are not affected by the difference in tape thickness between 0.1 mm and 0.2 mm. The samples have dominant (00l) peaks for the "2223" phase, together with only little peaks from the "2212" phase and others. This is consistent with the SEM observation (Fig. 5) that there exist highly c-axis-oriented microstructures. Although process 2 produces larger grains than process 1 at a fixed d, the lateral dimension of the grains decreases with decreasing d, independent of the fabrication process.

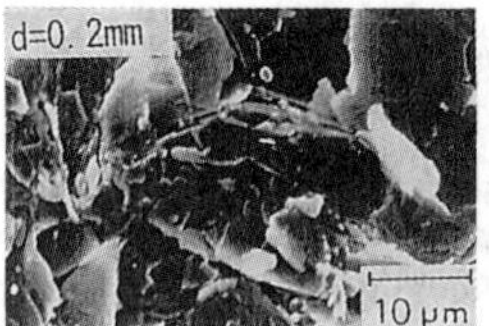

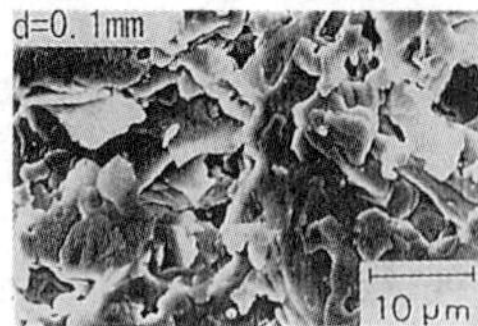

Fig.5.SEM images for a wide surface of ceramics core in screen-printed tape subjected to process 2.

3. RESULTS AND DISCUSSION

3.1. Thickness dependence of J_c at 77K and 0T

The thickness dependence of J_{c0} is shown in Fig. 6 for screen-printed "2223" tapes, subjected to processes 1 and 2 under $T_S=850\,^\circ$C and $t_S=100$ h. Note that the data points are taken as a maximum value of J_{c0} among several samples prepared under the same conditions. As can be seen, the behavior of J_{c0} with decreasing d is nearly independent of the process and characterized by the following two-stages classification: (1) a gradual increase of J_{c0} for 0.1 mm $\leq$ d $\leq$ 0.22 mm until showing a broad maximum at d=0.1 mm and (2) a gradual decrease of J_{c0} for d $\leq$ 0.1 mm after a maximum. It was confirmed by our preliminary study that the result shown in Fig. 6 is not affected by a 5°C deviation of T_S from the optimum value (850°C), hence it is considered to be an intrinsic behavior in screen-printed "2223" tapes.

At the first stage (0.1 mm $\leq$ d $\leq$ 0.22 mm), the densification of ceramics core, which is accompanied by the co-deformation between Ag substrate and ceramics, increases the geometrical form factors such as b/a and S_A/S_C, together with the density ρ, as shown in Fig. 4. An increase in J_{c0} with decreasing d at this stage can be ascribed to an improvement of the transport current path in ceramics core. Judging from the magnified view of transverse cross section (Fig. 3), no severe deformation on the Ag/ceramics interface due to a strong stress takes place.

At the second stage (d $\leq$ 0.10 mm), the densification progresses increasingly in ceramics core and enhances the factors b/a and S_A/S_C, together with the density ρ. Some stress acts on the Ag/ceramics interface and deforms it (Fig. 3), because of the difference in hardness between Ag and ceramics. In spite of an improvement of grain connectivity due to densification, the deformation at the interface should disturb the transport current flow and produces a strong self-magnetic-field, which may be locally far beyond the lower critical field H_{c1}. This is the reason why J_c drops in thinner tapes with d $\leq$ 0.10 mm. A broad maximum of J_c at d=0.1 mm is caused by a competition between the densification enhancing J_c and the disturbance in transport current path lowering J_c.

It is worthwhile mentioning that process 2 yields higher J_{c0} values than process 1. Judging from the behaviors of the factors b/a, S_A/S_C and the density ρ (Fig. 4), pressing is superior to rolling for enhancing the densification of ceramics core so that J_{c0} is increased.

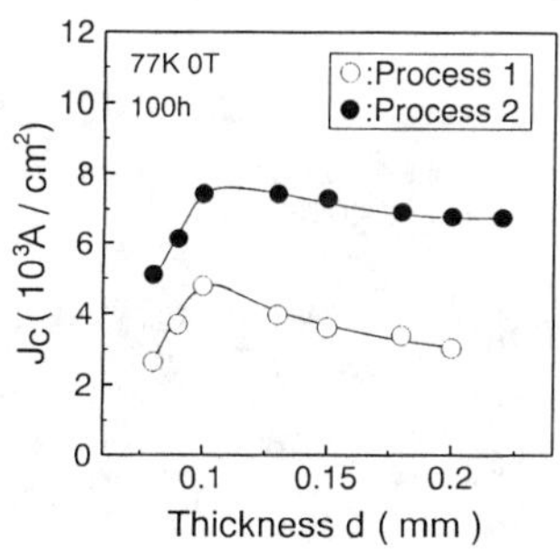

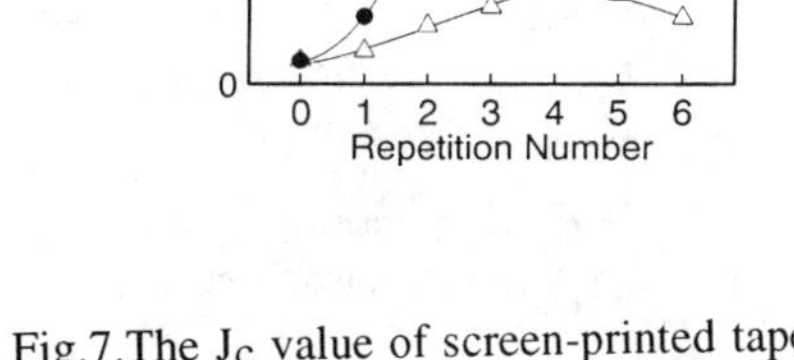

Fig.6. Thickness dependence of the J_c value at 77K and 0T for screen-printed tape subjected to processes 1 and 2.

Fig.7. The J_c value of screen-printed tape, plotted against a repetition number in process 3.

3.2. *Achievement of $J_c=1.5 \times 10^4$ A/cm^2 (77K, 0T) in process 3*

In an attempt to achieve higher J_{c0} values in screen-printed "2223" tapes, we arranged process 3 (see section 2.1) which is a modification of process 2 (i.e., a repetition of pressing and 25h-sintering), so as to progress the densification increasingly without causing severe distortion at the Ag/ceramics interface. The sintering temperature T_s was fixed to the optimum value (850 °C), and the sintering time t_s was changed between 0 h and 100 h by changing a repetition number n_p. We began process 3 when the tape thickness before the pre-sintering reached 0.15mm, by taking into account the fact that the distortion on the Ag/ceramics interface takes place in thinner tapes with d $\leq$ 0.10 mm and degrades J_{c0} (Fig. 6).

The J_{c0} value for process 3 is shown as a function of a repetition number n_p for pressing and sintering in Fig. 7. It is evident that J_{c0} strongly depends upon n_p. It increases monotonically with n_p until showing a maximum at n_p=4, (i.e., t_s=100h). To our knowledge, the maximum value of J_{c0}=1.5 $\times 10^4$ A/cm^2 is the highest record reported so far for screen-printed "2223" tapes. As temperature decreases from 77 to 23K, J_{c0} increases from 1.5×10^4 to 9.0×10^4 A/cm^2, almost linearly with temperature. This corresponds to an increase of critical current I_c from 10.0 to 60.0 A.

For discussion about the factors that are responsible for the achievement of J_{c0}=1.5 $\times 10^4$ A/cm^2, it is worthwhile mentioning that an increase of n_p progresses the densification of ceramics core increasingly. As can be seen from Fig. 8, an increase of n_p from 0 to 6 raises the aspect ratio b/a from ~400 to ~580 and the area ratio S_A/S_C from ~15 to ~22. According to the XRD patterns shown in Fig. 9, the densification of ceramics core due to an increase of n_p is accompanied by the enhancement of a volume fraction of the "2223" phase. A gradual decrease of J_{c0} for $n_p > 4$ after showing a maximum is likely to be ascribed to the evaporation of Bi or Pb. In addition to the XRD study, the SEM image (Fig. 10) for the sample with J_{c0}=1.5 $\times 10^4$ A/cm^2 shows a high degree of c-axis-alignment in microstructures [1]. Most platelet grains run nearly parallel to each other while slightly bending in places. An average misalignment angle between the grains is roughly estimated to 4-6°· The lateral dimension of the grains is approximately 20-25 μm. Accordingly, the achievement of J_{c0}=1.5 $\times 10^4$ A/cm^2 is attributed to both high degree of grain alignment in microstructures and densification of ceramics core, together with enhancement of a volume fraction of the "2223" phase. We note here that pressing under low pressure of P=0.5 GPa in process 3 causes insufficient densification of ceramics core and lowers J_{c0} to below 5 $\times 10^3$ A/cm^2 even when n_p=4 as shown in Fig.7.

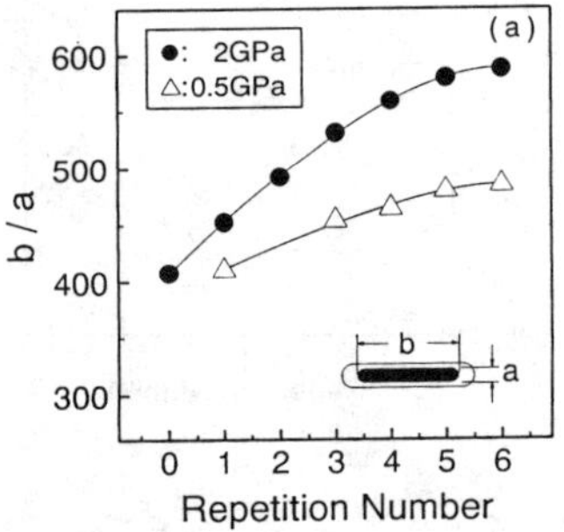

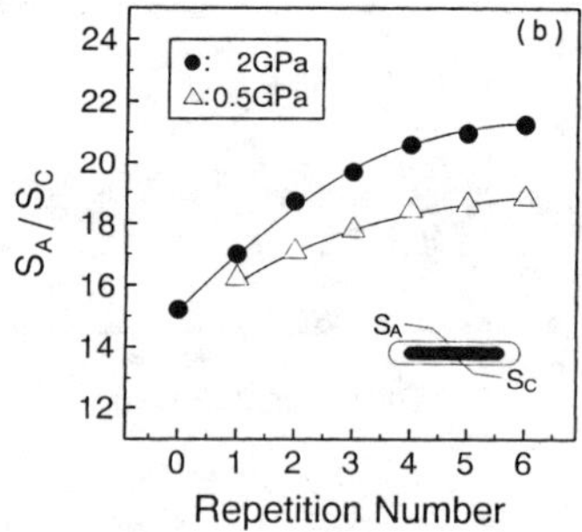

Fig.8.(a) Aspect ratio b/a and (b) area ratio S_A/S_C of screen-printed tape, plotted against a repetition number in process 3.

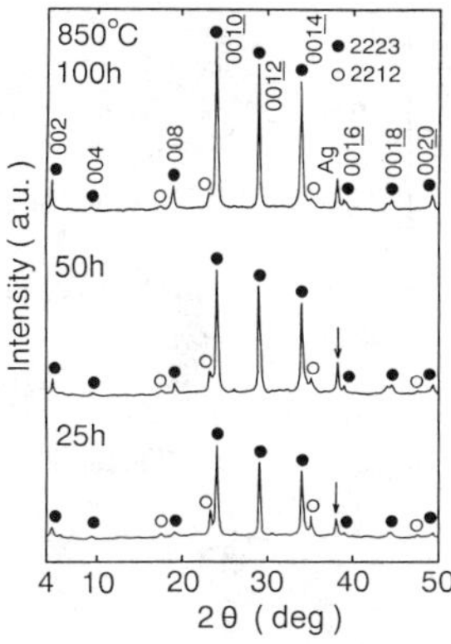

Fig.9.XRD patterns for a wide surface of ceramics core of screen-printed tape, as a parameter of a repetition number in process 3.

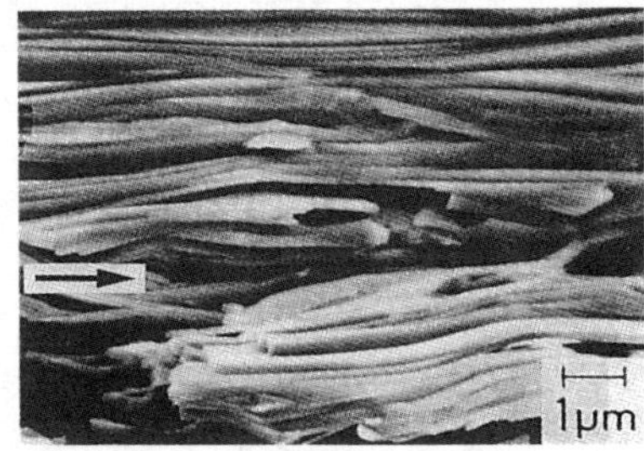

Fig.10.SEM image for a fractured surface of the screen-printed tape with $J_c=1.5 \times 10^4$ A/cm^2 (77K, 0T) [1]. An arrow shows a dislocation in microstructures.

In addition to a high degree of grain alignment, there is a dislocation in microstructures along the contact surface of the ceramics in fabrication process (see Fig. 10). This is the reason why, in spite of a better grain alignment, the J_{c0} value for screen-printed "2223" tape is much lower than that ($> 5 \times 10^4$ A/cm^2) for Ag-sheathed "2223" tapes [4].

Figure 11 shows the J_c anisotropy at 77K for the sample with $J_{c0}=1.5 \times 10^4$ A/cm^2. J_c is plotted against the angle θ between B and the tape normal (i.e. the c axis). As can be seen, J_c shows a sharp maximum (hereafter called $J_{c\perp}$) at $\theta=90°$ (i.e. B$\perp$c), decreases rapidly with a tilt of the field away from 90°, and shows deep minima (hereafter called $J_{c\parallel}$) at $\theta=0°$ and 180° (i.e. B$\parallel$c). An increase in B enhances the J_c anisotropy and sharpens the peak at $\theta=90°$. Both $J_{c\perp}$ and $J_{c\parallel}$ are derived from the J_c vs θ curves, and the result is shown in Fig. 12. As can be seen, $J_{c\parallel}$ shows a decay with B much faster than $J_{c\perp}$. As a result, the anisotropy ratio γ (defined as $J_{c\perp}/J_{c\parallel}$) is enhanced by an increase in B. The half-height angular width $\Delta\theta$, defined as the angular width corresponding to the value $(J_{c\perp}+ J_{c\parallel})/2$, is also shown in the insert of Fig. 12. As can be seen, there is a correlation between γ and $\Delta\theta$. Enhancement of γ, by an increase in B, is accompanied by the narrowing of $\Delta\theta$. It has been shown that the J_c anisotropy is related to the microstructure of the sample [3, 8]. In high fields above 0.5T, $\Delta\theta$ is likely to approach a misalignment angle (4-6°) between the grains.

3.3. A comparison with the Ag-sheathed "2223" tapes

It is well known [4,5] that the achievement of high J_{c0} values well above 1×10^4 A/cm^2 in Ag-sheathed "2223" tapes are attributed to both grain alignment in microstructures and densification of ceramics core. This situation is rather similar to screen-printed "2223" tapes. In an attempt to examine further the factors which dominate J_c, we will compare the thickness dependence of the J_{c0} value, geometrical factors (b/a and S_A/S_C), and density ρ between Ag-sheathed tapes and screen-printed tapes.

The thickness dependence of J_{c0} for Ag-sheathed tapes, which is characterized by a three-stages classification [5], is shown between 0.05 mm and 1.0 mm in Fig. 13. Note that the data points are taken as a maximum value of J_{c0} among several samples prepared under the same conditions. If we confine ourselves to the range below d $\leq$ 0.2 mm, the behavior of J_{c0} with decreasing d in Ag-sheathed tapes is characterized by a marked rise up to 0.1 mm, a sharp peak at d=0.1 mm and a rapid drop for d $\leq$0.1 mm after peaking. This is compared to the

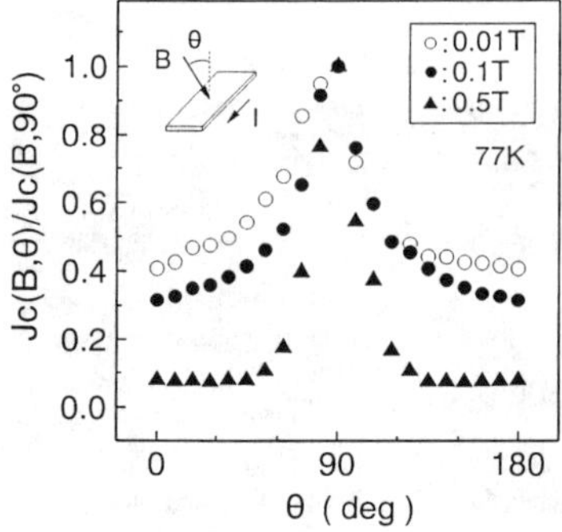

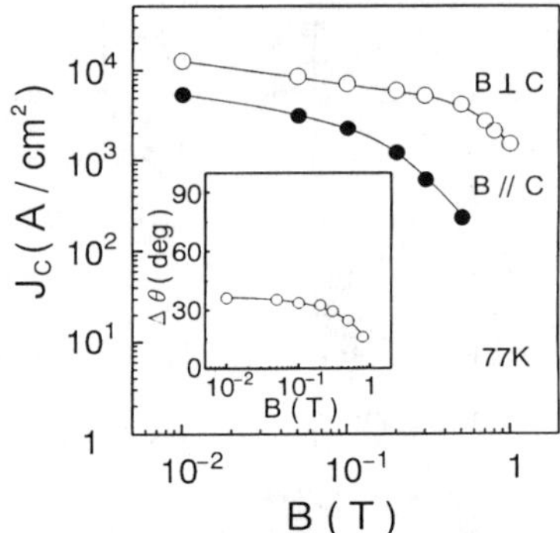

Fig.11.The J_c anisotropy at 77K for the screen-printed tape with J_c=1.5 ×10^4 A/cm^2 (77K, 0T) [8].

Fig.12.The J_c value at 77K for the screen-printed tape with J_c=1.5 ×10^4 A/cm^2 (77K, 0T) in both field configurations B ⊥ c and B ∥ c [1]. Also shown in the insert is the result for the half-height angular width $\Delta\theta$.

result of the screen-printed tapes (Fig. 6), which is characterized by a two-stages classification (see section 3.1). From the behaviors of the aspect ratio b/a and the density ρ shown in Fig.14, an increase of J_{c0} for 0.1 mm ≤ d ≤0.2 mm in the Ag-sheathed tapes is ascribed to the improvement of grain connectivity by densification of ceramics core. However, extreme densification causes severe distortion on Ag/ceramics interface and sticks "2223" platelets into an Ag sheath as shown in Fig. 15. Accordingly, a peak of J_{c0}= 1.5 ×10^4 A/cm^2 is explained by a competition between the improvement of grain connectivity and the disturbance in transport current path. We note here that the densification of ceramics core itself in Ag-sheathed tapes is followed by the co-deformation stage between Ag sheath and ceramics in thicker tapes with d ≥ 0.2 m. In this region, as shown in Fig. 13, both b/a and ρ show an insensitive variation with d, compared to thinner tapes with d ≤0.2 mm .

A comparison of the behaviors of both b/a and ρ in Figs. 4 and 14 leads us to the conjecture that the densification of ceramics core in Ag-sheathed "2223" tapes especially for d ≤ 0.1 mm is more significant than that in screen-printed "2223" tapes, which causes strong stress on the Ag/ceramics interface as shown in Fig. 15. This is the reason why J_{c0} in Ag-sheathed tapes exhibits a more significant decay with d in thinner tapes with d < 0.1 mm, compared to screen-printed tapes (Figs. 6 and 13).

The progress of densification in screen-printed tapes is likely to enhance J_{c0} in a straightforward way without causing severe deformation on the Ag/ceramics interface, which is different from the Ag-sheathed tapes. An application of pressure higher than 2 GPa, used in the present study, should promise further enhancement of J_{c0}. In addition, an improvement of the fabrication process so as to suppress a dislocation in microstructures (Fig. 10) would be needed. These should include some alterations of sintering conditions for process 3, and also the initial tape thickness (d=0.15 mm in the present study) at which process 3 is started.

4. Summary

The thickness dependence of J_c at 77K in screen-printed "2223" tapes with 0.06 mm ≤d≤ 0.22 mm is characterized by a two-stages classification, which is separated on a threshold value d=0.1 mm. As d decreases from 0.22 mm, J_c increases gradually up to 0.1 mm by the densification of ceramics core, while showing a gradual decrease by the distortion

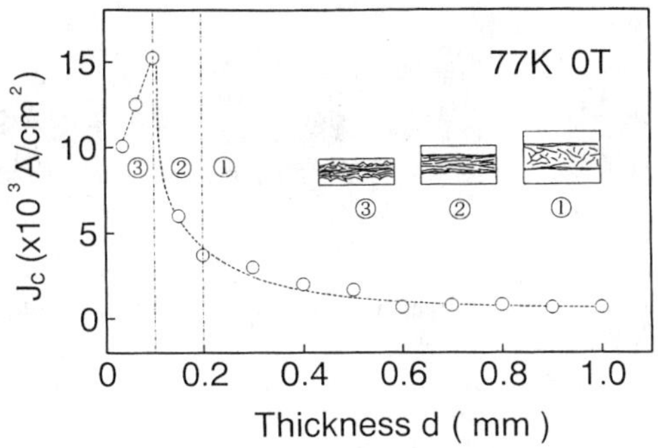

Fig.13.Thickness dependence of the J_C value at 77K and 0T for Ag-sheathed tape [5].

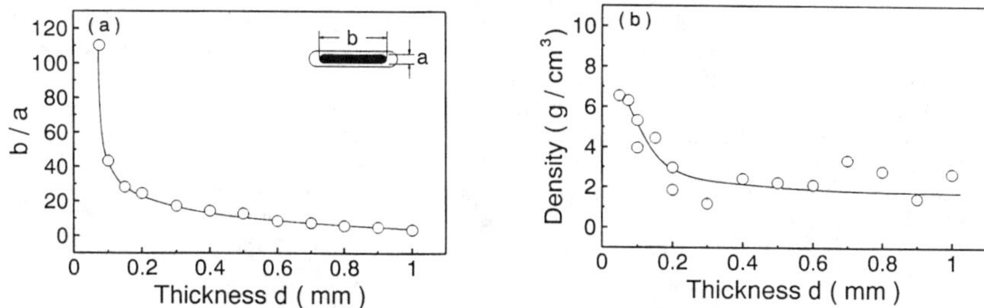

Fig.14.Thickness dependence of (a) aspect ratio b/a and (b) density ρ for Ag-sheathed tape [5].

of Ag/ceramics interface for $d \leq 0.1$ mm. Such behavior is similar to that for the Ag-sheathed "2223" tapes, but the degree is rather dull. The geometrical factors b/a, S_A/S_C and the density ρ in screen-printed tapes also show dull thickness dependence, compared to the Ag-sheathed tapes. These results, together with cross-sectional views around the Ag/ceramics interface, suggest that the magnitude of the stress acting on the Ag/ceramics interface during densification in the screen-printed tapes is weaker than that in the Ag-sheathed tapes.

It has been shown that a repetition of pressing and sintering is available not only for the Ag-sheathed tapes but also for the screen-printed tapes sandwiched between Ag substrates. Actually, four times repetition produces c-axis-oriented microstructures with high J_C values of 1.5×10^4 A/cm^2 (I_C=10.0A) at 77K and 0T, and 9.0×10^4 A/cm^2 (I_C=60.0A) at 23K and 0T. Incidentally the misalignment angle between the grains in microstructures is roughly estimated to 4-6°. In addition to a high degree of crystal alignment, there is a dislocation in microstructures along the contact surface of the ceramics for fabrication, which should cause weak links between the grains and lowers J_C. This is the reason why, in spite of a better grain alignment, the J_C value is much lower than that ($>5 \times 10^4$ A/cm^2 at 77K and 0T) for the Ag-sheathed tapes [4]. The suppression of the dislocation in microstructures, together with higher densification of ceramics core, is necessary for achievement of higher J_C values in the screen-printed tapes.

This work was supported by the AMADA Foundation (#AF-92006).

  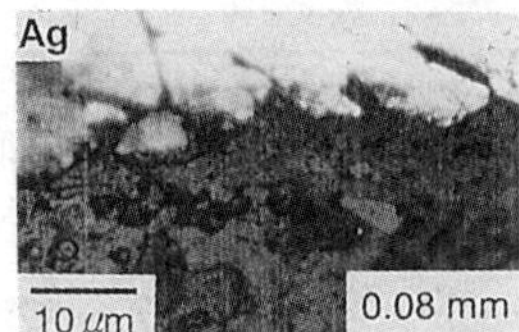

Fig.15.Magnified view for a polished surface near the Ag-ceramics interface in Ag-sheathed tape [5].

REFERENCES

[1] A.Oota, H.Matsui, M.Funakura, J.Iwaya and J.Maeda, Appl. Phys. Lett. 63 (1993) 243.
[2] H.Noji, W.Zhou, B.A.Glowacki and A.Oota, Appl. Phys. Lett. 63 (1993) 833.
[3] A.Oota, H.Matsui, M.Funakura, J.Iwaya and K.Mitsuyama, Physica C210 (1993) 489.
[4] K.Sato, T.Hikata, H.Mukai, M.Ueyama, N.Shibuta, T.Kato, T.Masuda, K.Nagata, K.Iwata and T.Mitsui, IEEE Trans. Mag. MAG-27 (1991) 1231.
[5] A.Oota, J.Iwaya, P.Songsak, T.Saigou and M.Funakura, Physica C214 (1993) 9.
[6] A.Oota, K.Ohba, A.Ishida, A.Kirihigashi, K.Iwasaki and H.Kuwajima, Jpn. J. Appl. Phys. 28 (1989) L1171.
[7] A.Oota, K.Ohba, A.Ishida, H.Noji, A.Kirihigashi, K.Iwasaki and H.Kuwajima, Jpn. J. Appl. Phys. 29 (1990) L262.
[8] A.Oota, M.Funakura, J.Iwaya, H.Matsui and K.Mitsuyama, Physica C212 (1993) 23.

PROGRESS IN THE DEVELOPMENT OF THE SILVER-ADDITION PROCESS FOR PREPARING TEXTURED "1223" TL-CA-BA-CU-OXIDE THICK FILMS

J.A. DeLuca, P.L. Karas, C.L. Briant, and J.E. Tkaczyk

GE Corporate R&D Center, Schenectady, N.Y. 12301

A. Goyal

Oak Ridge National Laboratory, Oak Ridge TN

Abstract

Progress in key areas is reported for the continued development of the silver-addition process for preparing textured "1223" Tl-Ca-Ba-Cu-oxide thick films; film homogeneity, film growth mechanism, scale-up, and flexible substrates. As a consequence of refinements in the precursor oxide film preparation and the development of a "flow" reactor, superconducting films on polycrystalline YSZ have been prepared that exhibit zero-field J_c values (77K) as high as 366,000 A/cm^2. Advantage has been taken of these process improvements to prepare a textured "1223" thick film with T_c=109K on a zirconia coated metal substrate. When placed in a magnetic field there is an initial drop in J_c, attributed to the presence of weak links. However, after this initial drop, the film exhibits the characteristic behavior expected of strongly linked "1223". The film growth mechanism underlying the silver-addition process is discussed in light of the results of microstructural studies of samples quenched at different times into the thallination process.

Processing of Long Lengths of Superconductors
Edited by U. Balachandran, E.W. Collings and A. Goyal
The Minerals, Metals & Materials Society, 1994

Introduction

Interest in the development of the "1223" Tl-Ca-Ba-Cu-oxide superconductor and other "single Tl-O layer" materials for electrical power applications has been stimulated by reports emphasizing not only their high superconducting transition temperatures but also their good "in-field" behavior at temperatures above 40 K [1-7]. Good performance of superconductors above 40 K in magnetic fields of 1-5 Tesla is especially attractive, because it raises the possibility of significant improvements in the efficiency, size, and weight of a wide variety of electrical equipment in which the superconducting elements are cooled by efficient and reliable cryogenic refrigerators. A significant challenge to achieving these goals is the fabrication of the long lengths of wires or tapes required for such applications; wires or tapes that it is generally agreed would contain an oxide superconductor in polycrystalline form. Toward this end we have pursued the development of a process suitable for the fabrication of "1223" Tl-Ca-Ba-Cu-oxide superconducting tape by the reaction of thallium oxide vapor with a silver-containing Ca-Ba-Cu-oxide precursor film [8-10]. In this paper we report on progress in several key areas of our process development efforts; film homogeneity, film growth mechanism, scale-up, and flexible substrates.

Experimental

Precursor oxide films having the cation stoichiometry, $Ca_2Ba_2Cu_3Ag_{0.37}$, were prepared by spraying an aqueous solution of the metal nitrates on to appropriate substrates which were held on a heating block maintained at 275C [8]. After the deposition process the samples were subjected to additional heat treatments both in air and oxygen as detailed below. The amount of material deposited was selected to yield a superconducting oxide film of ~3μm thickness. The precursor oxides were converted to superconducting "1223" films using our "two-step" thallination process [8]. In this process the precursor oxide film is placed (oxygen ambient) in one zone of a two-zone reactor and a boat filled with thallic oxide is placed in the second zone. The sample is heated to 860C. The source is heated first to 690C (step #1) and then to 745C (step #2). The reaction is allowed to proceed for 30 minutes with the source at 745C and then the furnace power is turned off and the system allowed to cool. The temperature/time schedules for both the sample and source can be found in Figure 2 of our earlier report [8]. Process or reactor modifications required for the studies reported in this work are described below.

Film Homogeneity

In an earlier study [9] we found evidence that the manner in which the spray deposited precursor oxide film is prepared affects the current transport characteristics of the final superconducting film. To aid us in understanding this effect we first prepared coatings which were heated to 400C on the deposition heater block. The coatings were removed from the substrates and used as samples for thermogravimetric analysis. We found that ~90% of the weight change upon heating occurred before the sample temperature reached 600C, and we observed inflections in the TGA plot between 500C and 600C that are associated with melting of the calcium and barium nitrate in the deposit. We performed a additional TGA analysis in which the sample temperature was programmed to rise to 500C, remain at 500C for one hour, and then continue to rise to 1000C. In this experiment we found that ~90% of the total weight change occurred during the heating at 500C, thus showing that film decomposition can be achieved without substantial melting of the nitrates.

232

In light of the TGA data, we prepared precursor oxide films on polycrystalline yttria-stabilized zirconia (poly-YSZ) using three deposition/decomposition schedules. **Schedule #1** - (a) the total amount of material required for an ~$3\mu m$ superconducting film was spray deposited at 275C, (b) the film was heated in air at 650C (5 minutes) while it was still on the deposition heating block, and (c) the film was placed directly into a combustion tube furnace at 500C (oxygen ambient), and the temperature was programmed to rise to 850C in 30 minutes at which point the furnace power was turned off and the sample allowed to cool under a flow of oxygen. **Schedule #2** - (a) 1/4 of the amount of material required for an ~$3\mu m$ superconducting film was spray deposited at 275C and the sample subjected to treatment-1b. After four such depositions/treatments the sample was subjected to treatment-1c. **Schedule #3** - (a) the total amount of material required for an ~$3\mu m$ superconducting film was spray deposited at 275C, (b) the sample was heated in air at 500C (10 minutes) while still on the spray heater block, (c) the sample was placed directly into a combustion tube furnace at 500C (oxygen ambient); the temperature was programmed to remain at 500C for an hour and then rise to 850C in 50C steps with a one hour dwell at temperature after each increment; the sample was allowed to cool under a flow of oxygen.

The precursor oxide films were converted to "1223" superconducting films using the apparatus and process schedule described above [8-11]. Also, as described previously, the "1223" samples were patterned with a four-segment 4mmx0.2mm test bridge. They were then annealed in oxygen at 600C [8,9] prior to a determination of the R vs T characteristics of the full bridges and the zero-field J_c values at 77 K, J_c(zf-77K), of the segments. The average superconducting transition temperatures (and standard deviation), T_c(SDEV), for the three groups were; **Schedule #1** - 106K(1), **Schedule #2** - 107K(2), **Schedule #3** - 107K(1). Following are listed the average segment J_c(zf-77K), the ratio of the standard deviation to the average, and the number of segments measured. **Schedule #1:** 16,000 A/cm^2-1.21-44. **Schedule #2:** 38,000 A/cm^2-0.54-56. **Schedule #3:** 40,000 A/cm^2-0.32-20. We examined a precursor oxide film prepared by each of the three schedules using a scanning electron microscope (Cambridge Stereoscan 240) equipped for energy dispersive spectroscopy (EDS-Tracor Northern spectrometer). SEM micrographs of these films are shown in Figure 1.

<u>Film Growth Mechanism</u>

To observe the evolution of the microstructure of our thick films we analyzed a series of samples each one of which had been brought to a different point in the process and then quenched by moving the reactor tube assembly to a position outside of the furnace. Samples were examined by X-ray diffraction, scanning electron microscopy equipped with energy dispersive spectroscopy, and transmission electron microscopy. For this report we will only summarize the results of these investigations. A more detailed description will be published at a later date.

The evolution of the microstructure is evident in the SEM micrographs shown in Figure 2. In our two-step process significant thallium oxide incorporation into the film and growth of the highly oriented "1223" phase occurs only after the thallium oxide boat temperature is raised to 745C. We have found that the thallium oxide is not incorporated in a uniform manner. Rather, it enters the film at isolated sites such as that indicated with an arrow in Figure 2a, which is an SEM micrograph of a sample quenched 5 minutes after the thallium oxide boat temperature reached 745C. Our analysis indicates that at these sites a liquid phase forms from which the "1223" phase grows. As more thallium oxide is incorporated,

A.   B.

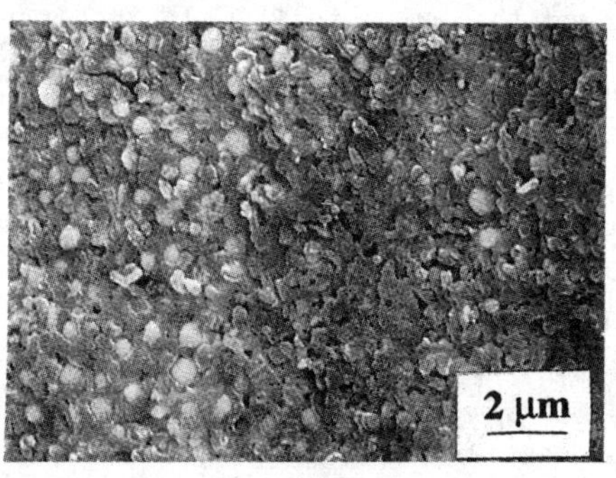

C.

Figure 1

SEM micrographs are shown for precursor oxide films prepared using (A) Schedule #1, (B) Schedule #2, and (C) Schedule #3 described in the text.

these sites grow in size until the entire sample is converted to the "1223" phase. This is evident in Figures 2b and 2c which are micrographs of samples quenched 6 and 42 minutes, respectively, after the thallium oxide boat temperature reached 745C. The data indicate that, although the thallium incorporation occurs at sites in areas that are predominately calcium oxide, barium and copper oxides and silver must also be present. We have found no evidence of a significant incorporation of silver into the "1223" phase. Rather, the silver is found only as isolated nodules in the final films.

Scale-Up

We refer to the two-zone reactor used in all of our studies to date [11] as a "static" reactor because of its closed alumina-tube geometry. In this reactor transport of the thallium oxide vapor from the thallic oxide boat to the sample, over a distance of ~20 cm, takes place via gaseous diffusion. One consequence of this design from the point of view of process scale-up is that the thallium oxide incorporation rate into the

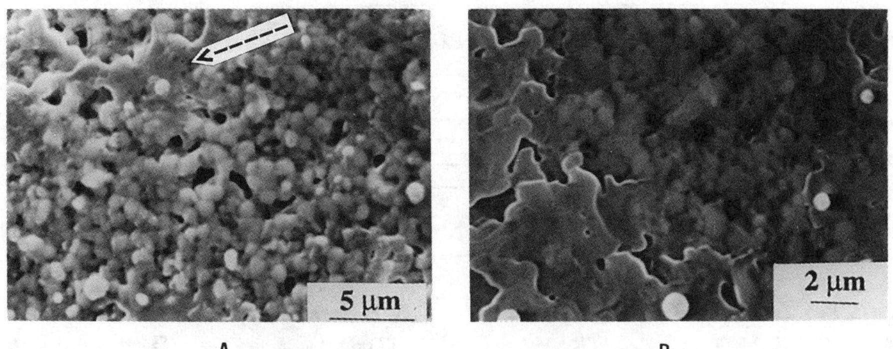

A.   B.

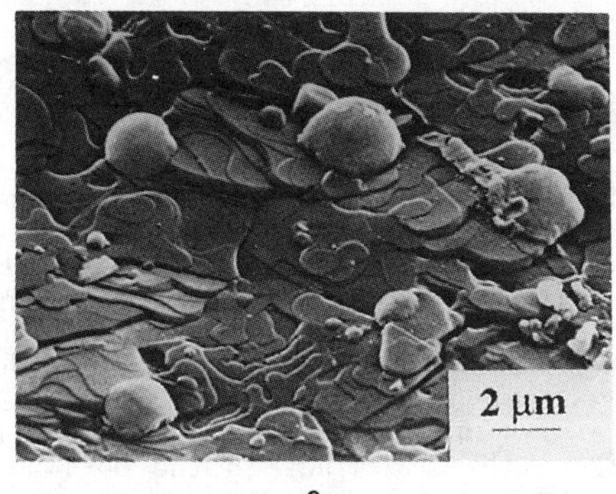

C.

Figure 2

SEM micrographs illustrating the evolution of the textured microstructure obtained in the silver-addition process. Shown are micrographs for samples quenched (a) 5 minutes, (b) 6 minutes, and (c) 42 minutes after the thallium oxide boat temperature attained 745C. The platelike "1223" growths are readily evident in (b) and (c). The arrow in (a) points to a growth front in its early stages of development.

film may be diffusion-transport limited. Another is that there will be a gradient in the thallium oxide partial pressure in the reactor tube which presents significant problems for obtaining samples with homogeneous properties over large areas. For example, in our "static" reactor we have evaluated the processing of three 12.5 cm long coated poly-YSZ substrates placed end to end in the sample zone (less than 3 degrees temperature difference across zone). Representing the cation stoichiometry of the film by X:2:2:3:0.37 (Tl:Ca:Ba:Cu:Ag), we have found in a typical experiment "X" values of 0.84, 0.70, and 0.43 for samples 1, 2, and 3 respectively (sample 1 was closest to the thallium oxide boat. For comparison we note that our optimum samples have compositions in which $0.75 \leq X \leq 0.90$

Because we believe that controlled transport of the thallium oxide vapor will be important in any scaled-up version of our process, we

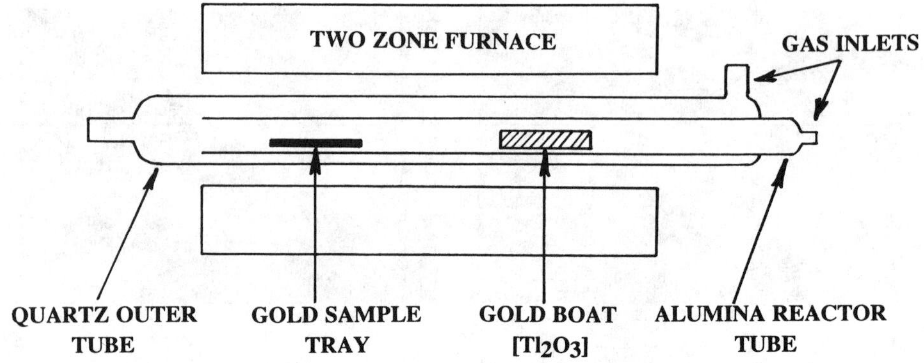

Figure 3

This diagram illustrates the essential features of the two-zone "flow" reactor.

initiated an evaluation of the preparation of films in the "flow" reactor shown schematically in Figure 3. We have used the same two-step process/process times described above with only minor modifications, namely; 1.) an oxygen flow is maintained both through the outer quartz tube (300 SCCM) and the alumina reactor tube (1.7 SCCM), and 2.) the thallium oxide boat temperature has been reduced to 730C in order to to obtain the desired thallium content in the films ($0.75 \leq X \leq 0.90$). Using these modifications, we can now obtain textured "1223" thick films (poly-YSZ/Schedule #2) with "X" values in the optimum range for each of three samples placed end to end in the sample zone. In Table 1 are listed the "segmented bridge" J_c(zf-77K) values obtained on four samples prepared in our flow reactor. The T_c values of the samples ranged from 104K to 107K.

Table 1

J_c(zf-77K) in A/cm^2

Sample	Segment 1	Segment 2	Segment 3	Segment 4
1	46,000	67,200	336,000	148,000
2	166,000	26,800	16,300	80,500
3	51,700	10,300	16,800	41,900
4	31,900	124,000	97,300	80,500

Flexible Substrates

A superconducting element for power applications must satisfy stringent electrical and mechanical performance requirements [12,13]. An important first step in meeting these requirements is the demonstration of good superconducting properties on a flexible substrate [14-16]. In our process the substrate material must resist chemical reaction at 860C with oxygen and thallium oxide vapor, as well as chemical and interdiffusion interactions with the precursor and superconducting layers. To minimize cracking of the films, the substrate should match closely the thermal expansion coefficient of the superconducting layer. Our demonstration that highly textured thick films can be obtained on a polycrystalline surface removes the requirement for a substrate that will support epitaxial growth.

However, it is likely that optimum texturing of the films will require that a flat substrate surface be maintained throughout the growth process. This presents additional challenges for metallic substrates [16].

For our first "flow reactor" experiments on forming "1223" thick films on metallic substrates we elected to use Fecralloy[R] [17], a high-temperature, oxidation-resistant iron-chromium-aluminum alloy. Earlier work in our laboratory [9] had shown that, although this alloy can withstand exposure to both oxygen and thallium oxide vapor at 860C, it reacts readily with the deposited film. In light of this, we have initiated work to evaluate the effectiveness of various barrier layers for minimizing this deleterious reaction. In the present case a sample of Fecralloy[R] (3.2x0.64x0.005cm) was coated by sputter deposition with an ~0.2μm thick layer of undoped zirconia. A amount of precursor oxide sufficient to form an ~3μm thick superconducting film was prepared using Schedule #3 described above. The sample was processed in the two-zone reactor and subsequently annealed in oxygen at 600C using the schedules previously reported [8,9]. Similar to the case of our samples on poly-YSZ substrates, x-ray diffraction analysis showed the film to consist primarily of strongly c-axis textured "1223" with only traces of an unidentified secondary phase. The sample exhibited a T_c=109K and J_c(zf-77K)=10,000 A/cm^2. The J_c vs field behavior at various temperatures is shown in Figure 4.

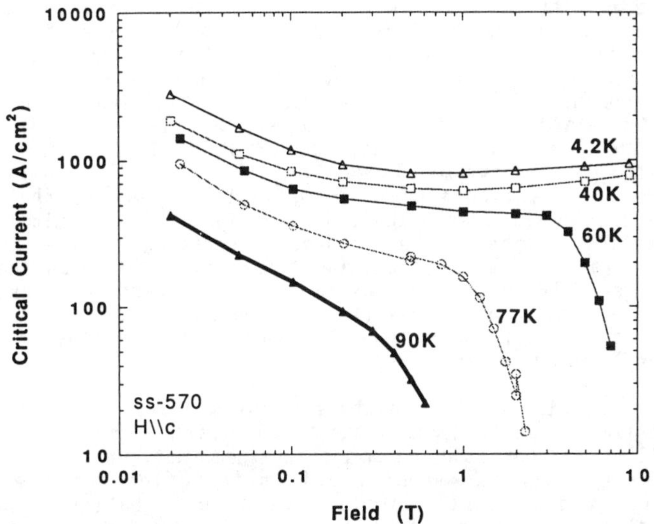

Figure 4

Critical current density vs. field at various temperatures for a textured "1223" film on zirconia coated Fecralloy[R] having J_c(zf-77K)=10,000 A/cm^2. The field was applied perpendicular to the substrate surface (parallel to c-axis).

Discussion

Although the three sample groups in our homogeneity study have similar T_c values, the precursor oxides prepared by Schedules #2 and #3 yielded superconducting films with significantly higher average J_c(zf-77K) values and less scatter in the segment values than those prepared from Schedule #1 precursor films. We believe that this behavior is related to the melting/decomposition characteristics of the calcium and barium nitrates. In the Schedule #1 precursors the rapid evolution of gas may cause some disruption of the film while the flow of the liquid nitrates before they decompose completely results in compositional inhomogeneities. The same processes occur during Schedule #2. However, the disruption by evolving gas is likely to be less severe and the effects of seggregation "averaged out" as the thinner layers are processed in sequence. In Schedule #3 the nitrates are decomposed slowly without melting, a situation expected to result in minimal disruption of the film and no post deposition seggregation. This interpretation is supported by the results of the microstructural characterization of precursor films. The SEM micrographs of the Schedule #3 and #2 precursors, shown in Figures 1a and 1b, respectively, reveal a fine-grained, uniform-appearing microstructure. The micrograph in Figure 1c of a Schedule #1 precursor has a coarser-grained, less uniform appearance. The EDS analysis of the Schedule #3 and #2 precursors revealed a relatively uniform chemical composition. On the other hand, the EDS analysis of the Schedule #1 precursor revealed a less homogeneous chemical composition and areas could be identified that were clearly either calcium or barium rich.

The results of our film growth mechanism studies are of particular interest. In our earlier work [8,9] we found that the silver addition is necessary to achieve the formation of a well interconnected and textured "1223" film via a process believed to involve a liquid phase assisted growth. Because silver exists as nodules in the final films and is found only in trace amounts in the "1223" phase, we believe that it plays an role only in the formation of the liquid phase. It is also important to note that, although all five metals must be present for the liquid to form, the liquid phase formation appears to initiate in regions that are calcium oxide rich. This suggests that, in addition to the chemical composition, the phase composition of the precursor is critical to the success of the process. That is, a "well calcined" precursor, in which the constituent oxides have fully reacted to form well defined compounds, may not give the desired results. This possibility should be taken into consideration in any attempt to adapt the silver-addition approach to other polycrystalline conductor fabrication processes.

Consistent with this picture of film growth are the results of a characterization of texture in our "1223" thick films from XRD azimuthal scans [18]. In that study it was determined that, on a macroscopic scale, the films exhibit excellent c-axis orientation but are essentially untextured with respect to basal plane orientation. However, on a microscopic scale, it was found that the films consist of an intergrowth of crystallite colonies within each of which there is a high degree of basal plane alignment. We believe that such colonies nucleate at sites, such as that indicated by the arrow in Figure 2a, where the liquid phase first forms as the thallium oxide vapor reacts with the precursor oxide films. Because the manner in which these colonies intergrow may determine the transport characteristics of the film [18], better control of this process may lead to significant improvements in the critical current densities obtainable.

The results of our first "flow-reactor" experiments with films on poly-YSZ substrates are very encouraging, especially when one considers the fact that in this initial set of samples are found three segments with J_c(zf-77K) values close to, and four segments with J_c(zf-77K) values well in excess of, the highest value (105,000 A/cm^2) ever measured by us in well over a hundred samples (four hundred segments) prepared in the "static" reactor. Additional work is required to understand why the J_c(zf-77K) values of the "flow" reactor samples are noticably higher than those prepared in the "static" reactor. However, we consider it likely that this is related, at least in part, to the presence of a more uniform thallium oxide partial pressure throughout the reactor which permits the "1223" phase to form under optimum equilibrium conditions at all positions on the film. It is apparent from the data in Table 1 that there is still a considerable segment to segment variation in J_c. Variations on the 200μm scale of the test bridge may be an unavoidable consequence of the film growth mechanism in the present process. Actually, we believe that the results obtained using this relatively narrow test bridge may present a misleading picture of the average properties over film dimensions of significance for practical conductors. To that end, we are presently assessing the suitablity of a wider test bridge.

In our first experiment with barrier-layer coated metallic substrates we have demonstrated the growth of a highly textured "1223" film. The T_c value of 109K, which is comparable to the highest values measured for us for films on poly-YSZ, indicates that the zirconia layer is effective in limiting film/substrate interactions. The J_c(zf-77K) value of 10,000 A/cm^2 measured for our film on Fecralloy^R:ZrO_2 is relatively low compared to our poly-YSZ samples, and the drop in J_c to ~1,000 A/cm^2 in weak fields indicates the presence of weak links. However, we find it most encouraging that the J_c vs field behavior after this initial low field drop indicates the presence of strongly linked current paths throughout our film.

Conclusion

The silver-addition process continues to show considerable potential for the production of superconducting tapes for use above 40K. Through refinements in the precursor deposition process and the development of a well controlled "flow-reactor", we have demonstrated significant improvements in the superconducting properties of "1223" thick films. The refinements reported here have permitted us to achieve an important milestone in the evolution of this technology, namely the preparation of a superconducting thick film, exhibiting characteristic "1223" strong linked behavior, on a flexible metal substrate.

Acknowledgments

This work was supported in part under Argonne National Laboratory Agreement No. 22242401 under W-31-109-ENG-38 (DOE), and Oak Ridge National Laboratory Cooperative Agreement 87X-SE934V.

References

1.   D.H. Kim, K.E. Grey, R.T. Kampwirth, J.C. Smith, D.S. Richeson, T.J. Marks, J.H. Kang, J. Talvacchio, and M. Eddy, Physica C **177**, (1991) 431.

2.   T. Kamo, T. Doi, A. Soeta, T. Yuasa, N. Inoue, K. Aihara, and S. Matsuda, Appl. Phys. Lett. **59(24)**, (1991) 3186.

3.   T. Sasoaka, A. Nomoto, M. Seido, T. Doi, and T. Kamo, Japn. J. Appl. Phys., **30**, (1991) L1868.

4.   S.P. Matsuda, T. Doi, A. Soeta, T. Yuasa, N. Inoue, K. Aihara, and T. Kamo, Appl. Phys. Lett., 59, (1991) 3186.

5.   M. Mittag, M. Roseberg, B. Himmerich, and H. Sabrowsky, Supercond. Sci. Technol. 4, (1991) 244.

6.   T. Doi, M. Okada, A. Soeta, T. Yuasa, K. Aihara, T. Kamo, and S. Matsuda, Physica C 183, (1991) 67.

7.   R.S. Liu, D.N. Zheng, J.W. Loram, K.A. Mirza, A.M. Campbell, and P.P. Edwards, Appl. Phys. Lett. 60(8), (1992) 1019.

8.   J.A. DeLuca, P.L. Karas, J.E. Tkaczyk, P.J. Bednarczyk, M.F. Garbauskas, C.L. Briant, and D.B. Sorensen, Physica C 205, (1993) 21.

9.   J.A. DeLuca, P.L. Karas, J.E. Tkaczyk, C.L. Briant, M.F. Garbauskas, and P.J. Bednarczyk, AIP Conference Proceedings 273 - "Superconductivity and Its Applications: Buffalo N.Y.-1992", 531 , American Institute of Physics, New York (1993).

10.   J.A. DeLuca, P.L. Karas, J.E. Tkaczyk, C.L. Briant, M.F. Garbauskas, and P.J. Bednarczyk, Mat. Res. Soc. Symp. Proc. Vol. 275, 669 (1992 Materials Research Society)

11.   J.A. DeLuca, M.F. Garbauskas, R.B. Bolon, J.G. McMullen, W.E. Balz, and P.L. Karas, J. Mater. Research 6, (1991) 1415.

12.   G. Ries, Cryogenics 33(6), (1993) 609.

13.   E.W. Collings, Cryogenics 28, (1988) 724.

14.   D.T. Shaw, MRS Bulletin Aug. 1992, 39.

15.   D. Jedamzik, GEC Jorunal of Research 8(2), (1990) 92.

16.   E. Narumi, L.W. Song, F. Yang, S. Patel, Y.H. Kao, and D.T. Shaw, Appl. Phys. Lett. 58(11), (1991) 1202.

17.   Available from Goodfellow Corp., Malvern Penn.

18.   D.M. Kroeger, A. Goyal, E.D. Specht, Z.L. Wang, J.E. Tkaczyk, and J.A. DeLuca, submitted to Applied Physics Letters.

THICK-FILM AND POWDER-IN-TUBE PROCESSING FOR

TL-OXIDE WIRE AND TAPE

R. D. BLAUGHER, R. N. BHATTACHARYA AND P. A. PARILLA
National Renewable Energy Laboratory
Golden, Colorado 80401-3393, U.S.A.

and

Z. F. REN AND J. H. WANG
State University of New York
Buffalo, New York 14214

ABSTRACT

The wire and tape processing of the Tl-oxides, compared to Bi-oxide, is not as well advanced due to a reduced level of effort. The Tl-oxide single layer compounds, moreover, offer high intrinsic pinning with potential for high field operation at 77K. The Tl-PIT processed to date has not achieved the expected transport properties due to poor inter-grain connectivity resulting from an inability to texture the superconducting core. On the other hand, thick-film processing methods, as offered by electrodeposition and spray pyrolysis have demonstrated the ability to obtain a highly textured microstructure with good in-field transport properties. This paper will review the progress in the Tl-oxide wire and tape processing for both PIT and thick-film techniques. The role of the powder precursor related to reactivity, mixing and particle size on Tl-PIT methods will be discussed. Electrodeposited wire or tape is advanced as a viable low cost alternative to powder-in-tube processing.

Processing of Long Lengths of Superconductors
Edited by U. Balachandran, E.W. Collings and A. Goyal
The Minerals, Metals & Materials Society, 1994

Introduction

The primary technical challenge that must be satisfied to permit usage of the high-T_c oxides in superconducting magnets or power related applications is the successful demonstration of a high field, high-current carrying wire or tape with acceptable mechanical properties. The development of technologically useful HTS wires or tapes has steadily progressed over the past few years. Most of this effort has concentrated on the $Bi_2Sr_2Ca_1Cu_2O_x$ (Bi-2212) and $Bi_2Sr_2Ca_2Cu_3O_x$ (Bi-2223) phases which show superconducting transitions at ~85 K and ~110 K, respectively. The processing approach for fabricating Bi-oxide wires and tapes has generally followed a powder-in-tube (PIT) method. The increase in Bi-PIT performance has been steady, with recent results for Bi-2223 at 5-7 x 10^4 A/cm^2 at 77 K, in zero field and over 10^4 A/cm^2 at 1.0 T. Bi-PIT processing has also produced long lengths, greater than 100m, with current densities demonstrated near 10^4 A/cm^2 at 77 K in zero field. The Bi-tape current densities are markedly improved at lower temperatures, showing near 10^5 A/cm^2 at 4.2 K in magnetic fields of 20 T which is relatively unchanged at 20 K.[1,2] The Bi-tape, however, shows pronounced degradation in current density at higher temperatures in magnetic fields parallel to the c direction, which is related to basic intrinsic limits to flux pinning for the Bi system at temperatures above 20 K.[1,2]

Tl-Oxide Superconductors

The Tl-based superconducting oxides present an attractive alternative to the Bi-oxides due to their high transition temperatures reaching 127 K. Polycrystalline development, moreover, appears more forgiving in this system than either the Y or Bi-oxides, with Tl-thin films reported with critical currents at 76 K of over 10^5 A/cm^2 with no evidence for weak-links in a magnetic field.[3] Tl compounds occur with both single- and double-layer Tl-O structures and multiple Cu-O layers from 1 to 5. Recent work by Kim, et al.[4] has argued that the single-layer Tl compounds should show improved coupling of the Cu-O planes resulting in improved magnetic performance at 77K. R. S. Liu, et. al.[5] has reported data on Tl-1223 that corroborates the Kim, et al. projection with over 10^5 A/cm^2 shown at 77K and 1.0 T. Nabatame et al. reported a H* or higher temperature irreversibility line for Tl-1223 that compares favorably at 77 K with the Y-123 performance[6]. The Tl single-layer compounds thus offer superior intrinsic magnetic field performance at 77 K compared to the Bi compounds and represent an excellent candidate for the development of HTS conductors operating near LN$_2$. The Tl-PIT results to date, however, have not been consistent with these expectations showing basic weak-link limitations in their transport properties. This weak-link problem is attributed to an inability to texture the superconducting core[7]. Thick-film processing on the other hand, as offered by electrodeposition and spray pyrolysis have demonstrated the ability to obtain a highly textured microstructure with excellent in-field transport properties in the difficult H ∥ c direction.[8,9,10] Recent studies on Tl-1223 thick films prepared by spray pyrolysis show a highly textured "plate-like" morphology is associated with the observed high transport currents. This behavior has been correlated to a variety of percolative models including a "brick-wall" structure, forgiving high angle twist or tilt boundaries, and a sub-grain network or "colony" of low angle boundaries.[11,12] These recent microstructural studies thus suggest that the earlier polycrystalline thin-film results of Ginley et al.[3] which showed a similar plate-like microstructure have an identical transport mechanism as proposed for the thick films. Since the thin-film results were obtained on film thicknesses well under a micron (0.3-0.7μm), the "brick-wall" model is less plausible, which favors the Kroger et. al.[12] sub-grain or colony microstructure as a more likely explanation. Kroger et. al.[12] observed the "brick" thickness for the ~3.0μm films was 0.5-2μm.

Powder-in-Tube Processing (PIT)

The development of wires and tapes based on the Tl-oxide superconductors has paralleled, to some degree, the methods followed for Bi and Y. The major processing approach for the Tl-oxides, analogous to Bi, has been the PIT technique. An oxide-based precursor powder is prepared and packed into a hollow metallic tube, which is usually silver (or Au-Pd) due to its relative inertness to the oxide and ability to permit oxygen diffusion. The composite tube is then swaged and drawn into a wire, and if desired subsequently rolled into a tape. The drawing and rolling operations are usually interspersed with intermediate heat treatments followed by a final reduction and final heat treatment in flowing oxygen. The combinations of mechanical deformations and heat treatments are crucial to forming the correct phase, promote texturing and obtain grain-to-grain interconnects for the superconductor.

The precursor powder preparation is critical to the PIT process, especially for the Tl system which has been shown to be more resistant to texturing than the Bi-oxide system. A variety of methods including mixed oxides, chemically precipitated powders, aerosols and spray pyrolysis have been investigated as PIT precursors both with and without Tl. In most cases, although the initially deposited precursor powders have the small particle sizes and high surfaces areas required for subsequent processing, these are lost to some degree during the steps required to add Tl and attain stoichiometric oxides suitable for PIT processing. The mixed-oxide approach uses the constituent oxide powders (except Tl_2O_3) which are normally mixed, sintered in flowing O_2, and reground into powder. An appropriate amount of Tl_2O_3 is then added, and the procedure is repeated to form a superconducting precursor. The sintered pellets are pulverized with particle sizes of a few microns and packed into either Ag or Au-alloy tubes followed by thermomechanical treatment. The alternative techniques for precursor preparation such as sol-gel, coprecipitation, aerosol, and so on provide more precise control on phase purity and stoichiometry, minimize preparative contamination, and offer better mixing and smaller size for the powder particulates. A chemical precipitation method developed by Sandia is being used to prepare precursor powders for the IGC/LANL Tl-PIT effort. The initial precursor is Ba-Ca-Cu-O with Tl_2O_3 (and Sr and Pb-oxides) subsequently added, mixed and reacted to form the desired Tl-compound. The high volatility and toxicity for Tl has prevented the inclusion of Tl in the initial precursor. This precursor approach should offer improved mixing and smaller particle size than the mixed oxides however the results to date have not surpassed earlier mixed oxide methods with typical J_c of 6.2 x 10^3 A/cm^2.[13]

Electrodeposition is an electrochemical process which allows (e. g., metals and oxides) to be deposited at the proper stoichiometry from a liquid electrolyte containing the ions of interest (e.g., Cu^{2+}, Tl^{1+}, Ba^{2+}, etc.) onto a conducting substrate. The ability to prepare homogeneous submicron precursor powders at the required stoichiometric ratios, including Tl, is one of the unique advantages of the electrodeposition method. The resulting precursor can then be heat treated and oxygenated to develop the oxide superconductor. The electrodeposition technique is simple, inexpensive, offers high deposition rates at many μm/min, and provides a highly reactive mixture on an atomic scale which reduces the reaction times to a few minutes. Complete details on this process have been previously published and should be referred to for a more complete description.[14,15,16]

TBCCO films were previously deposited using "dilute" molar concentrations at typical rates of $1\mu m/min.$[14,15,16] The deposition rate of each element under constant potential at steady state conditions can be represented by the following equation:

$$d[M^0]/dt = k_m[M^{n+}] \tag{1}$$

where k_m is the potential-dependent rate constant and $[M^{n+}]$ is the solution concentration of the metal ion. Since the deposition rate is directly proportional to the solution concentration, increasing the molar concentration will correspondingly increase the deposition rate to many $\mu m/min$ as required for a high rate powder process. The TBCCO deposit can be collected, filtered, rinsed, dried to a powder, and following pre-reaction can then be packed into a silver tube for PIT processing. As an alternative the precursor powder can be compacted into pellets, pre-reacted and then packed into a metallic tube for PIT processing.

If we assume a non-aqueous electrodeposition process the respective cations are reduced and deposited on the cathode as a "metallic" precursor.

$$\text{Example: } Ca^{+2} + 2e \rightarrow Ca^0$$

The inclusion of some water in the electrolyte which is introduced either by air exposure or through the waters of hydration in the component salts produces the following reaction at the cathode which applies to all of the cations, with Ca shown as an example,

$$\text{i. e. } Ca^{+2} + H_2O + 2e \rightarrow Ca(OH)_2$$

The hydroxides are subsequently converted to the oxides during the heat treatment and oxygenation to form the desired Tl-oxide compounds.

For the copper and barium hydroxides the typical reactions would be:

$$Cu(OH)_2 \rightarrow Cu_2O + H_2O \qquad \text{from } 130\text{-}270°C$$
$$Cu_2O + 1/2\, O_2 \rightarrow 2CuO \qquad \text{at } 850°C$$

and

$$Ba(OH)_2 \rightarrow BaO + H_2O \qquad \text{from } 560\text{-}700°C$$

TBCCO electrodeposited precursor powders were prepared as outlined above and studied to evaluate their suitability for PIT processing. SEM analysis of the powder showed spherical particulates of the order of $0.1\ \mu m$. Thermo-gravimetric and differential scanning calorimetry of the as-deposited precursor showed at least four different reactions from room temperature up to ~800°C which roughly corresponds with the four cation formulation. X-ray analysis of the precursor reacted at different temperatures showed the presence of oxides and hydroxides consistent with the above reactions. The subsequent reaction of this "oxygenated" precursor thus has sufficient oxygen to facilitate the development of the desired TBCCO crystalline structure. The electrodeposited precursor, furthermore, which allows deposition of all the components (including Tl) provides mixing of the as-deposited materials on an atomic scale which permits much shorter reaction times compared to mixed-oxides. Electrodeposited powder was prepared with the composition; $Tl_{1.1}Ba_2Ca_{.97}Cu_7O_{.8}$ with Ag added to promote nucleation, pressed into a pellet and reacted at 860°C x 30 min. Magnetization and x-ray analysis showed strong Tl-2223 development, T_c of 124K, and intrinsic J_c of ~10^5 A/cm^2 at

77K in zero field. The J_c at 77K dropped to essentially zero at 5kG which is consistent with the Tl-two-layer development.

Additional Tl-Ba-Ca-Cu-O and Ba-Ca-Cu-O electrodeposited precursors were prepared and processed as outlined in Ref. 17. In order to eliminate water formation during reaction the powders were usually pre-reacted at ~870°C x 5h and additional Tl_2O_3 added to compensate for the weight loss. Bi_2O_3 and $SrCO_3$ were also added to some of the precursors to enhance the Tl-1223 formation. Recent data by Ren and Wang[18] on Tl-Bi-Sr-Ba-Ca-Cu-O Ag sheathed tape showed enhanced Tl-1223 development with transport J_c at 77K in zero field of ~2 x 10^4 A/cm^2. The best results for PIT processed with the electrodeposited precursor were obtained with a three-step processing cycle as shown in Fig.1.

After introduction and packing of the powders, the Ag tube was swaged to 1.09mm and then rolled to 0.5mm. The tape was then annealed at 845°C x 10-15 min, rolled to 0.125mm and annealed at 825-845°C x 10-15 min, followed by pressing and a final anneal at 835-845°C x 5-8h. All of the anneals were conducted in a flowing O_2 atmosphere. SEM and x-ray investigation of the superconducting core showed a highly polycrystalline Tl-1223 development with no evidence of texturing. The highest J_c at 77K in zero field of ~10^4 A/cm^2 dropped to under 10^3 A/cm^2 at 500 G indicative of significant weak-link behavior.

Earlier data by Ren and Wang[17] on a 24m length of Tl-Pb-Ba-Sr-Ca-Cu-O Ag sheathed tape with the Tl-1223 structure showed transport J_c at 77K in zero field near 10^4 A/cm^2 for the entire length which also showed poor magnetic field performance.

The highest J_c results published to date for Tl-PIT silver sheathed tapes are due to Seido et al.[19], who used a variety of processing methods with their best results obtained using a phase change reaction from Tl-1212 starting powder to form Tl-1223. The J_c values for the Tl-1223 tape are notably better at fields above 2 T than the Tl-2223 or Bi-2223 tapes for both H ‖ c and H ⊥ c field orientations which is consistent with the improved magnetic performance suggested by Kim et al.[4]

The J_c-H dependence for the Tl-PIT data, nevertheless, showed evidence for weak link behavior with a precipitous drop in current density at a low field of 0.01T. The data, however, showed evidence for a strong linked component with a plateau extending to 4T The nature of the weak link behavior at low fields has not been established but is thought to be related to poor continuity or compositional variation at the grain boundaries. Microstructural investigation of Ag-sheathed Tl-1223 has generally shown random grain development with no evidence for precipitate at the grain boundaries.[19,20] Recent studies of the random grain development as seen by Seido and in virtually all Tl-PIT processed wire or tape concludes that the weak-links are related to a preponderance of high-angle boundaries.[7] Studies on Tl-2212 thin films shows a degradation in J_c for high-angle tilt boundaries similar to the earlier Dimos et. al studies on Y-123.[21,22] The realization of "useful" transport current at 77K in high magnetic fields for Tl-PIT thus requires a highly textured superconducting core which has not been observed to date.

Thick Film Methods

Various thick-film methods have been explored to fabricate a flexible Tl-based wire or tape. An early result by T. Goto et.al.[23] produced ~250μm filaments of Tl-1234 using a pre-fired powder of Tl-2223 mixed with an organic binder. Filaments up to 5m were extruded, cut into

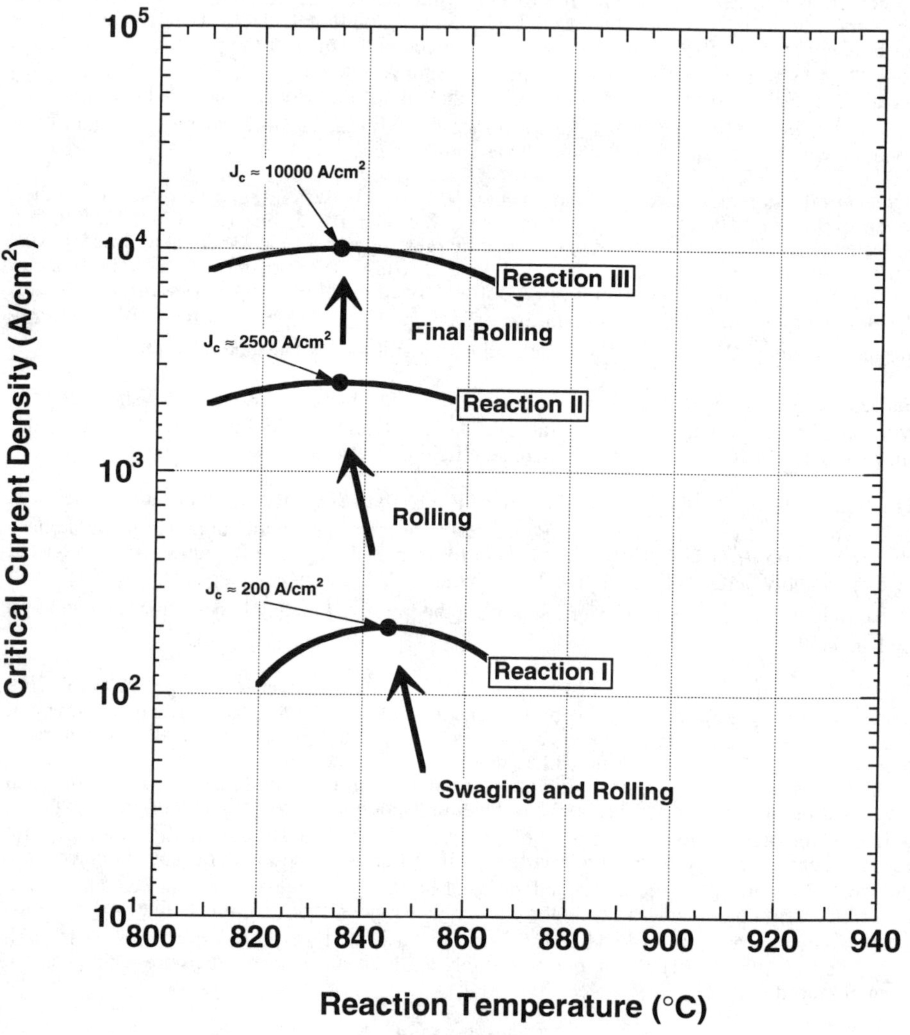

Figure 1. Influence of (PIT) thermomechanical processing on critical current density.

~10cm sections, pressed and fired at 500°C x 30 min. to volatilize the binder followed by heat treatment at temperatures up to 900°C. The best J_c was ~1.2 x10^4 A/cm^2 for pressed filaments reacted at 840°C x 20 min with Tl-1223 and Tl-1234 as the major phase development. Y. Yoshida, et. al.[24] used plasma spraying to produce ~100μm thick films of Tl-2223 on a Ag substrate. A Ba-Sr-Ca-Cu-O precursor was sprayed unto the Ag and reacted in Tl-vapor at 830°C for 5-100h. A J_c of ~4 x10^3 A/cm^2 was observed at 77K in zero field. The J_c dropped by a factor of 4 at 200G which is not as severe as seen for the Tl-PIT field dependence. The microstructure for both of these early studies was typical of a sintered development with mixed phases and indication of incomplete reaction. The results, nevertheless, were promising and with proper heat treatment to develop a "melted" morphology may have shown even better transport properties.

Thick-film deposition techniques using spray pyrolysis and electrodeposition have shown the most promise as an alternative to PIT processing. Recent data for both methods report critical current results at 77 K that surpass the Tl-PIT data and the best results for Bi-PIT processing. DeLuca, et al.[8] has reported the synthesis of Tl-Ba-Ca-Cu-oxide (TBCCO) thick films using a spray pyrolysis technique. An aqueous solution of the metal nitrates is sprayed onto heated (~300°C) substrates of yttria stabilized zirconia (YSZ) and is then oxidized to form a Ba-Ca-Cu-0 precursor. The precursor is then reacted in flowing O_2 in a two-zone furnace that allows control of the Tl vapor pressure to form the superconducting Tl-oxides. Initial results reported transport J_c at 77 K of ~30,000 A/cm^2 in zero field for films with a predominant formation of Tl-2223.[8] More recent data using the same technique reported transport J_c of 105,000 A/cm^2 at 77 K in zero field for 3.2-μm-thick films of single-layer Tl-1223 phase.[9] The field dependence for the Tl-1223 film was equally impressive, showing ~24,000 A/cm^2 at 1.0 T (77 K) for the field oriented parallel to the ab plane. The J_c for the more difficult pinning direction (H ‖ c) showed only modest degradation to near 10,000 A/cm^2 at 1.0 T (77 K). This data demonstrates that the single-layer Tl-1223 compound shows improved pinning and magnetic properties at 77K. Magnetization data by Liu et al.[5] and Komo et al.[25] on single-layer Tl-1223 show intrinsic intragranular J_c of 124,000 A/cm^2 at 77 K and 1.0 T and ~ 43,000 at 77 K and 1.0 T, respectively which supports the spray pyrolysis transport data.[9]

The electrodeposition of Tl-oxide thick films on a flexible conducting substrate also shows great promise as a technique for the fabrication of HTS wires or tapes. A multi-cation thick coating of many microns can be electrodeposited at the desired stoichiometry from an electrolyte solution onto a conducting substrate. Following deposition, the precursor is then reacted similar to the "pyrolized films" at temperatures from 800°-900°C to densify and completely oxygenate the film. The Tl stoichiometry is maintained using a two-zone system which provides an adjustable Tl vapor pressure during reaction.[8] A critical current density of 3.2x10^4 A/cm^2 at 77 K in zero field on Ag foil for electrodeposited TBCCO samples has been demonstrated when annealed using the two-zone thallation process to maintain Tl-stoichiometry.[10] The transport measurements for the electrodeposited TBCCO on Ag foil are higher at 77K in zero field than the Tl-PIT results published by Seido, et. al.[19]. These results on electrodeposited polycrystalline films of TBCCO are thus competitive with the best Tl-PIT results published to date . Moreover, since the film morphology and transition temperature (T_c) show predominant Tl-2223 development with T_c ~120 K, there is promise for even higher J_c values with further work to develop the single-layer composition. In Table I thin and thick film data are compared on various ceramic and metallic substrates. The epitaxial thin film data near 10^7 A/cm^2 is close to the predicted theoretical depairing limit for HTS films at 77K and represents an upper limit to the J_c for Tl-wire or tape. It is not unreasonable, however, that thick-film Tl-wire or tape can be eventually fabricated with J_c comparable with the "poly" thin film results i.e. in the mid 10^5 A/cm^2 range. Recent results by DeLuca et. al.[26] are approaching this value. An electrodeposited tape, moreover, has the potential to be a low cost process for the commercialization of HTS wire or tape compared to other techniques. The materials cost for the electrodeposition process is driven by the material consumed from the electrolyte during deposition which is basically the replenishment cost for the salts exclusive of the substrate. Adding on the cost of the substrate, which could be fairly inexpensive if we use a Ni base oxidation resistant alloy. The cost for the energy consumed during deposition and reaction is almost negligible. Since the process could be completely automated, the manpower or labor cost would be reduced to a monitoring function.

Conclusions

It is important to note that all of the spray pyrolysis and electrodeposited films are highly polycrystalline. The absence of weak link behavior in these films as discussed earlier is related to the "plate-like" morphology which provides a percolative path for current transport. These results combined with earlier results by Ginley et al.[7] of J_c ~6 x 10^5 A/cm^2 (77 K) on polycrystalline Tl-2223 thin films demonstrate that, with proper reaction, polycrystalline Tl-oxide films can be prepared with no evidence for weak link behavior. It is suggested that the grain boundaries for these non-weak-link Tl thin and thick films are in fact layered, analogous to the Bi-2212 development, and contain no stoichiometric variation or extraneous phases within the grain boundary that limit current transfer. Further microstructural studies on the Tl thick-films should resolve the exact mechanism i.e "brick-wall" or sub-grain morphology that is advantageous for high transport current. It is reasonable that the actual mechanism may be a combination of all of these proposed ideas which is consistent with the early intuitive argument that the Tl grain boundaries if processed correctly are more forgiving than the other oxide systems. The thick-film processing thus offers, at present, the most promising method for obtaining a textured polycrystalline development with the realization of the intrinsic Tl-single layer high field and high transport behavior. It is quite possible that processing will be developed to achieve similar high texturing for the Tl-PIT but the results to date have not presented any encouraging directions. The confined Ag-sheath for the PIT process compared to the "free-surface" reaction for the thick-film methods may be the limiting factor.

Table I Comparison of thick/thin film phase development and J_c for different reaction conditions and substrates

Source/ Method	J_c @ 77K, H(O) (A/cm^2 x 10^5)	X-ray Analysis	Annealing condition	Substrate
Thick film/metallic substrate				
GE/ spray pyrolysis	0.10	Mostly 1223	Two Zone Furnace: Sample Temperature, 860°C; Source Temperature, 735°C; Annealing Time, 30 min.	Ag foil
NREL/ electro-deposition	0.11	Mostly 1223	Two Zone Furnace: Sample Temperature, 855°C; Source Temperature, 740°C; Annealing Time, 26 min.	Ag foil
NREL/ electro-deposition	0.32	Mostly 2212 and 2223 + traces of 1223	Two Zone Furnace: Sample Temperature, 855°C; Source Temperature, 735°C; Annealing Time, 20 min.	Ag foil
Thick film/ceramic substrate				
NREL/ electro-deposition	0.56	Mostly 2212 and 2223	860°C x 10 min (O$_2$), in presence of TBCCO pellet (2223+2212).	SrTiO$_3$
GE/ spray pyrolysis	1.05	Mostly 1223	Two Zone Furnace: Sample Temperature, 860°C; Source Temperature, 735°C; Annealing Time, 30 min.	YSZ
Thin film/ceramic substrate				
SANDIA/ sequential e-beam	9.7 ("epi") 5.0 (poly)	2223	850°C x 10 min (air) + 750°C x 30 min (O$_2$) in presence of 2223 pellet	LaAlO$_3$
IBM/ sputtered	16.0("epi")	2223	Encapsulated growth: 830-860°C x 1-8 hours	LaAlO$_3$
STI/ laser ablated	10.6("epi")	2212	Two ZoneThallination and anneal	LaAlO$_3$
DUPONT/ *in-situ* sputtered	20.0("epi") (for 1.8m long x 10 μm wide)	2212	Two Zone Thallination and anneal	LaAlO$_3$

<u>Acknowledgement</u>

We would like to acknowledge the support from DOE contract No.DE-ACO2-83CH10093.

References

1. U. Balachandra, A.N. Iyer, P. Haldar, and L.R. Motowidlo, JOM, **45**(9), (1993) 54

2. K. Sato, "Processing and Properties of High-Tc Superconductors Vol.1 Bulk Materials" edited by S. Jin, World Scientific,Singapore, (1993) 1213.

3. D.S. Ginley, J.F. Kwak, E.L. Venturini, B. Morosin, and R.J. Baughman, Physica **C160** (1989) 42.

4. D.H. Kim, K.E. Gray, R.T. Kampwirth, J.C. Smith, D.S. Richeson, T.J. Marks, J.H. Kang and J. Talvacchio, Physica **C177** (1991) 431.

5. R. S. Liu, D. N. Zheng, J. W. Loram, K. A. Mirza, A. M. Campbell, and P. P. Edwards, Appl. Phys. Lett. **60** (1992) 1019.

6. T. Nabatame, J. Sato, Y Saito, K. Aihara, T, Kamo, and S. Matsuda, Physica C **193** (1992) 390

7. M. Okada, K. Tanaka, and T. Kamo, Jpn. J. Appl. Phys. **32** (1993) 2634.

8. J.A. Deluca, M.F. Garbauskas, R.B. Bolan, G. McMullen, W.E. Bolz, and P.L. Karas, J. Mater. Res. **6** (1991) 1415.

9. J.A. DeLuca, P.L. Karas, J.E. Tkaczyk, P. Bednarczyk, M.F. Garbauskas, C.L Briant, and D.B. Sorensen, Physica C **205,** (1993) 21

10. R.N. Bhattacharya, P.A. Parilla and R.D. Blaugher, Physica C **211**(1993) 475

11. D.J. Miller, J.G. Hu, J.D. Hettinger, K.E. Gray, J.E. Tkaczyk, J.A. DeLuca, P.L. Karas, J.A. Sutliff and M.F. Garbauskas, Appl. Phys. Lett. **63** (1993) 556

12. D.M. Kroeger, A. Goyal and Z.L. Wang, to be published

13. J.O. Willis, M.P. Maley, P.J. Kung, J.Y. Coulter, D.E. Peterson, P.G. Wahlbeck, J.F. Bingert, and D.S. Phillips, IEEE Trans. on Applied Superconductivity **3** (1993) 1219

14. R.N. Bhattacharya and R.D. Blaugher, "Electrodeposited TlBaCaCuO Superconductors", Tl-Based High Temperature Superconductors, M. Dekker, Inc., New York (1993) 279.

15. R.N. Bhattacharya, P.A. Parilla, R. Noufi, P. Arendt, and E. Elliott, J. Electrochemical Soc. **139** (1992) 67.

16. R.N. Bhattacharya, R. Noufi, L.L. Roybal, R.K. Ahrenkiel, P. Parilla, A. Mason, R.P. Hellmer, J.F. Kwak, and D.S. Ginley, J. Mater. Res. **6** (1991).

17. Z.F. Ren and J.H. Wang, Appl. Phys. Lett. **61** (1992) 1715

18. Z.F. Ren and J.H. Wang, to be published

19. M. Seido, J. Sato, T. Sasaoka, A Nomoto, T. Kamo and S. Matsuda, Proc. ISTEC/MRS International Workshop on Superconductivity, Hawaii (1992) 236

20. Y. Torii, H. Kugai, H. Takei, and K. Tada, Jpn. J. Appl. Phys. **29** (1990) L952.

21. A.H. Cardona, H. Suzuki, T. Yamshita, K.H. Young, and L.C. Bourne, Appl. Phys. Lett. **62**, (1993) 411

22. D. Dimos, P. Chaudhari, and J. Mannhart, Phys. Rev. B **41**, (1990) 4038

23. T. Goto and C. Yamaoka, Jpn. J. Appl. Phys. **29**, (1990) L1645

24. Y. Yoshida, T. Kanai, T. Kamo, T. Matsumoto, and R. Shiobara, Advances in Superconductivity III, Springer-Verlag, Tokyo (1991) 639

25. T. Kamo, T. Doi, A. Soeta, T. Yuasa, N. Inoue, K. Aihara, and S. Matsuda, Appl. Phys. Lett. **59** (1991) 3186.

26. J.A. DeLuca, P.L Karas, C.L. Briant, and J.E. Tkaczyk, TMS Materials Week, Pittsburgh, Pa. Oct.17-21,1993, to be published

THE PREPARATION OF PbBSCCO TAPES BY DYNAMIC MAGNETIC COMPACTION

AND INTERPRETATION OF THE I-V CURVES USING A NEW THEORETICAL MODEL

T. L. Francavilla, R. J. Soulen Jr., W. W. Fuller-Mora, P. Anderson
Naval Research Laboratory, 4555 Overlook Avenue SW, Washington DC 20375-5343

B. Chelluri, J. P. Barber, and T. Scholz
IAP Research Inc., 2763 Culver Ave, Dayton OH 45429-3723

A. K. Sarkar
University of Dayton, Research Institute, 300 College Park, Dayton OH 45469

D. H. Liebenberg
Office of Naval Research, 800 N. Quincy St., Arlington VA 22217-5660

Abstract

PbBSCCO conductors were prepared using the method of dynamic magnetic compaction. Their critical current properties were measured as a function of magnetic field at 77 K. In order to explore the utility of a new theoretical model, the resulting data were analyzed first by conventional methods using an E-field criterion of one micro volt per centimeter, and then in terms of a model developed by Soulen based upon the earlier theoretical work of Ambegaokar and Halperin. The latter involved computer fitting an equation derived from an adaptation of their model through the measured current voltage data points. The equation used to fit our data contained three fitting parameters: the normal state resistance, the critical current in the absence of thermal fluctuations, and a parameter related to the pinning potential. The fitting parameters were found to follow trends established in the measured physical properties of the wire.

Processing of Long Lengths of Superconductors
Edited by U. Balachandran, E.W. Collings and A. Goyal
The Minerals, Metals & Materials Society, 1994

Introduction

The discovery in 1987, of materials that are superconducting above liquid nitrogen temperatures initiated a flurry of intense research activity. Small scale devices such as narrow band filters, and Josephson junctions appeared rapidly after this initial discovery. The preparation of bulk materials in a technologically useful form such as wire that can be formed into magnets has taken a little longer.

Early attempts at preparing a bulk conductor resulted in sintered polycrystalline material that could carry current densities on the order of hundreds of A/cm^2 in magnetic fields of approximately 100 gauss. Subsequently highly textured bulk material was prepared and shown to be capable of carrying 10^4 A/cm^2 at 77 K in magnetic fields of several Tesla.(1). These early efforts suggested that the oxide superconductor YBCO was a potentially useful material if large specimens capable of this high current density could be prepared. In 1993, large samples of high quality YBaCuO typically 4 cm long, 0.5 cm wide, and 0.25 cm thick were shown to carry a total current of 393 A which is equivalent to a current density of 1871 A/cm^2. At a magnetic field of one Tesla, the sample could still carry a transport current of 229 A (1090 A/cm^2).(2) These samples were prepared by a continuous process that is capable of producing considerable lengths of rigid material with good superconducting properties.(3) These materials are potentially useful for rigid conductors such as high-current, low-loss bus bars or for magnetic bearings. Ultimately however, one would like to wind a magnet for motors or generators. Attempts to achieve this same level of current in a flexible conductor have not been as rapid, but some progress has been made.(4)

It is difficult to form YBaCuO into conductors with good current transport across grain boundaries. Consequently, most of the work on conductor development has been directed primarily toward the BSCCO compounds. Attempts to prepare a flexible conductor of the high T_C oxides have typically focused on the powder in tube method using the so called 2223 compound of $Bi_{2-x}Pb_xSr_2Ca_2Cu_3O_7$ or more commonly PbBSCCO. This material has properties such that drawing or rolling tends to result in good grain alignment. In conjunction with appropriate heat treatment, the result is a conductor with technologically useful critical current properties.

We have made use of dynamic magnetic compaction in an effort to increase the density of the superconductor core material of the composite. Three samples were prepared using this technique. Prior to the drawing or rolling process, the tubes were subjected to an intense transient magnetic field which resulted in radial compression of the silver tube and powder core. The composite was then heat treated to produce the final conductor which was subsequently studied, using both conventional analysis and a new theoretical model.

Recently, a new model was proposed by Soulen (5) that would account for the shape of the I-V curve, provide some physical insight to a process that was formerly empirical in nature, and suggest the possibility of an intrinsic definition of critical current to replace one that was arbitrary. This model, based on a paper by Ambegaokar and Halperin (A-H) (6), results in an equation which can be fitted to the data using three fitting parameters, the normal state resistance, the critical current in the absence of thermal fluctuations, and a parameter related to the magnitude of the pinning potential. This fitting process makes use of the entire I-V curve not just one point on the curve. Furthermore, once these parameters have been determined, the value of critical current for any empirical definition of critical current may be found. The data for one of the samples was also analyzed in this way in order to make a comparison with a more conventional analysis, and to look for correlation between the fitting parameters and experimental results.

Sample Preparation

PbBSCCO tapes with a silver sheath were fabricated using the powder in tube method, and dynamic magnetic compaction (DMC) was used to increase densification rather than the usual method of drawing and rolling. Rolling was still done however, to produce the final conductor.

254

Initially nitrate decomposed intermediate precursor powders containing only the 2212 phase were loaded into silver tubes, with nominal dimensions 6 mm OD and 4.5 mm ID whose ends were sealed with silver plugs. These were then subjected to DMC at IAP Research Inc. The densified tubes were subsequently rolled into tapes and sintered at temperatures between 840° C to 850° C for 72-100 hours and pressed. The heat treatment and pressing were repeated three times to achieve the required texturing and grain contact. These processing steps converted the initial 2212 phase to the desired 2223 phase.

A sketch of the experimental configuration used to produce DMC is shown in figure 1. A large current pulse is generated by the power supply and flows through the compactor coil in the direction shown. This generates an intense magnetic field which induces large opposing currents in the confining conductor which, in this case, is the silver tube. The opposing currents result in intense radial magnetic pressures that can be varied from zero to 150 Kg/in^2, with pulse durations of 10^{-3} to 10^{-6} seconds. The radial nature of the forces results in uniform compaction, with densities of up to 95% of the theoretical value.

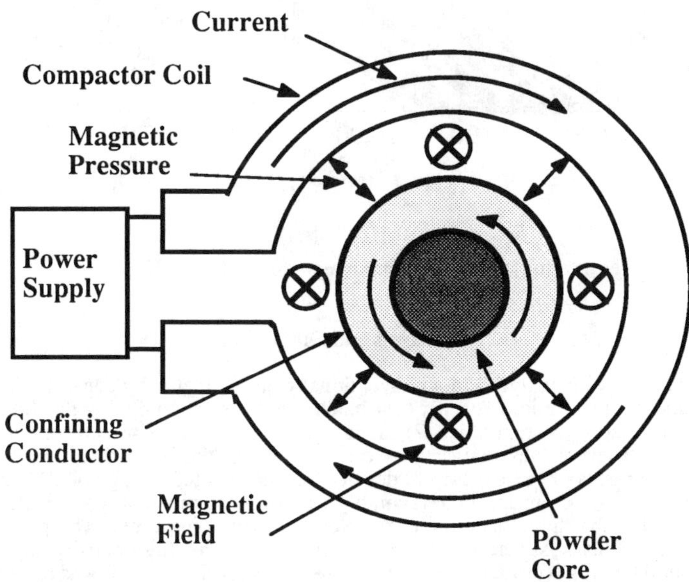

Figure 1. Schematic diagram of the dynamic magnetic compaction (DMC) process.

<u>Experiment</u>

The current-carrying capacity of these conductors was measured at liquid nitrogen temperatures in several different magnetic fields, the highest field being of sufficient intensity to drive the conductor almost completely into the normal state. Critical currents were measured as a function of applied magnetic field by generating current voltage (I-V) curves at each of these fields under computer control. The magnetic field was applied normal to the flat surface of the tape and normal to the current flow. Current from the power supply was incremented to a set value. After a delay of approximately one second, values of voltage and temperature were recorded, and the process repeated. Measurements were made for both increasing and decreasing current. The sample was immersed completely in liquid nitrogen, in order to minimize problems of local sample heating. An extra set of voltage leads at the current contacts allowed measurement of the dissipation present as current is transferred to the sample. A typical set of I-V curves at 77 K as a function of applied magnetic field is shown in figure 2.

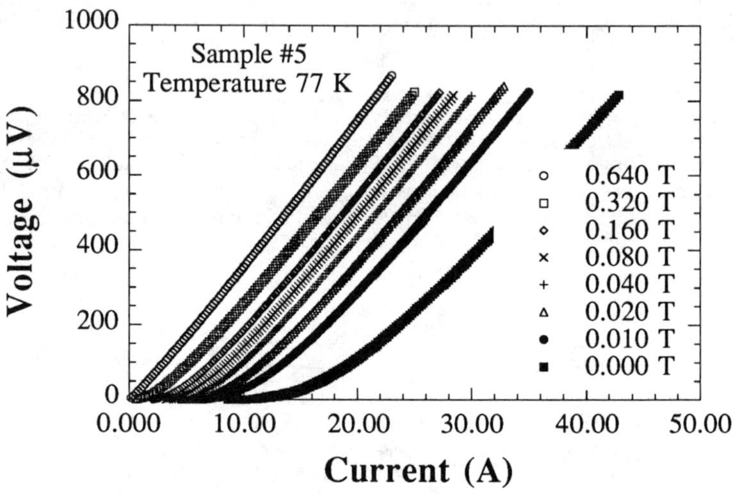

Figure 2 I-V curves as a function of external magnetic field at 77 K.

Analysis and Discussion

The resultant data were analyzed in a conventional fashion, that is by constructing the tangent to the I-V curve at the 1 micro volt per centimeter point and extrapolating this line to the point where it intersects the current axis. (7) This value was taken as the critical current at that particular field. These values, converted to critical current density, and plotted in figure 3, are the first measurements of critical current density for this new technology of DMC. The results of these measurements compare quite favorably with other commercially fabricated conductors measured at NRL that have been under development for some time. The process of DMC produces a high density superconducting core filament. This high density superconducting component in the soft silver matrix initially caused some problems as it was rolled into a tape. We believe that as the process improves, the current densities will also improve.

The one micro volt per centimeter criterion used to determine the critical current density is arbitrary but has been accepted by most investigators as a working definition and a means of comparison among conductors. As mentioned earlier in the paper however, this makes use of only one point on the curve, and does not give a physical basis for either understanding the shape of the I-V curve, or access to quantities such as pinning force that can be modified by varying the processing conditions. Ideally, one would like to have a physical model that utilizes all the information in the I-V curve. This model should have different parameters that can be associated with quantities that relate to material characteristics, which can be either measured or calculated. These parameters then could conceivably provide some guidance to wire processing other than empirical trial and error methods.

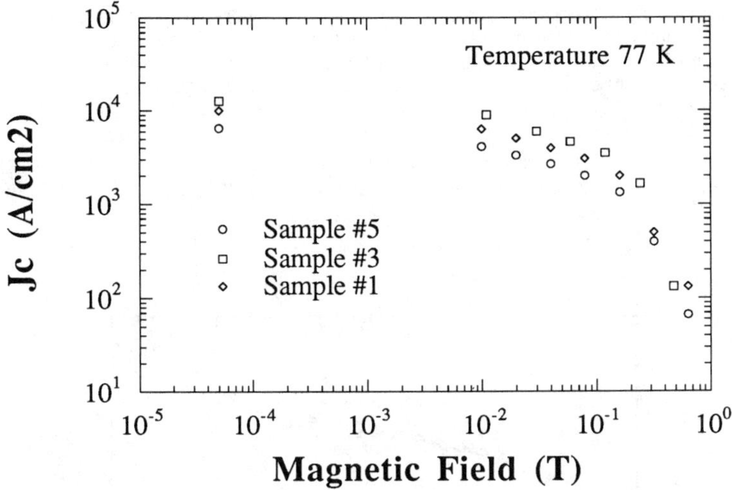

Figure 3. Plot of conventional critical current density as a function of magnetic field for the three IAP samples prepared by DMC. The area of the PbBSCCO core was used to compute critical current density

We have applied Soulen's adaptation of the A-H model to the above data. The model contains three fitting parameters, the normal state resistance of the superconducting material R_S, the critical current in the absence of thermal fluctuations I_c^*, and a quantity γ related to the pinning potential. Since the data obtained for all three samples were comparable, we selected one set of data (Sample 5), and fitted these data to the model. Soulen's model leads to the following equation

$$V = (I_c^* R_s) \frac{4\pi}{\gamma} (e^{4\pi\gamma x} - 1) \left\{ \left[\int_0^{2\pi} d\theta\, f(\theta) \right] \left[\int_0^{2\pi} \frac{d(\theta')}{f(\theta')} \right] + \int_0^{2\pi} d(\theta) \int_\theta^{2\pi} d(\theta') \frac{f(\theta)}{f(\theta')} \right\}^{-1}$$

where the function $f(\theta)$ is given by

$$f(\theta) = \exp\left\{ (\tfrac{\gamma}{2}) (\theta + \cos(\theta)) \right\} .$$

In order to fit these equations to the experimental data several steps were required. A baseline of zero voltage was established for the I-V curve. The entire I-V curve for both increasing and decreasing current was examined and the portion of the curve in the fully superconducting state was deleted, since the voltage in this region is zero and below the sensitivity of the measuring instrument. The motivation for deleting these data points was the recognition that the voltages observed in the fully superconducting state were due to the instrument fluctuations not the changing properties of the superconductor. This process used to prepare the data for curve fitting still being reviewed.

A complicating factor for these conductors is the fact that they are a composite consisting of a superconducting core surrounded by silver matrix. Consequently when the current exceeds the value at which dissipation begins to take place, the current is shared between the superconducting core and the silver matrix. Since the A-H theory applies only to the PbBSCCO core material and not the composite, the voltage attributed to the silver matrix was subtracted from the measured voltage. The remaining data was fitted to the above equations and values were obtained for the fitting parameters, R_S, I_C^*, and γ. This was done for the I-V curves of sample 5 at each of the magnetic field values shown in figure 3. An example of the fits obtained is shown in figure 4.

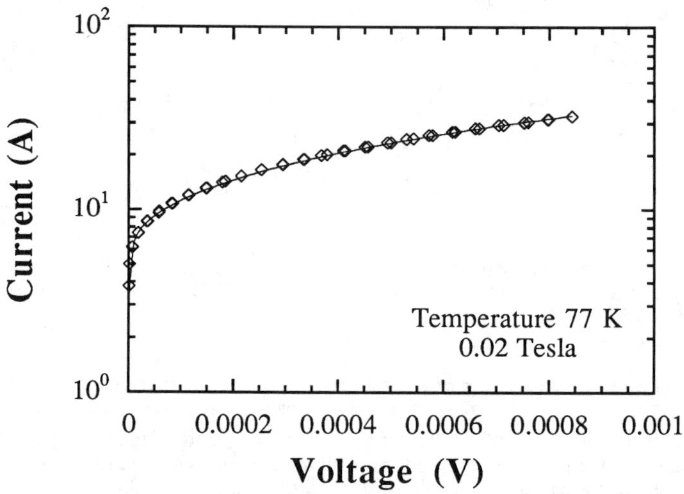

Figure 4. I-V curve fitted to the experimental data. Only one sixth of the approximately 300 data points are shown so that the fitted curve is visible.

The I-V curve can now be reproduced using this model once the three fitting parameters are obtained. One now has the advantage that the critical current for any arbitrary criterion can be determined. Thus a magnet designer would have more flexibility in engineering decisions when considering the alternatives for allowed conductor dissipation and corresponding cooling requirements.

We are currently examining the parameters themselves however, in an effort to determine if their utility extends beyond merely providing a means to reproduce a given I-V curve. For this study, we focused upon the trends of the fitting parameters with increasing magnetic field at a fixed temperature of 77 K.

We first examined the normal state resistance parameter R_S. In work that is currently in progress, other composites have been measured at currents taking them well into the normal state. These data, when fitted to the model, give reasonable values of normal state resistance. For sample number five however, the normal state resistance R_S does not in general give good agreement with the normal state resistance of the sample. We believe that this is due primarily to the lack of information in the high-current region of the curve where the voltage is due mostly to the normal state properties of the composite.

The critical current parameter in the absence of thermal fluctuations I_C^* should be greater than the more conventionally defined critical current but should decrease with increasing magnetic field. Furthermore, the parameter γ which is related to the pinning potential should have some positive value at zero magnetic field where the critical current is highest and decrease with increasing magnetic field as the superconducting properties are depressed. This trend is observed in figure 5, where J_c^* is I_c^* divided by the area of the PbBSCCO core.

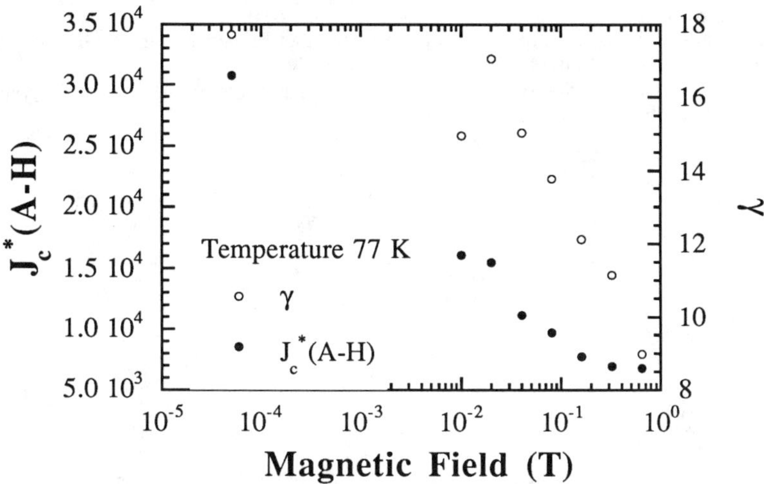

Figure 5. Fitting parameters I_c^* converted to J_c^*, and γ, plotted as a function of magnetic field at 77 K.

The general trends follow what one would expect but at this time it is not clear if the non-monotonic nature of the field dependence of these parameters is real. We expect questions of this sort to be resolved as experience is gained in applying this model. It may happen that these parameters can be used to characterize the quality of a superconductor at different temperatures and magnetic fields, becoming in a manner of speaking a figure of merit and a convenient way of distinguishing between superconductors. This analysis has been applied to a wide range of superconducting materials, so some of these issues are being addressed. Our future work will be focused on these questions.

Acknowledgments

The authors wish to express their appreciation to Dr. H. D. Dardy of the advanced concepts group NRL Code 5581 for access to their computer facilities.

References

1. J. W. Ekin, K. Salama, and V. Selvamanickam, "High Transport current density up to 30 T in bulk $YBa_2Cu_3O_7$ and the critical angle effect," Applied Physics Letters, 59 (3) (1991), 360-362.

2. T. L. Francavilla, R. L. Meng, P. Hor, C. W. Chu, J. W. Ekin and D. H. Liebenberg, "Magnetic field dependence of d.c. critical current for large samples of textured bulk Y-Ba-Cu-O," Cryogenics, 33 (3) (1993), 256-260.

3. R. L. Meng, C. Kinalidis, Y. Y. Sun, L. Gao, Y. K. Tao, P. H. Hor & C. W. Chu, "Manufacture of bulk superconducting $YBa_2Cu_3O_{7-\delta}$ by a continuous process," <u>Nature</u>, 345 (1990).

4. "Measurements of high temperature superconducting wires and coils: Liquid neon Facility," Buffalo Conference on Superconductivity and Applications, ed. H. S. Kwok, D. T. Shaw, and M. J. Naughton, (AIP Conference Proceedings No. 273, 1993) 455-461.

5. R. J. Soulen Jr., "Possible explanation for the shape of the I-V curves of superconductors and a more meaningful definition of I_c," <u>IEEE Transactions on Applied Superconductivity</u>, 3 (1) (1993), 1261-1264.

6. V. Ambegaokar and B. I. Halperin, "Voltage due to thermal noise in the dc Josephson Effect," Physical Review Letters, 22 (25) (1969) 1364-1366.

7. J. W. Ekin, "Offset criterion for determining superconductor current," <u>Applied Physics Letters</u>, 55 (9) (1989), 905-907.

MICROSTRUCTURE-PROPERTY RELATIONSHIPS IN BSCCO/Ag

TAPES FOR APPLICATION IN HIGH FIELD MAGNETS

Yusuf S. Hascicek, Steven W. Van Sciver

National High Magnetic Field Laboratory, Florida State University, Tallahassee, FL 32306

Leszek R. Motowidlo, Drew W. Hazelton, Pradeep Haldar

Intermagnetics General Corporation, Guilderland, NY 12084

Abstract

Microstructure of Bi-2223/Ag tapes, before and after bend strain experiments, were investigated by optical microscopy, SEM, EDS and WDS. Strain dependance of J_c at various temperatures for Bi-2223/Ag tapes produced by powder in tube process are presented. Critical strain for mono-core tapes varied between 0.25% and 0.6%. Microstructural results indicate that there exist some texturing and other phases. Radial cracks were observed in the longitudinal sections of bent tape conductors. "Wind and react" pancake coils built from these tapes were tested. $J_c(T)$ of these coils and that of short samples with corresponding bend strain are compared.

Processing of Long Lengths of Superconductors
Edited by U. Balachandran, E.W. Collings and A. Goyal
The Minerals, Metals & Materials Society, 1994

Critical current density of a superconductor, J_c, its dependence on temperature, applied magnetic field, mechanical strain and thermal cycling have to be well understood before any conductor can be utilized for magnet applications. All of these properties are microstructure sensitive. Therefore microstructure of the conductors need to be revealed and understood so that appropriate fabrication, processing and heat treatment conditions can be adopted to alter the microstructure which would in turn lead to a conductor with suitable superconducting and mechanical properties for practical applications.

Because of their high transition temperature (T_c) and very high upper critical field (B_{c2}), high temperature superconductors (HTS), especially $Bi_2Sr_2Ca_2Cu_3O_x/Ag$ (Bi-2223/Ag) composite conductors form a viable candidate for applications at high fields above 20 Tesla and at temperatures up to 20K. For the design of actual magnets it is imperative to precisely know the strain tolerance for the J_c of the HTS.

There have been a number of studies investigating tensile[1],[2], bend[1]-[5] strain dependance of Bi-2223/Ag conductors. Here we report on the temperature and strain dependence of critical current density of Bi-2223/Ag monocore tape conductors, and compare these with the performance of pancake coils built from similar conductors. Microstructure of the same is also investigated by optical and scanning electron microscopy (SEM), and by Energy Dispersive Spectroscopy (EDS) and Wavelength Dispersive Spectroscopy (WDS). Microstructure, critical current density and strain properties are related to better tailor a conductor for use in high field magnet applications.

Experimental

(BiPb)-2223/Ag superconducting tapes used in this investigation were produced by Intermagnetics General Corporation (IGC) employing the powder-in-tube process (PIT)[6]. The precursor for the PIT process are prereacted powders which are prepared with the nominal composition for the Bi-2223 phase. These powders are packed into silver tubes, swaged, drawn and then rolled to a final thickness of about 0.1 mm. Reduction ratios of about 15% per pass were employed in drawings to optimize the conductor properties. Two or three intermediate annealings are needed for the reduction process. All the heat treatments are performed between 830°C and 870°C for between 24 and 150 hours. The final dimensions of the tapes used in this study were 0.1mm x 6 mm with a superconductor core thickness of about 60 μm, and a silver to superconductor ratio of about 3:1. This process is used to produce long lengths of tape conductors, and by restacking to produce multiflamentary wires with practical critical current densities[7].

Transverse and longitudinal sections of the short tape samples used for J_c characterization were prepared for microstructural investigation by standard metallographic techniques. Also longitudinal sections of bent samples used for $J_c(\varepsilon,T)$ characterization were prepared and examined by using optical and scanning electron microscopy for possible cracks, crack morphology and distribution. Quantitative x-ray microanalysis were performed by EDS and WDS to identify chemical composition of individual grains in the longitudinal and transverse sections of these samples.

Measurement of the critical current density, defined in this paper, as the ratio of the critical current density to the superconductor crossection only, were carried out on short samples of Bi-2223/Ag tape conductors by using a computer controlled, four-wire DC pulse technique with a 1 μV/cm electric field criterion[8]. Critical current density of each short sample was measured as a function of temperature between 4.2K and T_c, and thermally cycled up to 20 times between 4.2K and room temperature under a helium atmosphere. These tapes were then bent around the largest step of our bend test rig at room temperature in a manner similar to winding a magnet. The bend test rig is an eight stepped cone with diameters between 10 mm and 42 mm. Each step has its own current and voltage leads. The sample was then cooled to 110K on the bend rig and J_c was measured as a function of temperature down to 4.2K, retaining the plastic and elastic strains just like one would have in a magnet built with react and wind approach. The same sample was then moved down to the next step in the bend rig and the process repeated until the smallest diameter (10 mm) were reached. The sample was warmed to room temperature in helium atmosphere between the steps.

Critical current density of pancake coils built from similar tape conductors were also measured in the same set up and over the same temperature range. A Hall probe was also placed in the center of the bore of the pancake coil for field measurement. These magnets were built by the wind-and-react technique. Two of these magnets were wound two Bi-2223/Ag tape conductors in hand, the third one was wound four conductors in hand.

<u>Results and Discussion</u>

Microstructural investigations by SEM, EDS, WDS and optical microscopy indicates that texturing, phase purity, uniformity of superconductor core were improved substantially compared with earlier results [9]. As seen in Figs.1, 2 and 3 the superconducting core is not quite uniform yet, secondary phases are still present as well as Bi-2223 and Bi-2212 phases. Fig. 1 shows a composite optical micrograph of a transverse section of a Bi-2223/Ag tape (0.1 mm

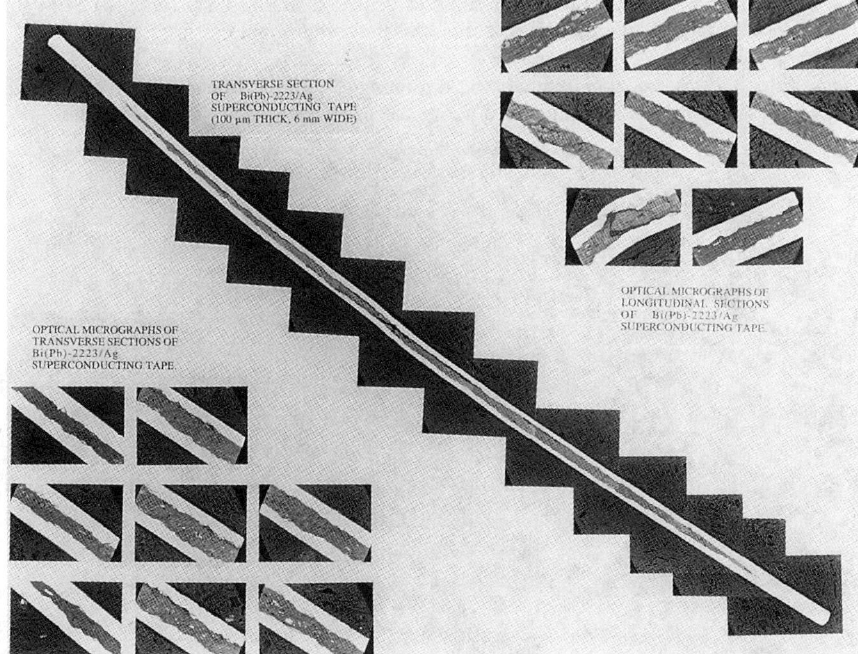

Fig. 1. Optical micrographs of transverse and longitudinal sections of one of the earlier Bi-2223/Ag tapes. The thickness of the tape is 0.1 mm.

thick) and several longitudinal (top right) and transverse (bottom left) optical micrographs. Fig. 2 a) shows an SEM micrograph of a longitudinal section of a Bi-2223/Ag tape bent to a surface strain of 1%. Fig. 2 b) shows another SEM micrograph of a crack from the same sample as in a). Dark gray platelike grains are Bi-2223, and roughly equiaxed, light gray grains are

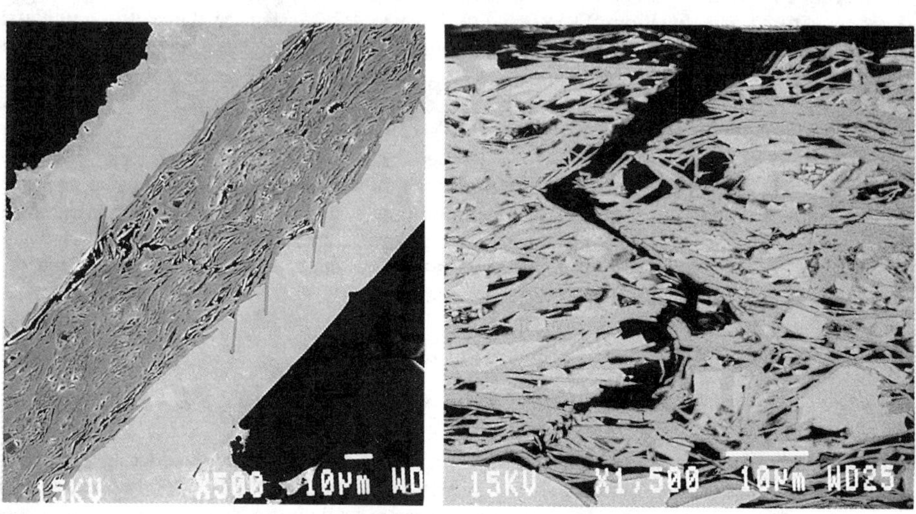

Fig. 2. a) SEM micrograph of one of the samples bent to a strain of 1% (left). b) SEM (back scattered electron image) micrograph of a crack in the same sample as in a) (right).

Bi-2212. Fig. 3 depicts a longitudinal crack running parallel to the superconducting/silver interface, but in the superconducting core not at the interface. This is observed quite

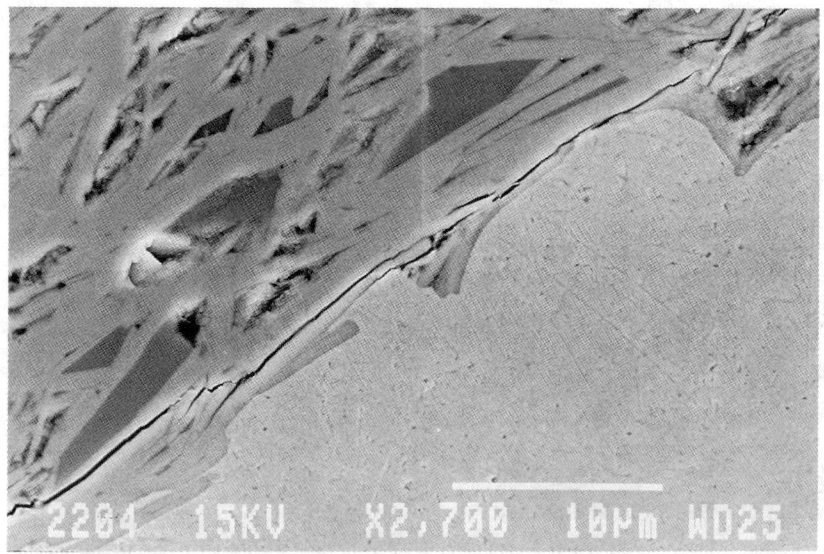

Fig. 3. SEM micrograph of Bi-2223/Ag interface of a bent sample. Note the longitudinal crack.

often, and is indicative of strong bonding at the interface. Despite the size and extent of several of these cracks, this particular sample retained about 5% of its zero strain J_c. It looks as if nonsuperconducting particles diverted propagation of the cracks from the expected growth direction. Visual inspection of this sample showed four distinct kinks at the silver surface after being bent around the first four largest steps of the bent rig.

Figure 4 depicts J_c versus temperature in zero external field for two short straight Bi-2223/Ag tape conductors and a pancake coil built from same type of conductor. All three were heat treated under the same conditions. The critical current density of the coil is over 60% of the short straight samples, which indicates reasonable uniformity of J_c over long lengths.

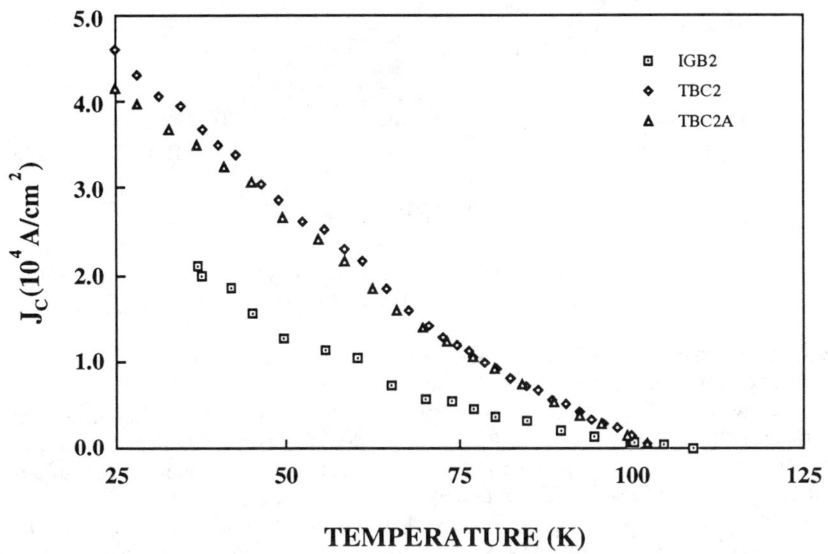

Fig. 4. J_c as a function of temperature for two short, straight Bi-2223/Ag tape samples and one of the single pancake coils (IGB2) wound from the same tape as the short samples and heat treated under the same conditions. The single pancake coil was wound with four tapes in hand, had an ID of 25.4 mm and 54 turns. The field produced at 4.2K was 0.225 Tesla (corresponding to an I_c of 185 A) at zero applied magnetic field.

Fig. 5 shows J_c versus temperature curves of one of the tapes, TBC2 as in Fig. 4, under various levels of bend strain as indicated, together with J_c versus temperature curve of the coil (IGB2). Innermost winding of the coil represents about 0.33% surface strain. J_c of the short sample under 0.33% is well comparable with that of the coil. This suggests that under favorable conditions "react and wind" and "wind and react" approaches to magnet fabrication from these conductors could give similar results. As it is seen from this figure general temperature dependence of the J_c does not change with strain but the whole curve is reduced by a common factor over the whole temperature range. This result suggests that there is no intrinsic strain dependence to J_c and also there is no additional strain introduced to the system due to the differential thermal expansion of the constituents of the bend rig. Strain dependence of critical current density at various temperatures deduced from these curves have the same shape when normalized to the zero strain critical current density (J_{c0}) [12]. It had been already

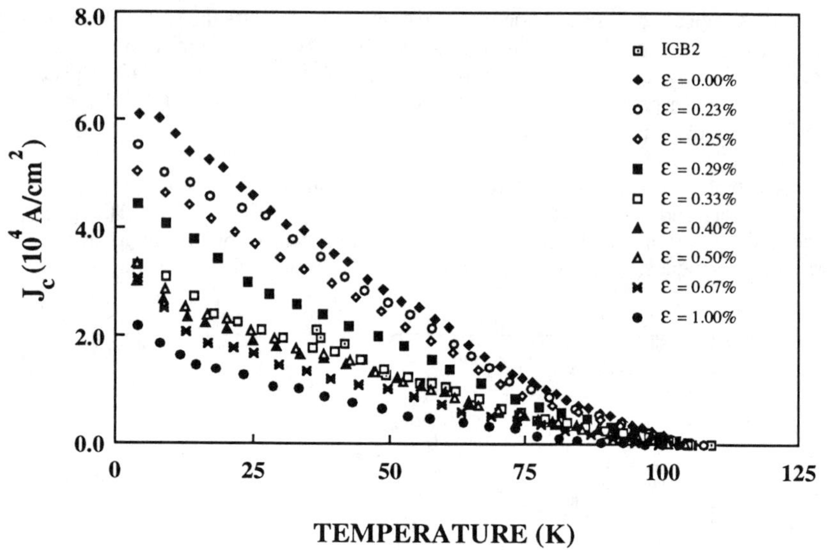

Fig. 5. Critical current density as a function of temperature for the pancake coil and one of the Bi-2223/Ag tape conductors (TBC2 as in Fig. 4) under various strains as marked.

established that thermal cycling under helium atmosphere has no effect on J_c in PIT Bi-2223 tapes[9]. Therefore, we attribute all the degradation in J_c to the strain in the tape due to bending. Figure 6 shows normalized critical current density, (J_c/J_∞), as a function of strain, ε, (strain at the surface of the tape) at 4.2K. Critical current density degrades with strain in a manner consistent with the models proposed by various workers[2]-[5]; J_c is independent of strain up to a critical strain, then starts falling precipitously to about $J_\infty/2$ followed by a rather slow decrease with increasing strain. Fig. 7 shows J_c versus temperature curves for five short Bi-2223/Ag tape conductors under 0.33% surface strain, together with J_c versus temperature curve of the coil as in Fig. 4. Note the very wide range of spread in the J_c of same Bi-2223/Ag tape conductor under same strain when bent after reaction. The wind and react coil data fall roughly in the middle of the spread. The data suggests that in the present state of magnet development, i.e. winding monocore Bi-2223/Ag conductors, it is safer to take the wind and react approach. In the present study, the critical strain was over a large range, between 0.25% to about 0.6%, which is quite high for a monocore Bi-2223/Ag tape produced by PIT. This result leads us to believe that there is room for improvement of strain tolerance of J_c even for monocore tapes.

<u>Conclusions</u>

Microstructural studies show there is increased texturing at the Bi-2223/Ag interface, and bonding at the same interface seems to be quite strong. Phase uniformity needs to be improved. The radial cracks are the main cause of degradation in J_c.

There seems to be no intrinsic strain dependence to the critical current density of HTS Bi-2223/Ag monocore tapes studied in this investigation. J_c starts falling once the surface strain of the sample reaches the local critical strain (varied between 0.25% and 0.6 in this study) probably due to the initiation of a crack. Upon further strain, J_c degrades either by formati

266

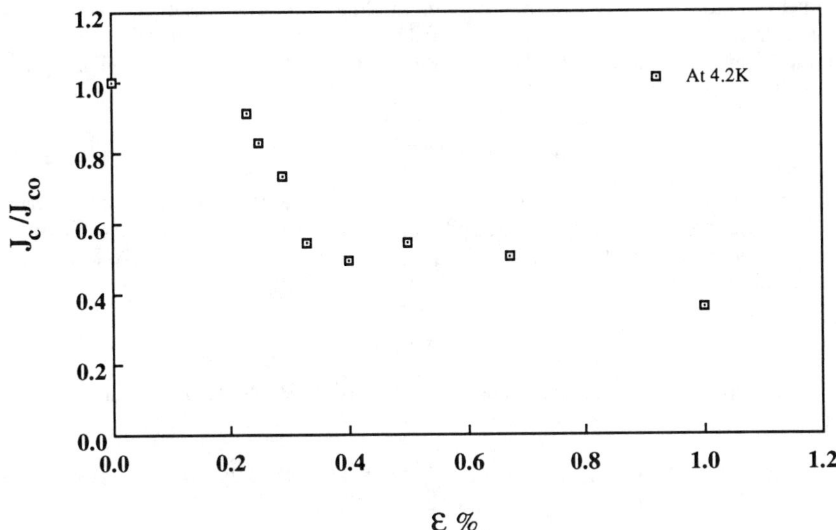

Fig. 6. Normalized critical current density versus strain (strain at the surface of the tape at room temperature) at 4.2K for the same sample, TBC2, as in Fig.5.

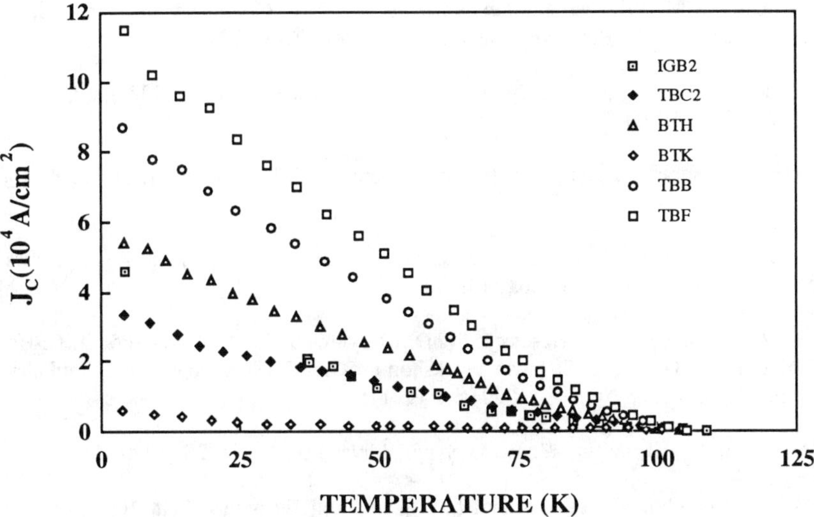

Fig. 7. J_c versus temperature of five similar Bi-2223/Ag tape conductors under 0.33% room temperature surface strain together with J_c versus temperature curve of the coil as in Fig. 4.

of new cracks or further propagation of the existing cracks or both. Although the wind-and-react technique is a more secure approach in magnet building from brittle conductors, comparison of the short straight sample, bent short sample and wind-and-react pancake coil J_c's show that the react-and-wind technique can also be employed for building magnets from PIT processed Bi-2223/Ag tape conductors.

Acknowledgement

The authors wish to thank several colleagues at FSU especially C. N. Chan and S. Gabor for careful data taking and handling, T. J. Fellers for assistance in SEM work, N. Fabre and S. Fearon for assistance in metallography. This project is supported by the State of Florida and the National Science Foundation through NSF Cooperative Grant No. DMR 9016241.

References

[1] J. W. Ekin, D. K. Finnemore, Qiang Li, J. Tenbrink. W. Carter, Appl. Phys. Lett. 61(7), 858 (1992).

[2] A. Otto, L. J. Masur, J. Gannon, E. Podtburg, D. Daly, G. J. Yurek and A. P. Malozimoff, IEEE Trans. Applied Superconductivity 3(1), 915 (1993).

[3] K. Sato, T. Hikata, H. Mukai, M. Ueyama, N. Shibuta, T. Kato, T. Masuda, M. Nagata, K. Iwata and T. Mitsui, IEEE Trans. Magn. 27(2), 1231 (1991).

[4] S. X. Dou, Y. C. Guo and H. K. Liu, Supercond. Sci. Technol. 6, 195 (1993).

[5] T. Kuroda, M. Yuyama, K. Itoh and H. Wada, Adv. Cryogenic Eng. Vol. 38, p. 1045, F. R. Fickett and R. P. Reed eds. Plenum Press, New York, 1992.

[6] L. R. Motowidlo, E. Gregory, P. Haldar, J. A. Rice and R. D. Blougher, Appl. Phys. Lett. 59(6), 736 (1991)

[7] U. Balachandran, A. N. Iyer, P. Haldar, and L. R. Motowidlo, J. of Metals 45(9), 54 (1993).

[8] Y. S. Hascicek, R. J. Kennedy, L. R. Testardi, H. Niculescu, P. J. Gielisse, Th. Leventouri, J. Appl. Phys. 69, 863 (1991)

[9] S. W. Van Sciver, Y. S. Hascieck, W. D. Markiewicz, L. R. Motowidlo, D. R. Hazelton, P. Haldar, Proc. The 5th U.S.-Japan Workshop on High T_c Superconductors, Tsukuba Japan, November 1992, edited by Kyoji Tachikawa p.115

[10] K. Sato, T. Hikata, and Y. Iwasa, Appl. Phys. Lett. 57, 1928 (1990)

[11] M. L. Hodgdon, R. Navarro and L. J. Campbell, Europhys. Lett. 16(7), 667 (1991)

[12] Y. S. Hascicek, S. Gabor, P. V. Shoaff Jr., H. W. Weijers, S. W. Van Sciver, P. Haldar, J. G. Hoehn Jr., L. R. Motowidlo, 13th Int. Magnet Technology Conf., September 20-24, 1993, Victoria, B.C. Canada. To be published in IEEE Trans. Magnetics.

LIMITATIONS FOR THE CRITICAL CURRENT DENSITY IN

MONOFILAMENTARY $(Bi,Pb)_2Sr_2Ca_2Cu_3O_x$ SILVER SHEATHED TAPES

B. Hensel, G. Grasso, A. Perin, and R. Flükiger

Université de Genève, DPMC
24, quai Ernest-Ansermet
1211 Genève 4
Switzerland

Abstract

It is shown that the main features of the temperature and field dependences of the critical current density j_c of silver sheathed $(Bi,Pb)_2Sr_2Ca_2Cu_3O_x$ (2223) monofilamentary tapes (prepared by the Powder In Tube method with intermediate pressing or rolling) can be interpreted in terms of a model that assumes the presence of 2212 in the path for the supercurrent, most probably in the form of layers of one or a few unit cells thickness that are intergrown at the frequent twist boundaries in the 2223 grains. Other main limiting factors are the pinning properties of the material and the texture of the filament.

This work has been supported by the Swiss National Science Foundation.

Processing of Long Lengths of Superconductors
Edited by U. Balachandran, E.W. Collings and A. Goyal
The Minerals, Metals & Materials Society, 1994

Introduction

The recent successes in the effort to increase the critical current density in silver sheathed, monofilamentary $(Bi,Pb)_2Sr_2Ca_2Cu_3O_x$ (2223) tapes [1,2] show that the material, as well as the prevailing fabrication technique (Powder In Tube, PIT) will allow the realization of technologically relevant projects such as power cables or high field magnets on the basis of High Temperature Superconductors (HTSC) [3].

The generally used measure for the 'quality' of HTSC tape and wire conductors is the critical current density $j_c(77K,B=0)$, i.e. at the temperature of liquid nitrogen and in zero external (i.e. in self) field. The interpretation of $j_c(77K,B=0)$, however, is by no means unambiguous as far as correlations between j_c and e.g. the microstructure or the phase composition are concerned. In many cases the temperature and field dependence $j_c(T,B)$ can provide much deeper insight.

The objective of the present work is to provide a simple scheme, that allows to 'qualify' a sample more profoundly than simply by the critical current density at 77K in self field. The developed method is succesfully tested on a selection of field depences $j_c(77K,B)$ (with the field oriented parallel to the tape surface) that were published by various authors in the last years. A correlation between the value of $j_c(77K,B=0)$ and the degree of degradation of j_c in magnetic fields at 77K is found.

One or a combination of the following effects are identified as the main origin for the strong limiting of the critical current density in magnetic fields: i) As can be seen from the microstructure in figure 1 the overall texture of the grains in the superconductor filament is not perfect and leads to an effective misalignment angle of the filament as a whole [4]; ii) The grain to grain connection is provided by low angle 'colony boundaries' [4,5] with a probably reduced supercurrent as compared to the value within a grain; iii) Intergrowths of one or few unitcells of 2212 occur at the twist boundaries that are characteristic for the 2223 colony structure [5].

The three mentioned effects lead to characteristic features in the measured field and temperature dependences.

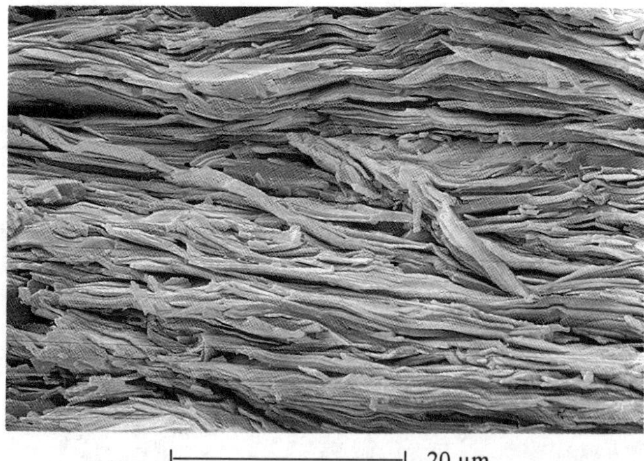

Figure 1 - Fracture surface (longitudinal) of a 'regular' 2223 tape prepared by the powder in tube method with intermediate uniaxial pressing ($j_c(77K)\approx20.000$ A/cm^2).

<u>Experimental</u>

The 2223 tapes that have been investigated in this work were prepared by the standard Powder In Tube method that is described in detail elsewhere [6,7]. The cation ratio was Bi:Pb:Sr:Ca:Cu = 1.72:0.34:1.83:1.97:3.13 throughout. The tapes have been rolled or uniaxially pressed between subsequent heat treatments. Critical current densities at 77K in self field of up to 15.000 A/cm^2 for the rolled tapes and up to 25.000 A/cm^2 for the pressed tapes were achieved. The short sample critical current was determined with a criterion of 1µV for a measured tape length of 1cm. For the measurement of the critical current as a function of temperature a maximum error due to sample self heating of 0.1-2.0K, depending on I_c, can be estimated.

<u>Results and Discussions</u>

Figure 2 shows the temperature dependence of the critical current $I_c(T)$ for a 'regular' (pressed) tape (j_c(77K,B=0)≈15.000 A/cm^2). $I_c(T)$ decreases linearly with temperature, except very near to T_c. The change in slope at T≈80K is commonly observed for a majority of the samples, but samples that show only a single slope over the whole temperature range are no exception. The difference of the data and an extrapolation of the 'high T' part (linear fit for T>80K) of the curve is shown in fig.2 as 'low T' part.

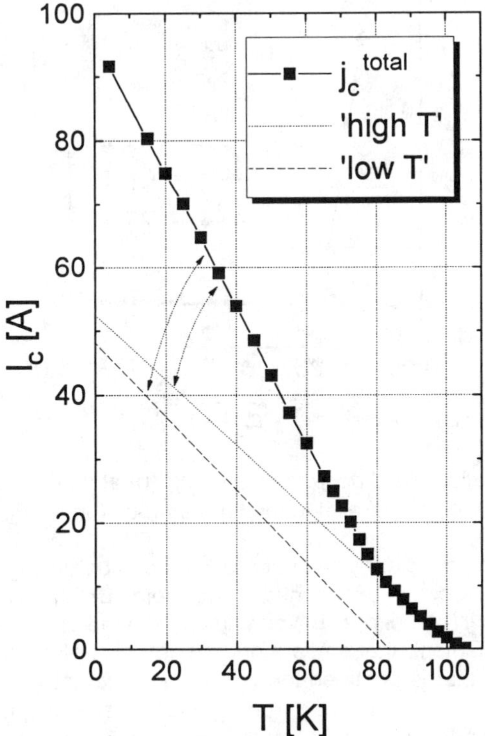

Figure 2 - Temperature dependence I_c(T,B=0) for a 'regular' (pressed) sample (j_c(77K,B=0)≈15.000 A/cm^2).

As the change in slope in different samples always occurs near 80K it is straightforward to associate it with the presence of a second superconducting phase, namely 2212, in the current path. The 2212 phase is distributed in intergrowths of one or a few unit cells 2212 at the frequent twist boundaries in the 2223 grains as has been shown by High Resolution Transmission Electron Microscopy (HRTEM) [5]. At temperatures below the critical temperature of 2212 ($\approx$80K) an enlargement of the effective cross section for the supercurrent occurs, thus enhancing I_c ('low T' curve). Figure 3 shows $j_c(B)/j_c(0)$ ($\underline{B}\perp c$) for the three temperatures T=65K, 77.35K, and 95K, i.e. below, near to, and above T_c of the 2212 phase.

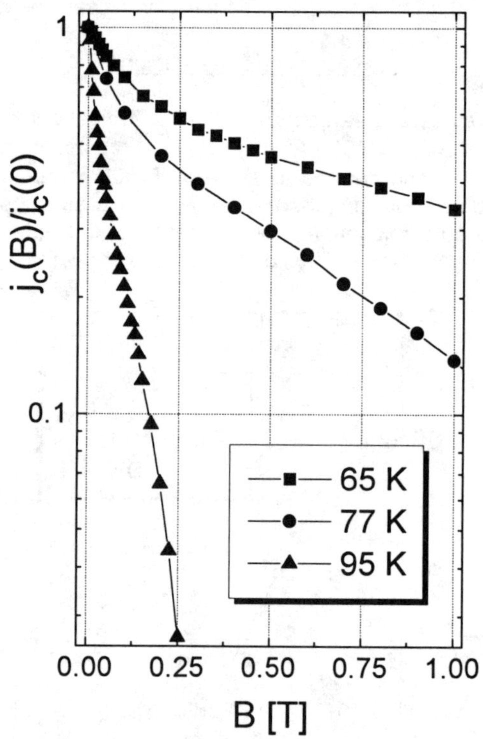

Figure 3 - Normalized field dependence $j_c(B)/j_c(0)$ at different temperatures for $\underline{B}$ oriented parallel to the tape surface (i.e. for $\underline{B}\perp c$).

For the lower temperatures clearly two field regions can be distinguished in which the critical current density decays exponentially with field (B<0.2T and B>0.2T at 65K). The form of the decay $I_c(B)\propto$ exp(-B/B*) originates from the above mentioned effective misalignment of the filament, that smears out the 'intrinsic plane pinning' peak in the anisotropy curve $j_c(\theta)$ that is obtained for 'perfect' samples at $\underline{B}\perp c$ [4]. Thus the field dependence $I_c(\underline{B}\perp c)$ can be calculated from $I_c(\underline{B}||c)$ by scaling the magnetic field as shown in fig. 4 and an effective misalignment angle $\phi_e \approx 8.5°$ can be obtained. Introducing two characteristic fields $B_1{}^*$ and $B_2{}^*$ together with two scaling factors c_1 and c_2 we can fit the field dependence ($\underline{B}\perp c$) with $j_c(B)=c_1\exp(-B/B_1{}^*)+c_2\exp(-B/B_2{}^*)$.

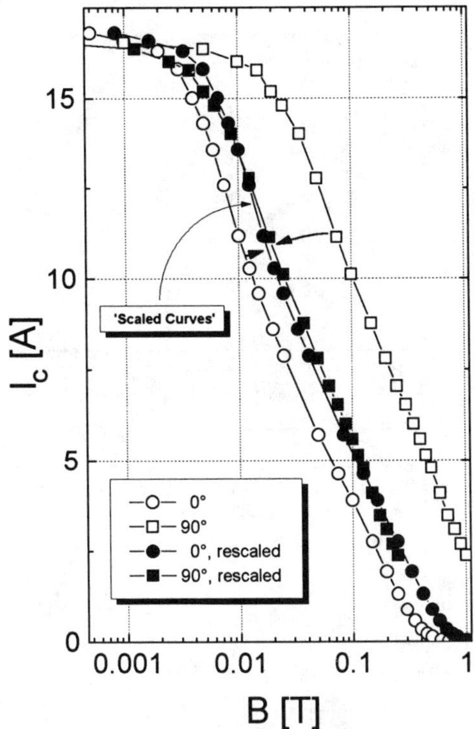

Figure 4 - Rescaling of $I_c(B)$. The two curves for the two field directions
($\underline{B}\|c$ and $\underline{B}\perp c$) can be superposed by rescaling $I_c(B)$.

As can be seen from figure 5 this function fits the data well. Excellent fits are generally obtained at temperatures T>20K. For lower temperatures a deviation from the above discussed behaviour is found as a result of the decreasing anisotropy of the critical current [8]. It should be pointed out that the field dependence is qualitatively the same for all investigated samples (i.e. for pressed and rolled samples), independent of the absolute value of $j_c(77K,B=0)$. Moreover, $I_c(B)$ is identical for pressed and rolled samples with the same critical current density at 77K in self field.
It should be noted that the fit with two exponentials is the result of a phenomenological approch to the problem, that finds its physical basis in the well known exponential field dependence of $j_c(\underline{B}\|c)$ in thin films and single crystals of 2212 (due to the above mentioned effective misalignment and the resulting absence of the 'intrinsic plane pinning' peak, $j_c(B,\theta)$ is effectively determined by a field component $B\|c$ at all angles).
One possibility for an association of the fit parameters to physical quantities might be, that c_1, $B_1{}^*$ and c_2, $B_2{}^*$ represent the contributions of current paths that are '2212 limited' (1) and '2223 pure' (2). This is only one (currently the most reasonable) of several possible interpretations and we have to defer the final answer to this question to a future publication [8]. Higher characteristic fields $B_1{}^*$ and $B_2{}^*$ result from an improved overall texture, as the effective misalignment angle ϕ_e will be smaller. This is of course only true if the intrinsic pinning properties of the material remain unchanged.

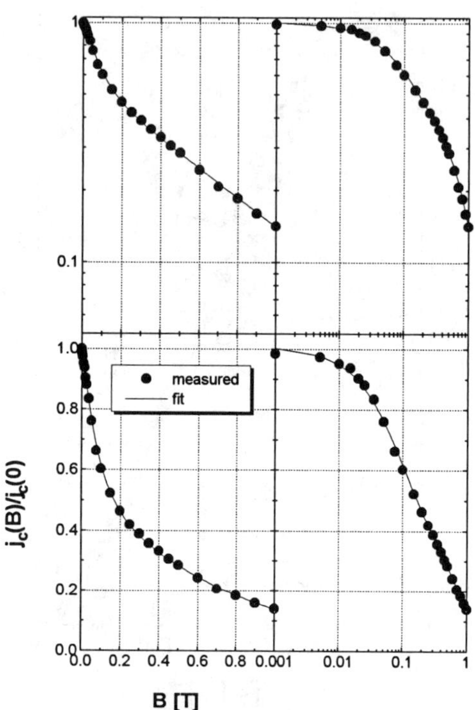

Figure 5 - Fit of the magnetic field dependence $j_c(\underline{B}\perp c)=c_1\exp(-B/B_1{}^*)+c_2\exp(-B/B_2{}^*)$.

An association of the steep slope in low fields (c_1, $B_1{}^*$) with 'weak links' in the current path can be ruled out, as no other experiment gives conclusive evidence for the presence of weak links in the supercurrent path in 2223 tapes.

In figure 6 we show a selection of magnetic field dependences $j_c(\underline{B}\perp c)$ of 2223 tapes reported by various authors. The above mentioned fit procedure is applied to the normalized curves and the results are shown in figure 7. Although there is a considerable scatter in the curves a general trend can be extracted from the data.

With increasing $j_c(77K,B=0)$ the slope of the exponential decay in magnetic fields is significantly reduced and the steep decay in low fields becomes less pronounced. Expressed in terms of the fit parameters: the characteristic fields $B_1{}^*$ and $B_2{}^*$ increase and c_1 decreases, whereas c_2 increases. Translated to material parameters this means that an improved texture and a reduced contribution of '2212 limited' current paths lead to an enhancement of j_c. This is in good agreement with the finding that the number of 2212 intergrowths is significantly lower in samples that have a high j_c, as compared to samples of poorer quality [9].

Once again it should be noted that this interpretation is not unambiguous and needs further investigation in order to have more evidence in favour or against this notion [8].

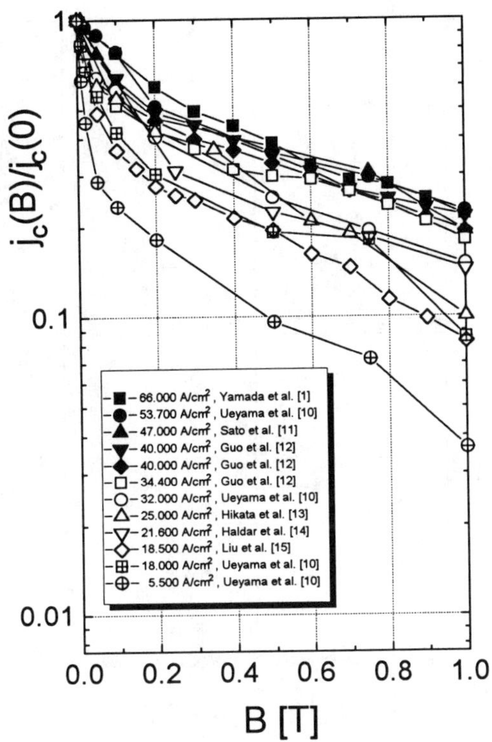

Figure 6 - Field dependences at 77K reported by various author for $\underline{B}$ oriented parallel to the tape surface.

<u>Conclusions</u>

The limiting mechanisms for the critical current density in silver sheathed 2223 tapes can be summarized as follows:

i) It is known from micro slice experiments that the supercurrent is not distributed homogeneously over the sample cross section [16] and there is evidence that only a very small part of the cross section carries the supercurrent.

ii) From High Resolution Transmission Electron Microscopy 2212 intergrowths in the 2223 grains are frequently found in samples with low critical current densities, whereas their number is significantly reduced in high j_c samples [9]. Although there is not yet a final understanding of the precise mechanism by which the intergrowths limit the critical current there is strong evidence that they do so.

iii) The magnetic field dependence of j_c also exhibits features that can be explained by the presence of two different phases in the path for the supercurrent, e.g. in the form of intergrowths.

iv) Another information that can be deduced from $j_c(\underline{B}\perp c)$ concerns the texture. The degree of texture of the grains in the path for the supercurrent can be quantified in the effective misalignment angle ϕ_e, that can be obtained from rescaling the two field dependences $j_c(\underline{B}\|c)$ and $j_c(\underline{B}\perp c)$.

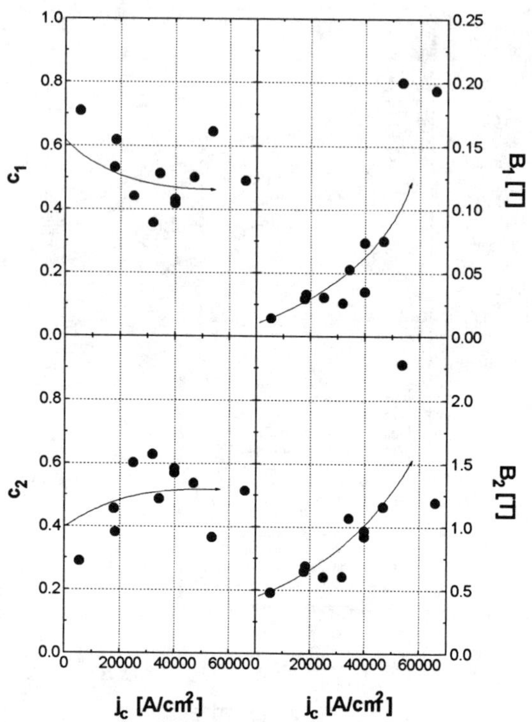

Figure 7 - Fit parameters obtained for the data in figure 6 (see text, the arriows are guides for the eye only).

Refined process parameters will lead to a further enhancement of the critical current density. Improved pinning properties will improve the performance of $j_c(B\|c)$. Simultaneously $j_c(\underline{B}\perp c)$ will also improve because of the scaling effect. An improved texture will reduce the effective misalignment angle and thus enhance the critical current in magnetic fields for the field orientation parallel to the broad tape surface, i.e. $j_c(\underline{B}\perp c)$. Finally an enlargement of the 'active' cross section for the supercurrent seems to be possible and will result in significantly higher critical current densities.

References

1. Y. Yamada et al., Proceedings of the Fifth International Symposium on Superconductivity ISS'92, Nov. 16-19 (1992) Kobe, Japan.

2. Q. Li et al., submitted to Physica C (1993).

3. P. Haldar et al., this conference.

4. B. Hensel et al., Physica C 205 (1993), 329.

5. Y. Feng et al., Physica C 192 (1992), 293.

6. R. Flükiger et al., Supercond. Sci. Technol. 5 (1992), 61.

7. G. Grasso et al., accepted by Physica C (1993).

8. B. Hensel et al., to be published.

9. A. Umezawa et al., submitted to Physica C (1993).

10. M. Ueyama et al., Jpn. J. Appl. Phys. 30 (1991), L1384.

11. K. Sato et al., Proceedings of the Applied Superconductivity Conference 1990, Snowmass Village, Colorado, Sept. 24-28, 1990.

12. Y.C. Guo, H.K. Liu, and S.X. Dou, Appl. Supercond. 1 (1993) 25.

13. T. Hikata et al., Cryogenics 30 (1990), 924.

14. P. Haldar et al., Appl. Phys. Lett. 60 (1992), 495.

15. H.K. Liu et al., accepted by Physica B, Proceedings of the 20th Conference on Low Temperature Physics, LT20 (1993).

16. D.C. Larbalestier, submitted to Nature (1993).

ARE THE GRAIN BOUNDARIES THE LIMITING FACTOR IN HIGH CURRENT
CONDUCTORS?

A.D. Caplin, S.M. Cassidy, L.F. Cohen, M.N. Cuthbert, J.R. Laverty and G.K. Perkins

Centre for High Temperature Superconductivity, Blackett Laboratory,
Imperial College, London SW7 2BZ, UK

and

S.X. Dou, Y.C. Guo and H.K. Liu
School of Materials Science and Engineering
University of New South Wales, Kensington, NSW 2033, Australia

Abstract

Detailed magnetic and electrical studies of high current BSCCO(2223) and (2212) phase
conductors show little indication of the grain boundary "weak-link" behaviour that occurs in
polycrystalline $YBa_2Cu_3O_7$. This conclusion is independent of whether a "brick-wall" model is
invoked to describe the grain connectivity, and it is further bolstered by circumstantial evidence
from irradiation experiments.

Work supported by the UK Science and Engineering Research Council, EC Brite-Euram
Contract BRE2 CT92 0229, Metals Manufactures Ltd, and the Commonwealth Department of
Industry, Technology and Commerce.

Processing of Long Lengths of Superconductors
Edited by U. Balachandran, E.W. Collings and A. Goyal
The Minerals, Metals & Materials Society, 1994

<u>Introduction</u>

The critical current (J_c) performance of conductors based on high temperature superconducting (HTS) materials continues to improve at an encouraging rate, and applications to magnets, etc. are within sight. However, the factors that control J_c in real conductors are far from clear, and are obscured by the complexity of the chemical phase diagram of these materials, their microstructure, and other aspects of their thermal and mechanical processing. Furthermore, many of these factors are interdependent, so that it is difficult for the materials scientist to prioritise better phase purity, cleaner grain boundaries, better texturing, or something else.

Here, we review the light that physical studies can upon this problem, and show that useful information can be obtained by going beyond the usual transport measurement of J_c.

<u>The Granularity Problem</u>

Soon after the discovery of HTS materials came the disappointment of the poor critical current performance of ordinary polycrystalline $YBa_2Cu_3O_7$. In that material, the supercurrent density $J_c(gb)$ that can flow *across* grain boundaries (Figure 1) is very much less than that which can flow *within* a grain, $J_c(gr)$.

A granular system is one in which the grains are weakly coupled to each other, so that the grains become superconducting essentially as isolated entities, and can sustain a large $J_c(gr)$. The strength of the *inter*granular coupling determines the size of $J_c(gb)$, and in a model system [1] can be made vanishingly small. Each of the grain-grain interconnections behaves as a Josephson Junction (JJ) or superconducting weak-link, whose behaviour is determined   by whether the intervening material is metallic or insulating, the temperature, the applied magnetic field (to which junctions are particularly sensitive), and its geometry [2]. Thus, a piece of polycrystalline HTS material can be considered as a network of Josephson Junctions, and this approach provided the early understanding of the observed behaviour [3,4]. More detailed experiments on bicrystalline thin films [5] demonstrated that rather small grain misorientation is enough to depress severely the critical current $J_c(gb)$ that can cross the boundary.

The key parameter in this granular model of HTS materials is the ratio $J_c(gb)/J_c(gr)$, which has

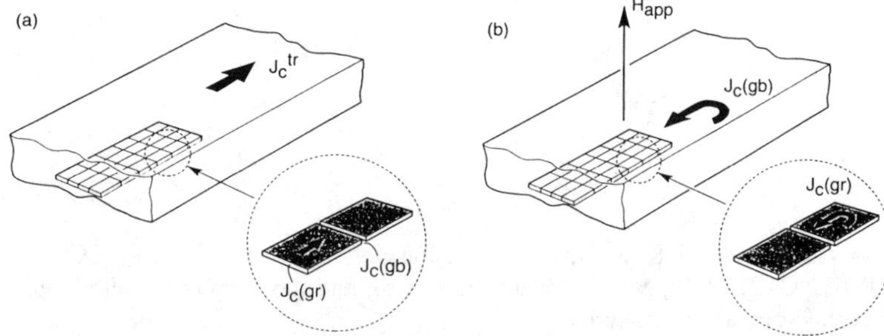

Figure 1   Schematic microstructure of a granular superconductor.
(a) In a transport measurement, the current that flows has to cross the grain boundaries, and so is usually limited by the grain boundary critical current density $J_c(gb)$. If the grains are very strongly coupled together, the limitation on J^{tr} may come instead from the *intra*granular critical current density $J_c(gr)$.
(b) In a magnetic measurement, screening currents flow: (i) on a macroscopic scale, crossing grain boundaries, and so involving $J_c(gb)$; (ii) simultaneously around each grain, involving $J_c(gr)$.

to be small compared with unity for the model to be applicable. Of course, both current densities are field- and temperature-dependent, so there may be regions in a field-temperature phase diagram where granularity dominates, but elsewhere the behaviour may be non-granular.

Why are grain boundaries (GB's) harmless in LTS, but so deleterious in $YBa_2Cu_3O_7$? A natural explanation is the shortness of the superconducting coherence length ξ, which is just a few nm in these oxides, whereas it is at least an order of magnitude larger in LTS materials. At a GB, the atomic structure is perturbed for perhaps 2 or 3 atomic layers, creating in HTS a region that is wide enough compared with ξ for it to be effectively non-superconducting, but in LTS too small compared with ξ to be noticed. If this explanation is correct, it should apply to *all* the HTS cuprate oxides, because they all have small ξ's, and they share common crystallographic features.

The excellent critical current performance of the Ag-based BSCCO materials, both the (2212) and the (2223) phases [6,7], which are extremely promising for applications, therefore comes as a welcome surprise. Although texturing is important in achieving high J_c's in the tapes, the c-axis alignment is generally no better than 5 or 10°, and there is little in-plane texturing. Thus, the *empirical* evidence is that here the GB's are much less of a problem than in $YBa_2Cu_3O_7$. An understanding of why that is the case should help not only to improve further the performance of the BSCCO conductors, but might offer also some insight as to how the GB problem can be overcome or circumvented in bulk $YBa_2Cu_3O_7$ and other HTS oxides.

Current-Voltage Characteristics

First consider a conductor that is macroscopically uniform, free of cracks and macroscopic voids. In a transport measurement, the current flow and voltage drop are along the length of the sample, and necessarily the current has to traverse the intergrain contacts. Thus, the measured transport current density, J_c^{tr}, is equal to $J_c(gb)$. The real material is certainly not fully dense, and the grains are in contact with each other over only a limited (and unknown) area (Figure 2); consequently, it is better to think in terms of an average intergrain critical current i_c. A simple model of well-textured platelet grains of side *2a* and thickness *2t* yields $J_c(gb) = i_c/4at$. A "brick-wall" model [8] has the same number density of contacts, and so yields an identical result.

In HTS materials, there is no sharply defined critical current, although it is often taken to be the current at some arbitrary electric field E, usually $\sim 10^{-4}$ V m^{-1}; therefore, we should drop the subscript c, and consider the measured J^{tr} as a function of E. Typical current voltage characteristics for a polycrystalline sample of $YBa_2Cu_3O_7$ are compared with those for a Ag-sheathed BSCCO(2223) tape in Figure 3. A clear qualitative difference between the two is

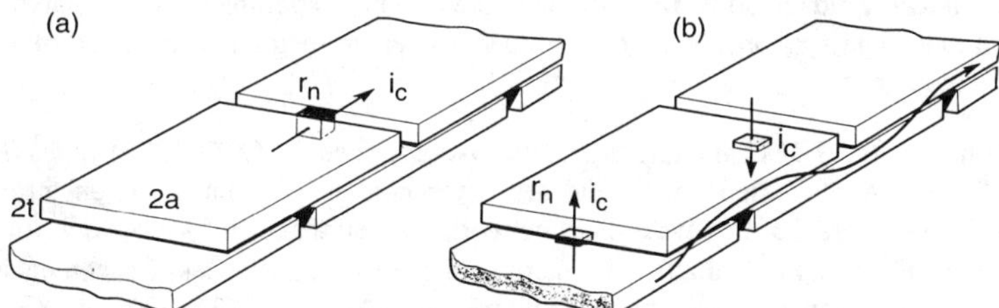

Figure 2   The contact between grains is over only some limited region, and is best described by its critical current i_c, and its normal state resistance r_n. The current transport may be (a) edge-to-edge of the plate-like grains, or (b) take a sinuous path involving face-to-face transport, as postulated in the "brick-wall" model.

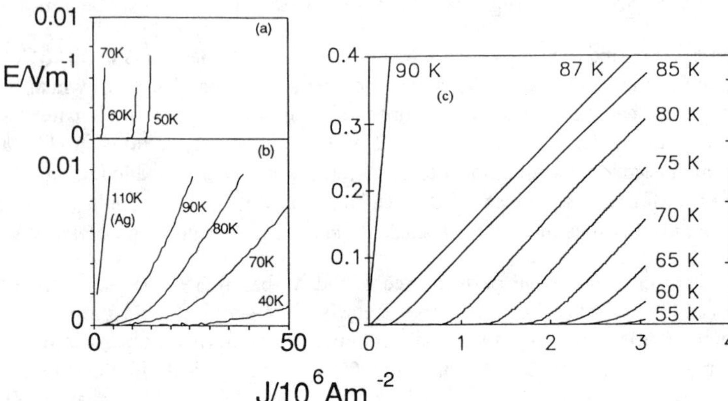

Figure 3   I-V characteristics in zero applied field of (a) polycrystalline YBa$_2$Cu$_3$O$_7$ rod. (b) Ag-sheathed BSCCO(2223) tape; the 110 K data represent the conductivity of the Ag sheath; at lower temperatures, the sheath contribution to the current has been subtracted. For purposes of comparison, the data are shown on identical scales in panels (a) and (b). In panel (c), the YBa$_2$Cu$_3$O$_7$ data are shown on the appropriate scales (changed by a factor of ~300 from panel (a)) to reveal the quasi-linear dependence of the characteristics at high currents.

apparent immediately: The former does show a fairly well-defined J_c, and at larger currents the dissipation can be described by a slope resistivity ρ' whose magnitude is almost independent of temperature and field (Figure 3(c)); this behaviour resembles that of a "classical" super-conductor-normal metal-superconductor (SNS) weak-link, and it is reasonable to associate ρ' with the dissipation at grain boundaries [9]. The analysis can be taken further by noting that the schematic microstructure of Figure 2 allows us to extract a slope resist*ance* per grain contact r_n, given by $\rho'/2t$. A fundamental parameter associated with any superconductive weak-link is the $i_c r_n$ product, which is therefore given by $2aJ_c(gb)\rho'$. This parameter involves only directly measured quantities, $J_c(gb)$ and ρ', and an average grain dimension; in particular, it is *independent* of the effective area of contact between grains, which is extremely difficult to estimate. For those reasons, it is a useful guide to the critical current performance.

As we showed some time ago [9], this analysis of the *I-V* characteristics of polycrystalline YBa$_2$Cu$_3$O$_7$ yields results that are consistent with experiments on isolated thin-film grain boundaries. The $i_c r_n$ products are ~ 100μV, similar in magnitude to those of single high angle grain boundaries.

In contrast, the I-V characteristics of the Ag-sheathed BSCCO(2223) tapes (Figure 3(b)) are strongly temperature-dependent, and are never convincingly linear. If an attempt is made to force-fit the data to the weak-link network model that works so well for polycrystalline YBa$_2$Cu$_3$O$_7$, the $i_c r_n$ products that emerge are of order 1μV or less (as can be seen in Figure 3, the I-V characteristics are much shallower in BSCCO(2223) than in YBa$_2$Cu$_3$O$_7$). This conclusion is independent of whether the "brick-wall" model is invoked.

Real conductors are presently far from macroscopically homogenous. Recent experiments [10] have shown that across the width of a tape, there can be considerable variation of J_c^{tr}; this variation is perhaps associated with greater or lesser texturing of the core material. Furthermore, the substantial reduction in J_c^{tr} that is reported invariably for long lengths (~10 m or more) compared with short sample (~10 mm) results is surely caused by macroscopic defects, cracks in particular. Thus, transport measurements on long lengths are indicative of the quality of process control, but are not at all informative on the *intrinsic* issue of whether the grain boundaries are imposing the fundamental limitation on the critical current.

Magnetization Loops

In response to an external applied magnetic field H_{app}, screening currents are set up, both within the grains, and on a macroscopic scale with currents crossing the grain-grain contacts (Figure 1(b)). These currents generate irreversible magnetic moments, whose sum Δm is measured by the magnetometer:

$$\Delta m = \Delta m(gr) + \Delta m(gb) \tag{1}$$

A vibrating sample magnetometer (VSM) is generally more convenient to use for these measurements, where it is helpful to monitor Δm as H_{app} is swept around a hysteresis loop; SQUID magnetometers are more sensitive, but are difficult to use in this mode.

For a tape sample with the field perpendicular to its face (i.e., parallel to the average c-direction of the textured grains), the magnetic moment can be written as:

$$\Delta m = (2\Omega/3) \, [f \, a \, J_c(gr) + \Lambda \, J_c(gb)] \tag{2}$$

where Ω is the sample volume, a is now the typical grain radius, Λ is the length scale on which the intergranular currents circulate [11], and f is the filling factor of the superconducting grains within the interior volume of the sheath, which for present purposes can be put equal to unity. In a well-connected tape, Λ should be equal to some average of the sample half-width and half-length. The separation of the screening currents into inter- and intragranular terms in equations (1) and (2) is valid strictly only if $J_c(gb) << J_c(gr)$, but suffices for the semi-quantitative analysis here. It is important to appreciate from equation (2) that the large ratio of Λ to a, typically 10^2 in tapes, weights strongly the *intergranular* contribution $\Delta m(gb)$ to the measured Δm of a tape sample, even if $J_c(gb)$ is considerably less than $J_c(gr)$.

A key issue is to determine the relative contributions of $\Delta m(gr)$ and $\Delta m(gb)$ to the measured Δm. A number of ways to do this have been developed:

First, in polycrystalline $YBa_2Cu_3O_7$, a well-defined "necking down" feature is observed in the magnetic moment-field (m-H) hysteresis loop [12] at fields ~mT. This is clearly caused by field suppression of $J_c(gb)$ by one or two orders of magnitude, and occurs at fields *below* the lower critical field of the grains themselves. Thus, m-H loops taken with a field amplitude of ~mT measure $J_c(gb)$ alone, and allow its field and temperature dependence to be followed. At higher applied fields, $J_c(gr)$ totally dominates Δm. Similar behaviour is visible in *poor quality* BSCCO (2223) material [13], but the necking-down feature is absent from high current BSCCO (2223) tapes (Figure 4).

Secondly, we have developed recently a quantitative analysis of magnetisation data, the length scale technique [11], that allows Λ in equation 2 to be measured straightforwardly. As long as no macroscopic defects are present, Λ should be of the order of the half-width of the tape.

The third, but destructive, approach is to compare directly the magnetic response of a tape sample with that of the separated superconducting grains extracted from it, which corresponds to $J_c(gr)$. Apart from some minor uncertainties (which introduce corrections of order unity),

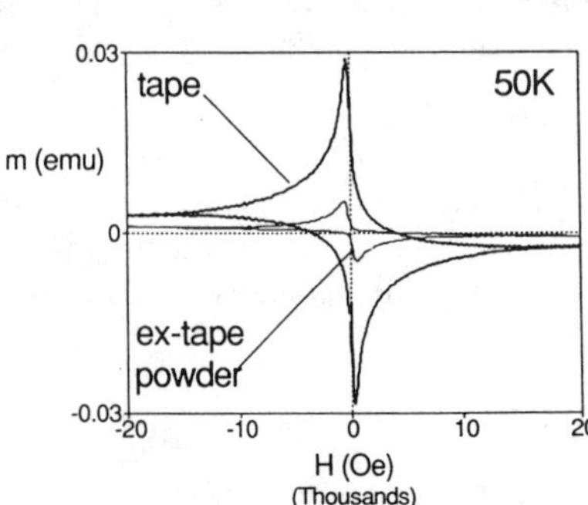

Figure 4.　m-H loops at 50 K for Ag-sheathed BSCCO(2223) tape sample 91402-2 (field perpendicular to plane of tape), and for the powder extracted from this sample (dispersed in wax). The small sloping background signals contain contributions from the Ag sheath and the wax, and also an instrumental term.

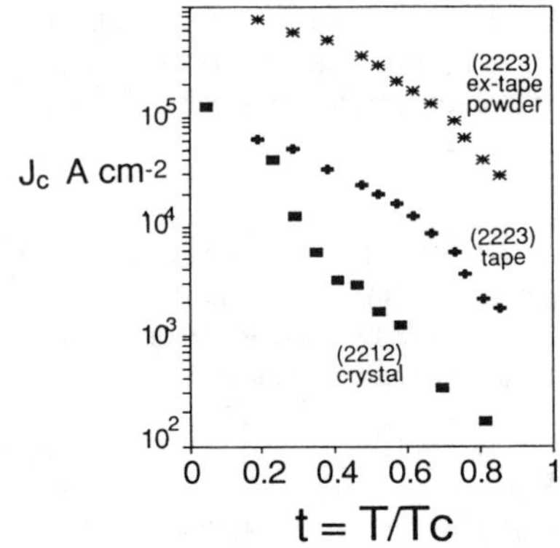

Figure 5.　Critical current densities obtained from the remanent moment of (a) BSCCO(2223) tape; (b) the randomly oriented powder extracted from that tape (assumed grain size 10 μm; (c) BSCCO(2212) single crystal.

such as the degree of texturing in the tape, and the random orientation of the powder, the difference between the two is an unambiguous measure of $J_c(gb)$. It is essential to measure $J_c(gr)$ of the BSCCO(2223) material extracted *from the tape itself*, rather than that fabricated by some other route, because the BSCCO(2223) phase is notoriously sensitive to the details of the processing, including temperature, mechanical treatment and atmosphere [14].

Figure 5 summarises the J_c data obtained from a BSCCO(2223) tape, using the techniques described above. Λ was always measured to be ~mm, but showed some tendency to diminish at high temperatures and high fields [15]. It would be useful to compare the tapes with a BSCCO(2223) single crystal were one available, but to our knowledge nobody has succeeded in growing such a crystal with reasonable quality. Instead, we make the comparison with a BSCCO(2212) crystal, and allow for the difference in transition temperatures. It is clear that the (2223) phase material extracted from the tape has a high $J_c(gr)$ compared with the (2212) crystal, and indeed is close to that of typical single crystal $YBa_2Cu_3O_7$. Although the BSCCO(2223) has slightly stronger inter-layer coupling than the (2212) phase, it is still quite two-dimensional; indeed, its micaceous structure is a key factor for successful texturing by mechanical processing. It would therefore be expected that flux penetrates as weakly-pinned "pancake" vortices [16]; consequently, the observed high $J_c(gr)$ requires understanding.

The fact that $J_c(gb)$ of the (2223) tape is an order of magnitude less than $J_c(gr)$ is not in itself an indication of 'weak-link' behaviour. It could be accounted for merely by the limited contact area between the grains within the tapes, as micrographs show that the texturing and packing of the present samples is far from perfect. It is noteworthy that the highest reported tape critical currents [17] are about an order of magnitude larger than that of the samples described here, and so are of about the same size as our $J_c(gr)$.

Comparison Between Magnetic & Transport Data

In a magnetic measurement, the induced shielding currents are normal to the applied field and in the plane of the tape, and so should be compared with transport measurements with the magnetic field applied *normal* to the tape surface. This is of course the "poor" direction for these conductors, in which J_c drops far faster with field than when the field is in the plane of the tape. The anisotropy between magnetic field normal and parallel to the tape becomes more severe at high temperature and high fields.

Again, we consider first a macroscopically uniform sample. Then, the transport current and the currents associated with *inter*granular magnetic shielding currents should be the same (Figure 1), and both should correspond to $J(gb)$. However, as Figure 3 shows, $J(gb)$ is a rather smoothly varying function of the electric field E in these tapes. The electric field in magnetic measurements arises from induction associated with the sweep rate of the applied field [18], and is several orders of magnitude smaller than in a transport measurement. Thus, because of very different voltage criteria, it is to be expected that magnetic measurements should yield a significantly lower value of the "critical current density" $J_c(gb)$ than do transport measurements. A general feature of the E-J characteristics in these materials is that they become steeper as the temperature is reduced, in the sense that if they are approximated by $E \propto J^n$, n becomes larger as T decreases. Consequently, the discrepancy in $J_c(gb)$ between the magnetic and transport measurements becomes greater at high temperatures.

It is better to use the two experimental approaches to generate complementary E-J characteristics, although this is time-consuming. We have some data of this kind on BSCCO(2212) ribbons (Figure 6), where we see that the agreement is quite good at low applied magnetic fields, but the discrepancies become substantial at higher fields [19]. These suggest that

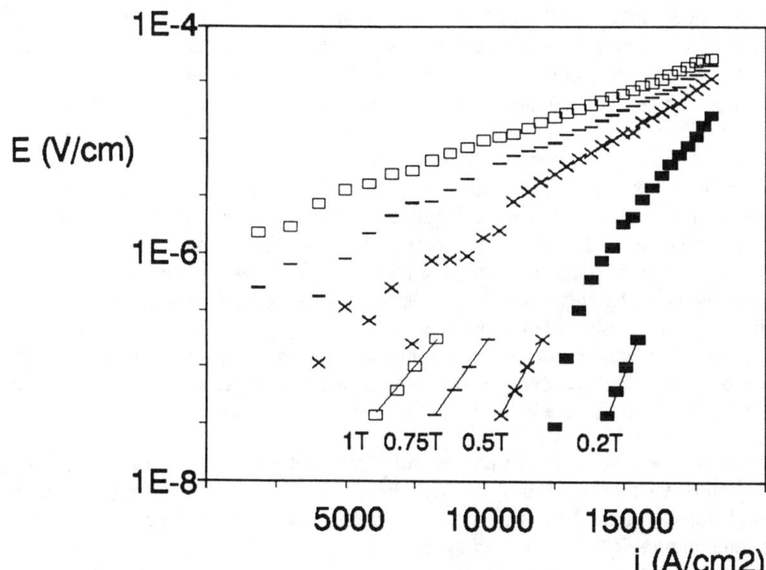

Figure 6   I-V characteristics of a (2212) ribbon at 25 K for different fields applied normal to the face of the ribbon [19]. At high electric fields, the data are from transport measurements, at low fields from magnetic studies.

macroscopic inhomogeneities are significant.

Transport monitors the strongest current path between the two voltage probes, the critical current of that path being limited by its weakest point. Thus, if a crack extends across the whole width of the conductor, the transport J_C would be zero. In contrast, the magnetic measurement looks at the strongest closed current loops, weighted by their area. The samples used in a magnetometer are generally no more than 10 mm long and usually 2 to 3 mm wide; the length scale parameter Λ is influenced primarily by the smaller dimension, so that the same crack would have little effect on Λ, and the magnetic J_C could be large. Thus, macroscopic defects affect the magnetic moment only when they are sufficiently closely spaced to reduce the typical scale of the pattern of screening currents to less than that of the geometric dimensions (width or length) of the sample.

Of course, these macroscopic defects need not be perfectly insulating; they might simply be regions of different stoichiometry or of poorer texturing, in which $J_{c(gb)}$ is reduced. Also, they could just be regions of enhanced sensitivity to temperature or magnetic field.

Finally, we must bear in mind that these conductors are fragile and easily damaged by handling. Also, the Lorentz forces involved in a transport measurement in field are not negligible. The most useful comparison is between transport and magnetic data on the *same* sample, best taken in that order, because the contactless magnetic measurement is less likely to be damaging. A valuable cross-check is to perform an initial magnetic measurement, so as to monitor the damage induced during the transport measurements. Naturally, this sequence is a time-consuming one, and systematic exhaustive studies of this kind have really only just commenced.

<u>Other Evidence</u>

An interesting recent study of BSCCO(2212) has investigated the role of microstructure on transport and magnetic behaviour [20]. Melt-cast rods of this phase contain bundles of plate-like grains. Within any one bundle, the grains are well-textured with c-axes parallel, and are typically 10 μm across and 1 μm thick. The bundles have size ~100 μm, and random orientation with respect to each other. Elschner et al. have explored the scale of current flow by successively grinding the material down into particles that are first much larger than the bundles, then smaller than the bundles but larger than the grains, and finally as small as possible; at each stage they have measured the magnetic moment. Recalling (equation 2) that for a sample of given volume, the moment is proportional to the length scale of current flow, the latter can be inferred from the dependence of moment on particle size (the length scale technique [11] cannot be used for rods, because it depends upon having a sample geometry of large demagnetising factor). Their conclusions are that the contacts between bundles are *not* limiting the critical current, but that the limitation is imposed between the grains within a bundle; transport current measurements are consistent with this view. The contacts between bundles involve *high angle* GB's, so these result support the idea that in this material, the latter do *not* behave as weak-links.

A second, circumstantial piece of evidence comes from the irradiation studies carried out by Kummeth et al. [21] on BSCCO(2223) tapes, in which they have explored in detail the impact on J_c^{tr} of the columnar defects created by high energy (2.6 GeV) heavy ions. Over most of the range of field and temperature, J_c^{tr} is *enhanced* by damage. Since the suppression of J_c at grain boundaries is associated with the presence of non-ideal material, and the depression of the super-conducting order parameter, it is reasonable to expect irradiation to reduce $J_{c(gb)}$. On the other hand, irradiation of this kind certainly enhances $J_{c(gr)}$ in (2212) single crystals [22], so the effects observed in the tapes are consistent with $J_{c(gr)}$ being the controlling factor in them too.

Conclusions

Detailed magnetic and transport studies undertaken on the *same* sample of conductor allow the distinction to be made between the *intra*granular critical current density $J_{c(gr)}$, the local *inter*granular critical current density $J_{c(gb)}$ associated with "well-formed" material, and the long-length J_c as limited by macroscopic defects.

Such studies show that the grain boundaries in these BSCCO materials are qualitatively and quantitatively much more favourable than those in YBCO; this conclusion is reinforced by the circumstantial evidence from irradiation experiments. In many circumstances, it may well be that $J_{c(gr)}$ is the limiting factor, rather than (as had previously been assumed) $J_{c(gb)}$; this is likely to influence the strategy needed to enhance further the performance of these conductors.

The intragranular critical current density of the BSCCO(2223) material, *as formed in the tape process*, appears to be much higher than might have been anticipated for a phase that is strongly anisotropic and two-dimensional. The nature of the flux pinning in this material deserves detailed further attention.

If the contrasting grain boundary behaviour of the BSCCO phases and $YBa_2Cu_3O_7$ can be understood at a microstructural level, there might be some hope of improving the behaviour of the grain boundaries in the latter material. Being more three-dimensional, $YBa_2Cu_3O_7$ is attractive in other respects for the fabrication of long lengths of conductor, and solution of the grain boundary problem would remove a major impediment to its use.

References

1   P. England, F. Goldie and A.D. Caplin, "Electrical transport and magnetisation measurements on a 3D disordered superconductor-normal-metal composite", <u>J. Phys.F: Met. Phys.</u> 17 (1987), 447-458.

2   A. Barone, <u>Josephson Effect - Achievements and Trends</u> (Singapore: World Scientific, 1986)

3   J.R. Clem, "Granular and superconducting glass properties of high temperature superconductors", <u>Physica C</u> 153-155 (1988), 50-55.

4   S.E. Male, J. Chilton and A.D. Caplin, "The correlation of critical current and magnetisation in $YBa_2Cu_3O_7$", <u>Physica C</u> 153-155 (1988), 1483-1484.

5   D. Dimos, P. Chaudhari and J. Mannhart, "Superconducting transport properties of grain boundarise in $YBa_2Cu_3O_7$ bicrystals" <u>Phys. Rev</u>. B 41 (1990), 4038- 4049.

6   K. Heine, J. Tenbrink and M. Thöner, "High-field critical current densities in BSCCO (2212)/Ag wires", <u>Appl. Phys. Lett</u>., 55 (1991) 2441-2443.

7   M. P. Maley, "Overview of the status of and prospects for high Tc wire and tape development", <u>J. Appl. Phys</u>., 70 (1991), 6189-6193.

8   L.N. Bulaevskii et al., "Model for the low-temperature transport of Bi-based high temperature superconductors", <u>Phys. Rev. B</u>., 45 (1992) 2545-2548.

9   S.S. Bungre et al., "Are classical weak-link models adequate to explain the current-voltage characteristics in bulk $YBa_2Cu_3O_7$?", <u>Nature</u>, 341 (1989), 725-727.

10   D. Larbalestier et al., "Local measurements of the I-V characteristics in polycrystalline BSCCO", Science (submitted).

11   M. A. Angadi et al., "Non-destructive determination of the current-carrying length scale in superconducting crystals and thin films", Physica C 177 (1991), 479-486.

12   S. E. Male et al., "Critical currents, magnetisation and netweork models of polycrystalline YBa$_2$Cu$_3$O$_7$", Supercond. Sci. Technol. 2 (1989), 9-16.

13   J.R. Laverty, A.D. Caplin and S.E. Male, "High critical currents in bulk Pb-doped BSCCO", Physica C, 162-164 (1989), 1165-1166.

14   D. Larbalestier and M.P. Maley, "Conductors from superconductors", MRS Bulletin, (Aug 1993), 50-56.

15   S.M. Cassidy et al., "High critical currents in Ag-BSCCO(2223) tapes: Are the grain boundaries really weak-links?", J. Alloys & Compounds, 195 (1993), 503-506.

16   J.R. Clem, "Fundamentals of vortices in the high temperature superconductors", Supercond. Sci. Technol. 5 (1992), S33-40.

17   K. Sato et al., "Bismuth superconducting wires and their applications", Cryogenics 33 (1993), 243-246.

18   A.D. Caplin et al., "Mapping the E-J-B surface in high temperature superconductors", Supercond. Sci. Technol. (submitted).

19   M. Dhallé et al., "Field, temperature and angle dependence of the critical current density in BSCCO(2212) ribbons", J. of Superconductivity, in press.

20   S. Elschner et al., "Influence of granularity on the critical current density in melt-cast processed BSCCO(2212)", Supercond. Sci. Technol. 6 (1993), 413-420.

21   P. Kummeth et al., "The influence of columnar defects on the critical current density in BSCCO(2223) tapes", J. of Superconductivity, in press.

22   W. Gerhäuser et al., "Correlation of flux line pinning and irradiation damage in Bi-2212 single crystals after 0.5 GeV iodine irradiation", Physica C, 185-189 (1991), 2339-2340.

THE PATH FOR LONG RANGE CONDUCTION IN HIGH J_c

$TlBa_2Ca_2Cu_3O_{8+x}$ SPRAY-PYROLYZED DEPOSITS*

D. M. Kroeger, A. Goyal, E. D. Specht, Z. L. Wang

Metals and Ceramics Division
Oak Ridge National Laboratory
P. O. Box 2008
Oak Ridge, Tennessee 37831

and

J. E. Tkaczyk, J. A. Sutliff, and J. A. DeLuca

General Electric Corporate R&D
P. O. Box 8
Schenectady, New York 12301

Abstract

Grain boundary misorientations and local texture in polycrystalline $TlBa_2Ca_2Cu_3O_{8+x}$ deposits prepared by thallination of spray-pyrolyzed precursor deposits on yttria-stabilized zirconia have been determined from transmission electron microscopy, electron backscatter diffraction patterns and x-ray diffraction. The deposits were polycrystalline, had small grains, and excellent c-axis alignment. The deposits contained colonies of grains with similar but not identical a-axis orientations. Most grain boundaries within a colony have small misorientation angles and should not be weak links. It is proposed that long range current flow occurs through a percolative network of small angle grain boundaries at colony intersections.

*Research sponsored by the U.S. Department of Energy Office of Advanced Utility Concepts - Superconducting Technology Program under contract DE-AC05-84OR21400 with Martin Marietta Energy Systems, Inc.

Processing of Long Lengths of Superconductors
Edited by U. Balachandran, E.W. Collings and A. Goyal
The Minerals, Metals & Materials Society, 1994

Deposits of $TBa_2Ca_2Cu_3O_{8+x}$, prepared by vapor phase thallination of spray-pyrolyzed precursor deposits on polycrystalline yttria-stabilized zirconia substrates, have been shown to have excellent critical current density behavior in high fields at high temperatures (1,2). The deposits are polycrystalline and have small grains with excellent c-axis alignment. Resistively measured critical current density, J_C, as high as 3×10^5 A/cm^2 at 77 K in zero field has been obtained, and at 60 K values remain well above 10^4 A/cm^2 in a 1T field applied parallel to the c-axis. Furthermore, studies of deposits containing columnar defects resulting from heavy-ion irradiation have shown that with defect densities the irreversibility field at 77K is in the range reported for $YBa_2Cu_3O_7$ (3). These results indicate the great promise of Tl-1223 deposits for high field conductor applications, at high temperatures.

Typically, J_C vs H curves for H parallel to the c-axis decrease by a factor of 3-5 at 0.01T and at higher fields exhibit an extended plateau. This field dependence indicates the sample contains parallel weakly and strongly linked current paths. In this paper we report extensive characterization studies directed toward determining what aspects of the microstructure lead to the strongly linked paths. Transmission electron microscopy, electron back scatter diffraction and x-ray microdiffraction were used to define colonies of grains with similar but not identical a-axis orientations. Within a colony, c-axis tilt grain boundaries have small misorientation angles and thus are not expected to be weak links. It is proposed that long-range current flow is through a percolative network of small angle boundaries at colony intersections.

Experimental

Samples were prepared according to procedures described previously (1,3). Film thicknesses were approximately 3 μm. Zero field J_C's at 77 K were measured with an electric field criterion of 1 μv/cm on a bridge pattern etched from the film. The bridge dimensions were 4 x 0.2 mm. There were five voltage connections, one at each of the current tabs and three at 1 mm intervals along the bridge (points A-E of the inset in Fig. 6), permitting measurement of J_C in each of four 1 mm segments. As described previously (1), J_C varies by as much as a factor of three or four from segment to segment. Microstructural information regarding grain morphology and grain boundary chemistry and misorientation was obtained using scanning and transmission electron microscopy (SEM and TEM). Local texture information was inferred from x-ray diffraction (XRD) measurements, TEM, and electron back scatter diffraction patterns (EBSP). This last technique provides Kikuchi diffraction patterns formed by back scattered electrons in a scanning electron microscope. Spatial resolution was ~1000 Å. Since grain sizes in the basal plane were in the range 3-10 μm, these patterns could be analyzed to obtain the orientations of individual grains at the surface of the film.

Results

Figure 1 shows a TEM image of a cross-section specimen in which both twist boundaries (parallel to the basal plane) and tilt boundaries (approximately perpendicular to the basal plane) are seen. Alignment of the c-axis perpendicular to the substrate surface is seen to be good. SEM images of fracture surfaces and EBSP measurements to be discussed below indicate that basal plane dimensions of grains are in the range of 3 to 10 μm. TEM images indicate that in the c-direction grain thicknesses tend to range from 0.5 to 2 μm. Aspect ratios are thus ~10, intermediate between those in Y123 and Bi-Sr-Ca-Cu-O. The average thickness of the deposit is ~3 μm, so that in most areas it is only a few grains thick. Examination of a number of images suggests that perpendicular boundaries tend, in association with short segments of parallel boundaries, to extend through the thickness of the deposit (see arrows in Fig. 1). Composition determinations at edge-on twist boundaries from energy dispersive electron spectroscopy indicate thallium enrichment compared to the bulk. High resolution images suggest that this non-stoichiometry results from the nature of the grain termination at the boundary rather than a grain boundary phase. Thallium enrichment at c-axis twist boundaries is significantly enhanced by a post-thallination treatment in oxygen at 600°C which has been found to increase J_c by a factor of 3-4.

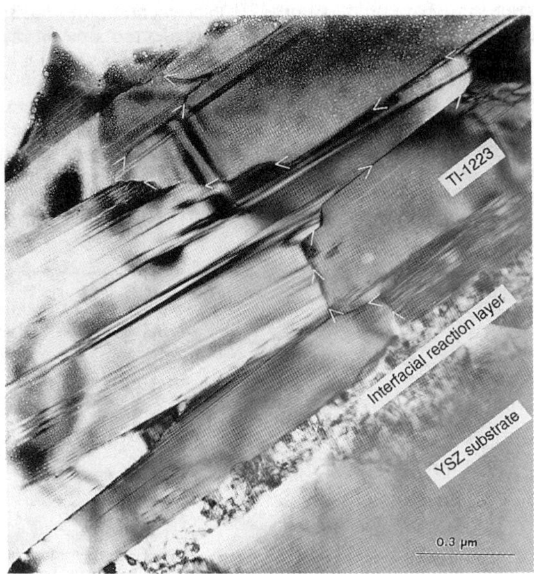

Figure 1 - TEM cross-section image showing the substrate, a thin reaction layer, and tilt (perpendicular to the substrate) and twist (parallel to the substrate) grain boundaries. As indicated by the arrows, tilt boundaries tend, in combination with short segments of twist boundary, to extend over several grains. The thin area in this view covers about half the sample thickness.

Misorientation angles for 20 grain boundaries were derived from convergent beam electron diffraction patterns for grains on either side of the boundary. Of eleven c-axis twist boundaries, ten had misorientation angles less than 10° and five of nine tilt boundaries had misorientation angles under 15°. This is a small sample, but the results do not appear to be consistent with a previous conclusion from EBSP determinations of the orientations of a large number of grains at widely separated positions on the sample surface that the basal plane orientations are random (4). Taken together, the two results suggest the presence of local texture, in which adjacent grains tend to have similar orientations even though macroscopically all orientations are present in similar numbers.

The presence of local texture was confirmed by XRD azimuthal scans using large and small x-ray beams. Figure 2 shows two XRD azimuthal scans of the same deposit using a large beam which illuminated approximately half the sample including one of the 4 x 8 mm current tabs and a small beam which was ~0.5 mm in diameter. The positions of source and detector, and the axis of rotation of the sample are indicated. The detector was positioned to detect a non-(00ℓ) peak. Since the deposit has good c-axis alignment the intensity at a given angle ϕ is proportional to the fraction of the deposit area illuminated by the beam which had some specific basal plane orientation. Thus, the ϕ-scans of Fig. 2 are, effectively, distribution plots for the directions of the a-axes of grains relative to a fixed arbitrary angle. Figure 2(a) indicates the intensity of the (103) reflection as a function of rotation about the film normal (i.e. the c-axis) using the wide beam. The intensity distribution consists of several peaks superimposed on an average intensity which is well above background. These data suggest that basal plane orientations of grains are largely random, but that there is a tendency for clustering about several specific directions. This result is consistent with previous EBSP measurements of the orientations of a large number of grains at random locations. However, a very different intensity distribution was obtained when a narrow x-ray beam was used, indicating that locally grains are far from randomly oriented. Figure 2(b) shows an azimuthal scan of the (10$\underline{10}$) reflection obtained with a narrow beam which had a projected area on the sample of ~0.5 x 1 mm. The intensity distribution shows two strong peaks with a low intensity (near background) at all other angles, indicating that basal plane orientations for grains within the illuminated area cluster strongly about only two directions. For approximately twenty such local scans obtained on separate areas of two different deposits, the number of peaks ranged from 1 to 5. These results indicate that, although over the whole deposit many basal plane orientations occur, there is a high degree of local texture indicative of the presence of colonies of grains with similar orientations.

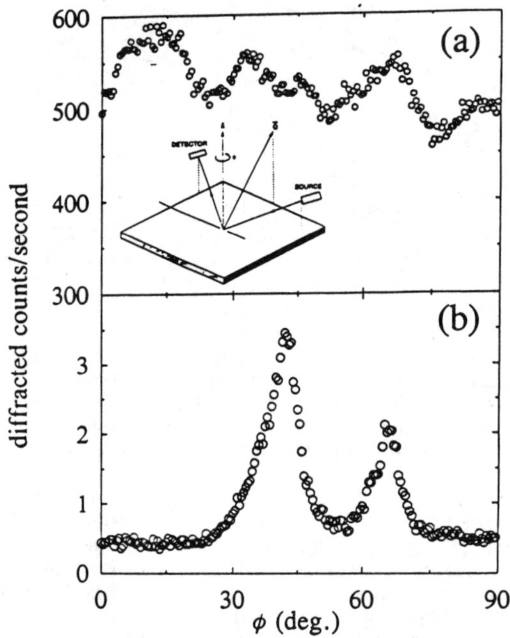

Figure 2 - Intensity distribution for x-ray diffraction azimuthal scans using (a) wide, and (b) narrow x-ray beams.

The presence of colonies of grains with similar a-axis orientations was confirmed using electron back scatter diffraction. Diffraction patterns were obtained from a large number of points within a rectangular area on the sample surface, which had been lightly polished to remove silver and secondary phase particles. While observing the diffraction pattern the electron beam was moved in a line over the sample surface. When a shift in the pattern was observed the new pattern was recorded and analyzed to obtain the local crystalline orientation. Since the c-axes are very well aligned, the orientation can be represented by a single arrow in the a-direction. In Fig. 3 the a-axis directions are plotted for a 40 x 120 μm area. The beam was moved along parallel lines separated by 12 μm to obtain data over the rectangular grid. An arrow is plotted only where a shift in diffraction pattern was seen, i.e., at a grain boundary. The spacing between arrows is roughly indicative of grain size, although surface irregularities prevented obtaining patterns in some areas, and therefore some shifts in orientation were missed. Clearly this area is within a single colony. The total spread in a-axis orientation is ~15-20°. Figure 4 shows a similar map over a somewhat larger area of the same sample. In this case an irregularly shaped colony intersection can be discerned clearly. The average a-axis orientations of the two colonies differ by ~35-40°. Note that the misorientation across the colony intersection is not constant. Figure 5 shows an EBSP map of another colony intersection. In this case there appears to be a gradual change of orientation across the intersection, and hence the colony boundary is not easily discerned.

The variation in the number of colonies and their orientations over the length of the patterned bridge was obtained from a series of XRD azimuthal scans taken at 0.5 mm intervals. The results are shown in Fig. 6 The angular positions of XRD peaks from plots obtained using a narrow beam, as in Fig. 2(b), are plotted as a function of position along the gauge length. The measured J_c's for each of the four sections are indicated. The projected area of the x-ray beam is approximately 0.5 x 1 mm, so there is substantial overlap of areas sampled for adjacent measurements, as indicated by the horizontal bars. Vertical bars indicate the full width at half maximum of the peaks, and thus are a measure of the spread in a-axis orientations. The size of the circle indicates peak intensity, which was influenced not only by the relative number of grains with that orientation, but also by the amount of deposit area sampled, as determined by proximity to current and voltage tabs. The number of a-axis orientations varied from 1 to 3 over the gauge length. Although particular a-axis orientations, presumably indicative of colonies of grains of similar orientation, arise and disappear as a function of position, it appears that there is usually a small angle boundary connection between adjacent areas. The results for this particular sample suggest that within the gauge length there may be only three colonies, one of which extends over the whole length. Alternatively, there may be a larger number of colonies, with a tendency for colony orientations to be correlated locally.

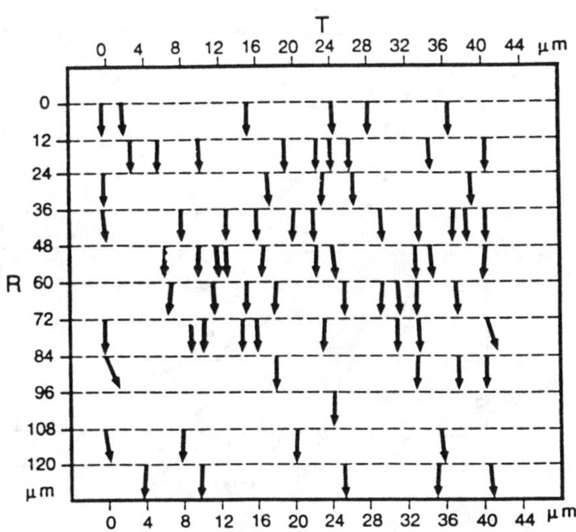

Figure 3 - EBSP measurements indicating the structure within a single colony. The arrows represent the direction of the a-axis of the grains. The c-axis is normal to the substrate.

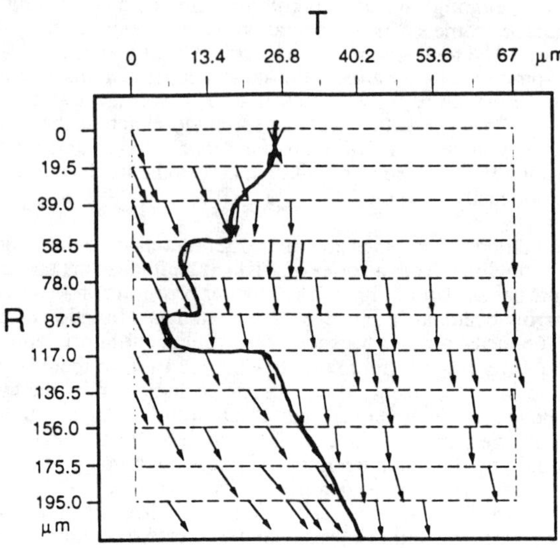

Figure 4 - EBSP measurements indicating the structure at a colony intersection. Many high angle boundaries are formed at the intersection.

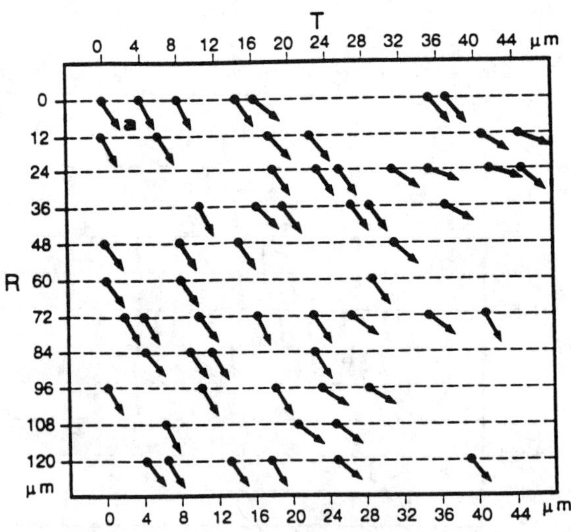

Figure 5 - EBSP measurements of another colony intersection. A gradual change of orientation across the intersection from one colony to another can be seen.

294

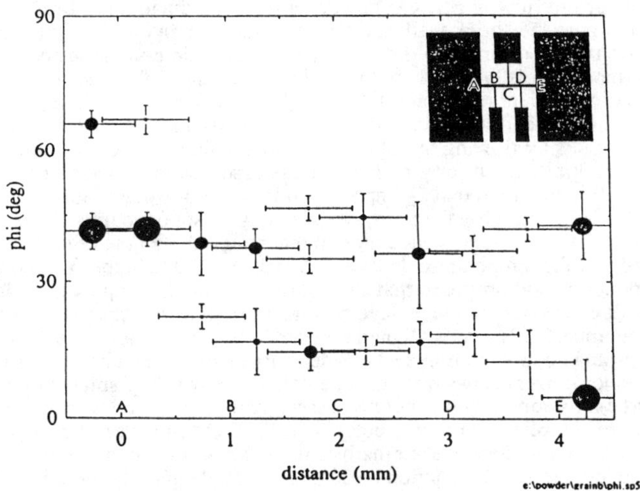

Figure 6 - Angular positions of peaks in x-ray diffraction azimuthal scan intensity distributions obtained using a narrow beam [as in Fig. 1(b)] as a function of position along the gauge length. The bridge length A-E (see inset is 4 mm and the segments A-B, B-C, etc. are 1 mm. The current pads are 4 x 8 mm. The size of the symbol indicates peak intensity, vertical bars indicate the peaks FWHM, and horizontal bars indicate the dimension (1 mm) of the beam, and thus the overlap of area sampled.

It appears that many colonies are not much smaller than the beam diameter. Smaller beam diameters require a high intensity x-ray source. Azimuthal scans along the length of the bridge have been obtained using a 100 µm diameter beam at the National Synchrotron Light Source at Brookhaven National Laboratory. The results indicate that low current density bridge segments are associated with large misorientation angles at colony intersections (11).

Discussion

Results of XRD azimuthal scans using small beams and EBSP determinations of the orientations of individual grains indicate the presence of colonies of grains with similar basal plane orientations. Colony dimensions in the plane appear to be in the range of 0.1 mm to 1 mm. Colonies have irregular shapes and therefore large areas of intersection. We note that these colonies are different from those in BSCCO, which consist of stacks of c-axis aligned grains which are connected by twist boundaries. In these Tl-1223 deposits colony intersections are made up of a large number of c-axis tilt grain boundaries. Within a colony most grain boundaries have small misorientation angles and thus should not be weak links. We have proposed (5) that long range current transfer occurs through a percolative network of small angle grain boundaries at colony intersections. The number of such small angle grain boundaries should increase with the spread in orientation of grains within a colony and with any tendency for adjacent colonies to have similar orientations as might result from a local bias of colony orientation. The FWHM of peaks in XRD azimuthal scans tends to be ~10-20°, indicative of a similar spread in a-axis orientation. Since Tl-1223 is tetragonal the maximum misorientation angle between c-axis aligned grains is 45°. Thus, if colonies are randomly oriented and each colony intersects several other colonies, it is probable that a significant fraction of intersection area will have overlapping grain orientation distributions.

Strongly-linked current flow in BSCCO powder-in-tube conductors has been discussed primarily in relation to the "brick wall" model for conduction in c-axis aligned materials (6-8). Conduction in the c-direction through c-axis twist grain boundaries is an essential element of that model. It is believed, on the basis of the results of Dimos et al., that c-axis tilt grain boundaries with large misorientation angles ($\geq 10°$) are weak links. In an idealized brick wall model of the microstructure of a c-axis aligned material, current may avoid these weak links by flowing through c-axis twist boundaries which may also be Josephson junction limited, but for grains with large aspect ratios, have much larger areas. Umezawa et al. (9) have found that J_C improves in Bi-2223 powder-in-tube conductor as the number of Bi-2212 intergrowths decreases at c-axis twist boundaries. This evidence supports the importance of c-axis conduction in that material. On the other hand, Hensel et al. (10) found that the temperature dependence of J_C in Bi-2223 tapes is inconsistent with c-axis conduction and proposed that current flows through complex boundaries (with both twist and tilt components) which have relatively small c-axis misalignment and occur frequently in the imperfectly c-axis aligned powder-in-tube materials. We have presented a model of long-range conduction in Tl-1223 deposits in which the above controversy does not arise, since current flows through a percolative network of small angle c-axis tilt boundaries, and c-axis conduction is not necessary. Although the relevance of these results to the more complicated microstructures in BSCCO conductors remains to be determined, it is clear that there are alternatives to the brick wall model of the microstructure which permit conduction through small angle grain boundaries, and that macroscopic determinations of texture are not adequate for delineating such structures.

References

1. J. A. DeLuca, P. L. Karas, J. E. Tkaczyk, P. J. Bednarczyk, M. F. Garbauskas, C. L. Briant, and D. B. Sorensen, "The Preparation of "1223" Tl-Ca-Ba-Cu-Oxide Superconducting Films via the Reaction of Silver-Containing Spray Deposited Ca-Ba-Cu-Oxide with Thallium Oxide Vapor," Physica C 205 (1993), 21-31.

2. J. E. Tkaczyk, J. A. DeLuca, P. L. Karas, P. J. Bednarczyk, M. F. Garbauskas, R. H. Arendt, K. W. Lay, and J. S. Moodera, "Transport Critical Currents in Spray Pyrolyzed Films of $TlBa_2Ca_2Cu_3O_x$ on Polycrystalline Zirconia Substrates," Appl. Phys. Lett. 61(5) (1993), 610-612.

3. J. E. Tkaczyk, J. A. DeLuca, P. L. Karas, P. J. Bednarczyk, D. K. Christen, C. E. Klabunde, and H. R. Kerchner, "Enchanced Transport Critical Current at High Fields After Heavy Ion Irradiation of Textured $TlBa_2Ca_2Cu_3O_x$ Thick Films," Appl. Phys. Lett. 62(23) (1993), 3031-3033.

4. D. J. Miller, J. G. Hu, J. D. Hettinger, K. E. Gray, J. E. Tkaczyk, J. DeLuca, P. L. Karas, J. A. Sutliff, and M. F. Garbauskas, "Brick-Wall Structure in Polycrystalline $TlBa_2Ca_2Cu_3O_x$ Thick Films with High Critical current," Appl. Phys. Lett. 63(4) (1993), 556-558.

5. D. M. Kroeger, A. Goyal, E. D. Specht, Z. L. Wang, J. E. Tkaczyk, J. A. Sutliff, and J. A. DeLuca, "Locak Texture and Percolative Paths for Long-Range Conduction in High Critical Current Density $TlBa_2Ca_2Cu_3O_{8+x}$ Deposits" (Appl. Phys. Lett., 1993).

6. A. P. Malozemoff, "Critical Current Density of High Temperature Superconductor," High Temperature Superconducting Compounds II, eds. S. H. Whang, A. DasGupta and R. B. Laikowitz (TMS Publications, Warrendale, PA, 1990), 3.

7. L. N. Bulaevskii, L. L. Daemen, M. P. Maley, and J. Y. Coulter, "What Limits the Critical Current in High T_C Superconducting Tapes?" (Phys. Rev. B.).

8. J. Mannhart and C. C. Tsuei, "Limits of the Critical Current Density of Polycrystalline High-Temperature Superconductors Based on the Current Transport Properties of Single Grain Boundaries," Z. Phys. B - Condensed Matter 77 (1989), 53-59.

9. A. Umezawa, Y. Feng, H. S. Edelman, D. C. Larbalestier, Y. S. Sung, E. E. Hellstrom, and S. Fleshler, "Electromagnetic Granularity, Critical Current Density, and Low-T_C Phase Formation at the Grain Boundaries in $(Bi, Pb)_2Sr_2Ca_2Cu_3O_x$ Silver-Sheathed Tapes," Physica C 198 (1992), 261-272.

10. B. Hensel, J.-C. Grivel, A. Jeremie, A. Perin, A. Pollini, and R. Flükiger, "A Model for the Critical Current in $(Bi, Pb)_2Sr_2Ca_2Cu_3O_x$," Physica C 205 (1993), 329-337.

11. E. D. Specht, A. Goyal, D. M. Kroeger, J. E. Tkaczyk, and J. A. DeLuca, "X-ray Microdiffraction Analysis of MM-Scale Orientational Correlation in Tl-1223 High-Critical -Current, High Temperature Superconducting Thick Films," (Physica C, 1993).

THE EFFECT OF PROCESSING ON THE MICROSTRUCTURE OF

Ag-SHEATHED Bi-2223 WIRES

D.J. Miller, J.G. Hu, P. Kostic
Materials Science Division,
Argonne National Laboratory
Argonne, Illinois 60439

U. Balachandran
Energy Technology Division
Argonne National Laboratory
Argonne, Illinois 60439

P. Haldar
Intermagnetics General Corp.
Guilderland, New York 12084

Abstract

The effect of processing on the microstructure of long-length $(Bi,Pb)_2Sr_2Ca_2Cu_3O_x$ (Bi-2223) wires has been evaluated using scanning and transmission electron microscopy. In particular, these studies have focused on the effect of the final heat treatment conditions on the microstructure of the wires and the corresponding variations in critical currents (I_c). It is found that variations in I_c for segments cut from various points along the length of a wire can be correlated with variations in the volume fraction and nature of second phases. Similar variations are observed between wires processed under different final annealing conditions. A common feature found in all of the wire segments with lower I_c values was the presence of the $Pb_3Sr_2Ca_2CuO_x$ phase (Pb-3221). Microchemical analysis suggests that the formation of this phase leads to variations in the composition of the Bi-2223 grains. As a result, the wires which exhibit the Pb-3221 phase consist of an assemblage of grains with a range of transition temperatures based on composition. Thus, in these segments, the current carrying capacity of individual grains at 77K varies from very good to poor, and a significant reduction in the overall I_c of the segment is observed.

Research at ANL and part of the work at IGC is supported by the U.S. Department of Energy, Divisions of Basic Energy Sciences-Materials Sciences, and Energy Efficiency and Renewable Energy, as part of a program to develop electric power technology, under Contract W-31-109-Eng-38.

Processing of Long Lengths of Superconductors
Edited by U. Balachandran, E.W. Collings and A. Goyal
The Minerals, Metals & Materials Society, 1994

Introduction

The development of reliable, long-length conductors based on $(Bi,Pb)_2Sr_2Ca_2Cu_3O_x$ (Bi-2223) superconductors has been pursued by a large number of investigators as a key to commercial development of these materials [1]. The oxide-powder-in-tube technique (OPIT) is an efficient method by which to produce such lengths of conductors. However, although sufficiently high critical currents (I_c) and critical current densities (J_c) have been attained in short length samples [2,3], it has been more difficult to achieve similar properties over long lengths [4-7]. At the current stage of development, it appears that variations from point to point along the length of these wires play a crucial role in limiting the development of these wires for practical applications. Thus, it is desirable to identify the features which limit current densities in long length wires.

The basic mechanism of current transport in wires is still not clearly understood, but it is clear that a high degree of texture is required for most of the proposed transport mechanisms [8,9], and this has been confirmed by experiment [10]. Some macroscopic features which disrupt aligned growth have been identified as factors which may lead to variations in critical currents in long lengths. For example, severe variations in core thickness ("sausaging") have been observed in rolled samples [11,12]. Such sausaging creates a constriction for current flow, frequently destroys the alignment of the superconducting grains, and introduces a further difficulty in defining the appropriate cross-sectional area used for calculation of J_c. In addition, the presence of second phases has been reported to disrupt aligned growth, leading to decreased critical currents [11], although some investigators have reported a trend towards less disruptive effects due to alignment of second phase particles as a function of processing conditions [13]. In addition to these somewhat more macroscopic features, microscopic features may also influence critical currents. For example, the presence of a few layers of Bi-2212 along [001] twist grain boundaries has been correlated with reduced J_c in Bi-2223 wires [14]. The purpose of this work has been to correlate the microstructure with measured properties for segments cut from long wires in order to identify the macroscopic and microscopic features which limit current densities in long length wires.

Experiment

Long-length (≈ 70 meter) wires of the Bi-2223 superconductor were prepared using conventional OPIT processing techniques. Partially reacted precursor powders of a lead-added Bi-2223 composition were loaded into Ag tubes, drawn and rolled with an intermediate heat treatment, and then given a final anneal in air for ≈ 50 hours. The two wires examined in this work were processed under identical conditions except that one wire was given a final anneal at 835°C (wire A) while the other was annealed at 840°C (wire B). In addition, a number of short segments of as-rolled wires were annealed at various temperatures to study phase development. The critical current at 77 K and zero field was measured for several sections of each wire which were cut from various points along the length of the wires. Microstructural characterization was then carried out by scanning electron microscopy (SEM) on each measured segment for both wires. The wires were prepared for SEM studies by standard metallographic techniques using alcohol- or oil-based slurries to avoid contact with water. Microstructural characterization by transmission electron microscopy (TEM) was also performed on some of the samples. These studies were carried out on the segments of each wire which exhibited the highest and lowest I_c values. Samples were prepared by standard cross-section techniques including ion milling at liquid nitrogen temperature. Microscopy was performed in a Philips CM-30 microscope equipped with an EDAX thin-window detector for energy dispersive spectroscopy (EDS). Compositional analysis based on EDS spectra was performed using semi-quantitative routines.

Results and Discussion

Figure 1 shows a plot of I_c measured for each segment as a function of arbitrary distance along the wire. For reference, an *average* I_c was determined for each wire from the I_c values of the various segments, although this value has no physical significance since the properties of a long length are expected to be dominated by the segment with the lowest I_c. Two features are of particular interest in this plot. Firstly, there is a significant variation in I_c from point to point within a given wire. In comparison with the average value, it is seen that for each wire there are a number of segments which fall below or near the average value and a few segments which lie significantly above the average. Thus, these segments can be grouped into two categories: one of "typical" samples, most closely associated with I_c values expected across a long length of wire, and "exceptional" samples which exhibit significantly higher I_c values. Secondly, there is a distinct difference in the average I_c. Even neglecting the "exceptional" segments, the "typical" I_c values for wire A are significantly better than for wire B. The microstructural work performed in this study was aimed at understanding the basis for these variations from point to point within a given wire as well as the differences between the wires.

The differences in I_c from point to point within a given wire can be explained in part on the basis of differences in the relative volume fractions of second phase and Bi-2223. For example, Fig. 2 shows backscattered SEM images from two segments of wire B, one with an I_c value of ≈ 16 A, the other with an exceptionally high I_c near 20 A. Although lower magnification imaging revealed some variations in core thickness, no sign of severe sausaging was observed in any of these wires. However, there are distinct differences in the volume fraction of second phases, most of which appear dark in these images. In particular, the segment with the lower I_c exhibits a few large particles which appear to limit the cross-sectional area of superconductor available to carry current, similar to the type of variation reported previously [15]. It is likely that these variations result from subtle inhomogeneities in composition during initial packing of the tube. Variations in core thickness have been linked with packing density [16], so it is not unreasonable to assume similar factors could lead to

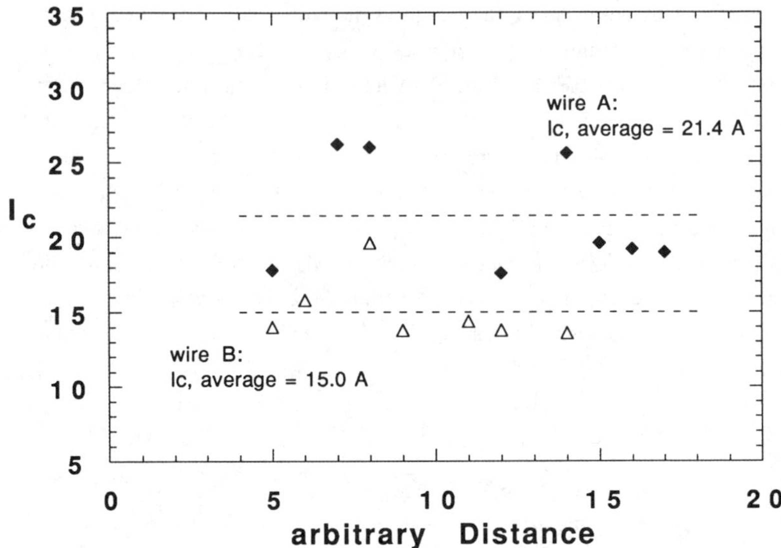

Figure 1:   Plot of I_c (77 K, zero field) measured for segments cut from Bi-2223 wires as a function of arbitrary distance along the wire.

compositional variations. These inhomogeneities may become more distinct and exaggerated by mechanical working. For example, most of the second phases observed in these images are CuO, CaO, $(Ca,Sr)_{14}Cu_{24}O_{41}$ (14-24 phase), and Ca_2CuO_3. Since none of the second phases observed are rich in Bi, it is clear that the composition in the portion of the wire seen in Fig. 2a contains relatively less Bi than the region seen in 2b. In wire A, differences in second phase fraction from segment to segment were more difficult to correlate with I_c values and, as will be described in more detail below, the formation of a $Pb_3Sr_2Ca_2CuO_x$ phase (Pb-3221) seems to play a key role.

The subtle differences in phase assemblage leading to variations in I_c from point to point in

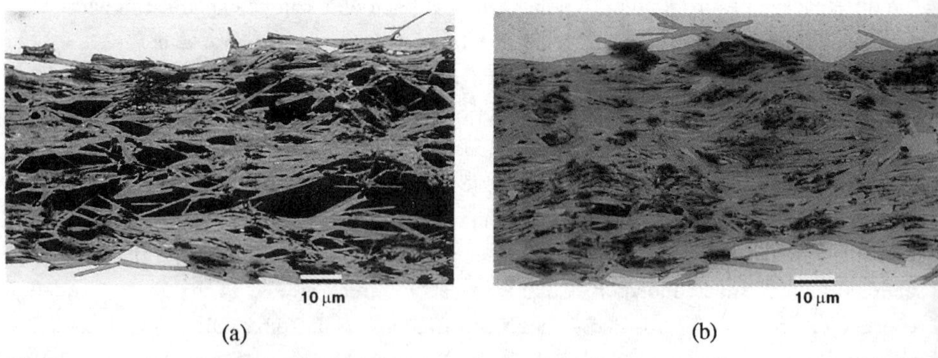

Figure 2   Backscattered electron SEM micrographs of two segments from wire B with $I_{c,segment}$ of (a) $\approx$16 A and (b) $\approx$20 A

wire A were similar to the differences observed between segments from wire A and wire B. The most striking difference in microstructure between wire A and wire B was the presence of regions of material which appeared to have decomposed. This is illustrated in Fig. 3 in which the microstructure of the segment of wire A with the highest I_c is compared to that of the lowest I_c segment of wire B. As seen in the micrographs, the segment from wire B exhibits a large fraction of these decomposed regions while the segment of wire A shows few if any of these regions. In comparisons between the two wires as well as from point to point within a given wire, it was found that the "exceptional" segments of wire A showed very few of these decomposed regions, the more typical segments from wire A and the exceptional segment from wire B showed some of these regions, and the typical segments from wire B exhibited large fractions of these regions. Thus, a correlation can be made with the presence of these decomposed regions and I_c.

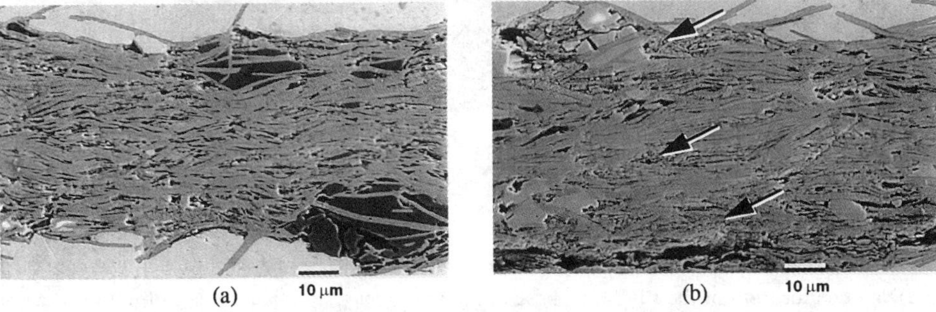

Figure 3   Secondary electron SEM micrographs of segments from (a) wire A with $I_{c,\,segment}$ $\approx$ 26 A and of (b) wire B with $I_{c,\,segment}$ $\approx$ 14 A. Some of the regions which appear to have undergone some decomposition are indicated by arrows in (b).

In order to more fully understand the nature of the decomposed regions, TEM was employed to examine the best and worst segments from each wire. Each wire was observed to consist of colonies of Bi-2223 grains with second phase particles interspersed between them. A typical colony structure is shown in Fig. 4 for a segment from wire A with the highest I_c. These TEM samples also served to confirm the effect of second phase volume fraction on I_c. As an artifact of TEM sample preparation, preferential ion milling of second phases tended to yield an enhanced view of the superconducting portions of the core. In the exceptional wires with little second phase, broad, continuous colonies were frequently observed. In contrast, for the wires with higher fractions of second phase, fewer, narrower colonies were observed. In general, the presence of second phase particles was not observed to have severely disrupted colony growth.

Examination of the wires by TEM did allow the presence of a Pb-3221 phase to be established. The Pb-3221 phase has been identified as a hexagonal crystal with a = 0.9919 nm, b = 0.3471 nm and a reported composition of $Pb_{3.4}Bi_{0.33}Sr_{2.6}Ca_{2.3}Cu_{1.0}$ [17]. An example of this phase is seen in Fig. 5 in which a large particle is present between two colonies. The identity of this

Figure 4   TEM bright-field micrograph of a typical colony structure observed in all of the wire segments.

phase was established by selected area electron diffraction patterns which can be indexed consistently according to this phase, as seen in Fig. 6, and by compositional analysis based on EDS. The location of the Pb-3221 phase did not suggest a significant role in limiting I_c. For example, there was no sign that this phase segregated to grain boundaries or resulted in an accumulation of impurities in those locations. While the Pb-3221 phase was frequently observed to grow into Bi-2223 grains from adjacent Ca_2PbO_4 grains, most of the samples exhibited many small, discrete particles. It is believed that the decomposed regions observed in SEM are associated with the presence of the Pb-3221 phase. For example, Fig. 7 shows a particularly large Pb-3221 grain which is clearly associated with the surrounding decomposed regions. In addition, TEM suggested a correlation between the volume fraction of Pb-3221 and I_c. The lowest I_c segment showed the highest volume fraction of Pb-3221 while the best segment showed virtually none of this phase. This is the same correlation with I_c noted for the fraction of decomposed regions observed by SEM. Based on TEM observations, it is not clear what mechanism leads to the appearance of the decomposed regions. As noted above, although TEM generally revealed many small Pb-3221 particles, there was no evidence of any severe decomposition reaction associated with these particles, as is clear from Fig. 5. It is believed that the appearance of the decomposed regions in SEM samples is due to a reaction during polishing. However, it is concluded that the reaction that leads to formation of the Pb-3221 phase indicates the onset of decomposition since this phase was most commonly observed in the wire annealed at 840°C while significantly less was found in the wire annealed at 835°C. The significant impact of these differences based on only slight variations in processing

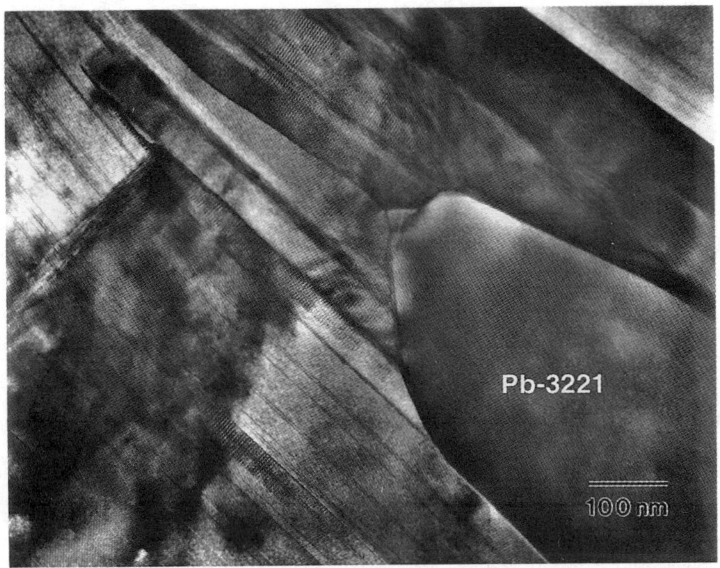

Figure 5   TEM bright-field micrograph showing a Pb-3221 particle within a Bi-2223 colony.

temperature underscores the importance of precise control of processing parameters during processing. This is a particularly critical issue for long length wires which pose a more difficult control problem due to the larger size of these samples.

The formation of the Pb-3221 phase did have an effect on the composition of the superconducting grains. EDS measurements were performed on approximately ten Bi-2223 grains from each sample examined by TEM in order to measure their composition. The grains were selected at random without regard to their proximity to Pb-3221 particles. The results of these measurements are summarized in Table 1 in which it can be seen that the average composition of the Bi-2223 grains in the best segment examined is significantly different from that measured for the other segments which showed progressively more of the Pb-3221 phase.

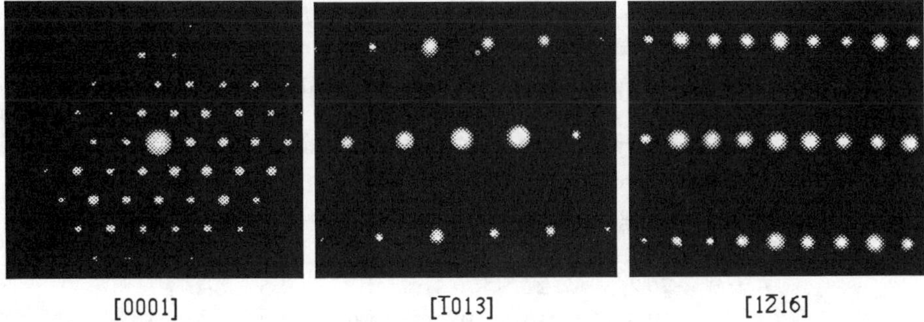

Figure 6   Selected area electron diffraction patterns from the Pb-3221 phase, indexed as indicated.

Figure 7   Secondary electron SEM image of a segment from wire A. A large grain of the Pb-3221 phase surrounded by decomposed material is visible near the center.

It is expected that a larger sample size would lead to differences in average composition between the other three wires as well. The variations in composition of the Bi-2223 grains between the segments are expected to lead to some differences in superconducting behavior from grain to grain. For example, changes in cation content in Bi-2212 have been shown to lead to significant differences in transition temperature.[18,19] In order to explore the potential effect of these compositional variations, resistance and magnetization measurements were performed as a function of temperature on the samples from each wire with the highest and lowest I_c values. The resistance data is shown in Fig. 8. The general behavior is that samples with lower I_c values show a broadened transition with a correspondingly lower $T_{c, R=0}$. However, the resistance data only yields information for the very best conduction path based on the measurement current. During an I_c measurements, a sample must carry the optimum current in *all* possible conduction paths to maximize the total current. The magnetization data provides a measure of the superconducting behavior of all the grains in the sample and thus gives a measure of the bulk behavior. Shown in Fig. 9, the magnetization data reflects an increasingly broadened transition as a function of decreasing I_c. The implications of such a broadened transition on I_c are significant in that it suggests that there is a wide variation in the

Table 1   Cation stoichiometry for the Bi-2223 grains in the highest and lowest I_c segments from wire A and wire B.

wire	$I_{c, segment}$	Bi	Pb	Sr	Ca	Cu
A	26.2	2.2	0.4	2.2	1.5	2.7
A	17.6	1.8	0.3	1.9	1.9	2.9
B	19.6	1.8	0.3	2.0	1.9	2.9
B	13.6	1.8	0.3	1.9	1.9	3.0

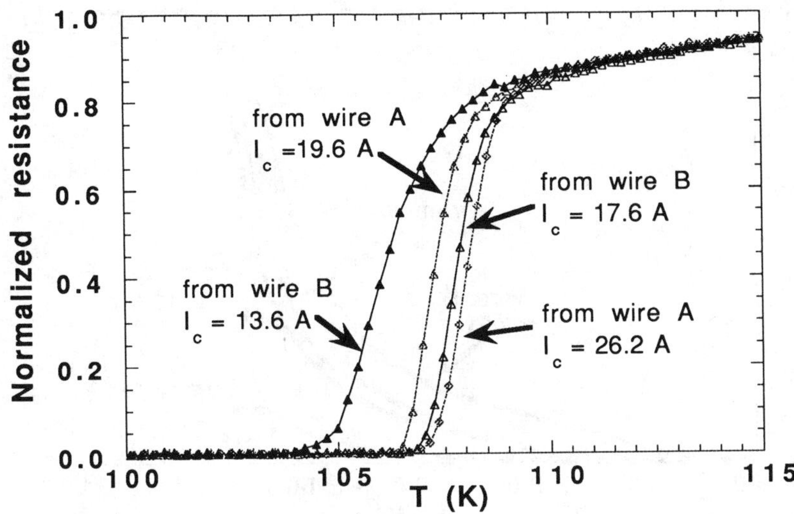

Figure 8   Normalized resistance versus temperature for the highest and lowest I_c segments from wire A and wire B.

conduction capacity of individual grains at a given temperature since the J_c of a sample tends to be proportional to the difference between the transition temperature and the measurement temperature. For example, in the sample from wire B with an I_c of 13.6 A, it is seen that there are a large fraction of grains which are not superconducting at the measurement temperature of 77 K and thus can carry no superconducting current. Similarly, there is a large number of grains which have transition temperatures only slightly above the measurement temperature and will carry a correspondingly lower current compared to other grains with transition temperatures which are much higher. In contrast a large fraction of grains will be far below T_c for the best segment of wire A during measurement at 77 K. It should be noted that these measurements have not been normalized based on the volume of the sample and therefore do not reflect the absolute volume of superconducting material, only the relative fraction of the total superconducting volume. Thus, differences in superconducting volume may also play a role and can be used to account for the smaller variations in I_c between wires which show an intermediate transition.

The relationship between the volume fraction and composition of secondary phases which lead to changes in the Bi-2223 grains suggests a change in phase equilibrium as a function of processing conditions. Variations in these features likely result from the interrelated effects of composition and annealing temperature. As noted previously, small variations in composition can be seen from point to point along the length of a wire. It is expected that changes in phase relations may result from these variations alone but such an effect may be enhanced at higher processing temperatures for which the compositional range of the superconducting phase tends

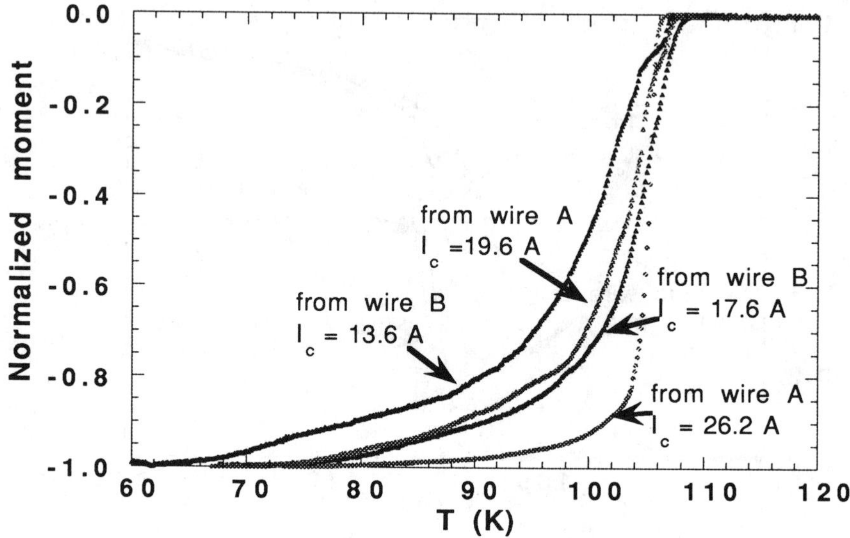

Figure 9   Normalized magnetic moment versus temperature for the highest and lowest I_c segments from wire A and wire B.

to decrease. Thus, all the segments of wire B, processed at 840°C, were found to contain the Pb-3221 phase, although the volume fraction of that phase varied from segment to segment, similar to the variations observed in the fraction of alkaline earth cuprates. In contrast, only the low I_c segments from wire A, processed at 835°C, showed a significant fraction of this phase. At this lower processing temperature, only the more extreme compositional fluctuations may have led to the formation of Pb-3221, leaving most of the segments relatively devoid of this phase.

Conclusions

One microstructural basis for variations in I_c from point to point along the length of lead-doped Bi-2223 wires has been found to be related to the volume fraction of second phases as well as the specific reactions which lead to the formation of those impurities. Specifically, the presence of a Pb-3221 phase has been found to be a key marker which indicates a shift in equilibrium. The reaction which results in the formation of the Pb-3221 phase also produces a change in composition of the Bi-2223 grains which results in degraded superconducting properties of these grains and therefore limits I_c in the segment. The significant variations noted in microstructure and properties as a result of only small differences in processing temperature between the various samples examined in this work underscores the importance of precise control of processing conditions, particularly for long length wires.

References

1   see, for example, *IEEE Trans. Appl. Superconductivity* 3 (1) Part II: Superconducting Materials (1993).
2   Q. Li, et al., "Thermomechanical processes of Ag-sheathed Bi-2223 tapes" (Paper H12.26 presented at the 1992 Fall MRS meeting, Boston, MA, 3 December 1992).
3   K. Sato, et al., *IEEE Trans.* MAG-27 (1991) 1231.
4   G. Riley, Jr., W.L. Carter, *Am. Cer. Soc. Bull.* 72 [7] (1993) 91.
5   E.E. Hellstrom, *MRS Bulletin* 17 (8) 45 (1992).
6   T. Kitamura, T. Hasegawa, H. Ogiwara, *IEEE Trans. Appl. Superconductivity* 3 (1) (1993) 939.
7   H. Mukai, et al., "Multifilamentary Bismuth (2223) Multilayer-wound Conductors for Power Transmission Lines" (Paper presented at the 1992 Spring MRS meeting, San Francisco, CA, April 27-May1, 1992).
8   B. Hensel, et al. *Physica C 205* (1993) 329.
9   L.N. Bulaevskii, et al., *Phys. Rev. B* 45 [5] (1992) 2545.
10 N. Enomoto, et al., *Jpn. J. Appl. Phys.* 29 (1990) L447.
11 Y. Feng, et al., *Physica C* 192 (1992) 293.
12 L.R. Motowidlo, et al. *Appl. Phys. Lett.* 59 [6] (1991) 736.
13 J.S. Luo, et al., manuscript submitted to *J. Mat. Res.* (1993).
14 A. Umezawa, et al., *Physica C* 198 (1992) 261.
15 T. Kato, et al., *MRS Bulletin* 17 [8] (1992) 52.
16 Y. Yamada, B. Obst, R. Flükiger, *Supercond. Sci. Tech.* 4 (1991) 165.
17 S.X. Dou, et al., *Supercond. Sci. Tech.* 4 (1991) 203.
18 T.G. Holesinger, et al., *Physica C* 202 (1992) 109.
19 P. Majewski, H-L. Su, B. Hettich, *Advanced Materials* .

ANALYTICAL REPRESENTATIONS OF DEFORMATION TEXTURE GRADIENTS IN CERAMIC SUPERCONDUCTORS

*Krishna Rajan *, Debes Bhattacharyya† and Wei Gao†§*

**Rensselaer Polytechnic Institute
Troy,N.Y. USA &
†University of Auckland
Auckland , New Zealand*

Abstract

A new approach to represent crystallographic texture associated with deformation of ceramic superconductors is presented. This approach differs from the classical Eulerian projection in a number of ways, the most notable being the representation of misorientation data by a single three dimensional vector known as the Rodrigues-Frank vector. In this paper we outline a procedure to derive orientation distribution functions in Rodrigues-Frank space. This is applied as a simulation technique to assess *local* crystallographic misalignments associated with deformation induced textures in bulk ceramic superconductors which are composed of grains macroscopically aligned on *average* along the c-axis. The application of these simulation studies to continuum mechanics descriptions of deformation processing of long lengths of superconducting ceramics is also discussed.

*   *Department of Materials Engineering*
†   *Department of Mechanical Engineering*
†§ *Department of Chemical and Materials Engineering*

Processing of Long Lengths of Superconductors
Edited by U. Balachandran, E.W. Collings and A. Goyal
The Minerals, Metals & Materials Society, 1994

Introduction

The deformation processing of long lengths of ceramic superconductors poses a significant challenge in developing controlled microstructures with well defined crystallographic features. It is widely accepted that important correlations exist between the macroscopic orientation or texture of a polycrystalline aggregate in these materials and the current carrying capabilities of high temperature superconductors. An important challenge in developing correlations between critical current behavior (J_c) and the macroscopic texture of long lengths of superconductors is to be able to capture the dimensional length scale of both the microstructure and that of the "weak links" controlling J_c behavior. A first step in making these correlations is to be able to assess the local variations in texture associated with grain *specific* data that lead to the macroscopic texture. In this paper we will outline an approach to characterize texture such that the local variations in grain orientations can be related to the macrotexture in a statistically meaningful way. We shall utilize this approach to simulate possible variations in local texture associated with variations in J_c as measured in experimental studies on $(BiPb)_2Sr_2Ca_2Cu_3O_y$ subjected to deformation processing.

Theoretical Background

A central theme to our study is the utilization of the so called Rodrigues-Frank space orientation mappings. Apart from representing grain orientation and grain boundary misorientation using the usual stereographic projection analysis, it is proposed use a crystallographic representation originally proposed nearly a century ago by the mathematician Rodrigues and recently popularized by Frank. Unlike conventional two dimensional crystallographic representations of grain orientation, Rodrigues-Frank representations (henceforth referred to as R-F space) involve a three dimensional framework where vectors represent axis of misorientation between grains rather than simply the normal vector to a single grain and the magnitude of the vector represents the magnitude of the angular misorientation . The use of R-F space analysis is a very useful and powerful means of representing texture variations in a given view of microstructure, as has been demonstrated by recent studies [1-3].

In R-F representations, unlike conventional stereographic representations of texture, *single* vectors represent *misorientations* between two grains. The axis of this misorientation vector represents the equivalent axis of rotation between two different crystal orientations and the magnitude of this R-F vector represents the magnitude of this rotation, such as to achieve a coincidence in the lattice frame of reference. Unlike the two dimensional stereographic projections where the vectors of crystallographic plane normals are represented, R-F vectors are projected in a three dimensional format. The exact shape of this format, or "fundamental zone" is dependent on the crystal symmetry and is derived by the use of what is termed "quaternion theory". It is beyond the scope of this article to review the group theoretical derivations of this concept, but suffice it to say that for cubic symmetry for example, the fundamental zone is that of a cube with truncated corners. The three dimensional fundamental zone may be considered as analogous to the two dimensional stereographic projection.

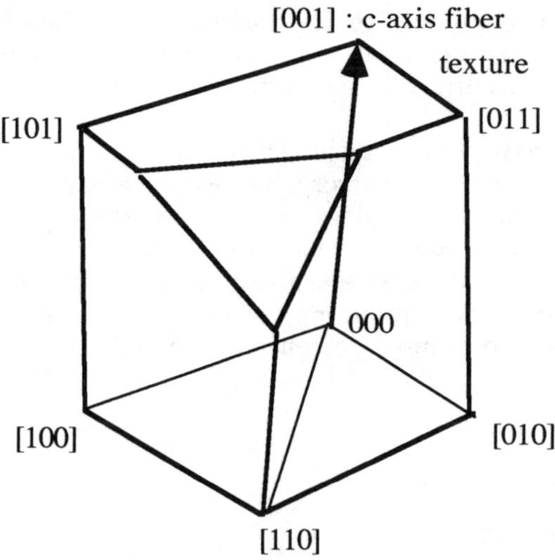

Figure 1 - A schematic of a portion of the fundamental zone in Rodrigues-Frank space for cubic symmetry. The R-F vector is defined in this fundamental zone as:

$$\vec{R} = \vec{l}\tan\frac{\theta}{2} \; ; \text{ where } \vec{l} = \text{ misorientation axis and } \theta = \text{ rotation}$$

angle.

Simulation of Texture Gradients

In this section we shall outline a procedure for calculating the orientation distribution functions (ODFs) in the framework of the Rodrigues-Frank space. The calculation of an ODF permits a quantitative assessment of the preferential variations in local crystal orientation. The R-F space map reveals orientations simultaneously using 3-D space, as do Euler maps. The major advantage with R-F space is the rectilinear geometry that is conducive to the visualization of texture components in terms of rotation. Another advantage of using R-F space is that the space is virtually "homochoric", which means that there will be uniform distribution of points in the fundamental zone for a material with no preferred orientation. This property of spatial uniformity is convenient from a computational standpoint, since the orientation density (points/orientation space) between subdivisions of the fundamental zone may be compared directly when assessing the texture components. This facilitates the calculation of the ODF purely based on the frequency of occurrence of points in orientation space.

For the purposes of this simulation, we are assuming that the BSSCO superconductor has a macroscopic texture aligned along the c axis (i.e.. [001]). When viewed along the c axis, the crystal symmetry may be considered to be pseudo-cubic as $a \sim b$. Thus for the purposes of this simulation study, the R-F fundamental zone associated with cubic symmetry will be used. The question we wish to focus on is what will be the distribution of local misalignments within the a/b plane which will lead to a given macroscopic texture (Figure 2).

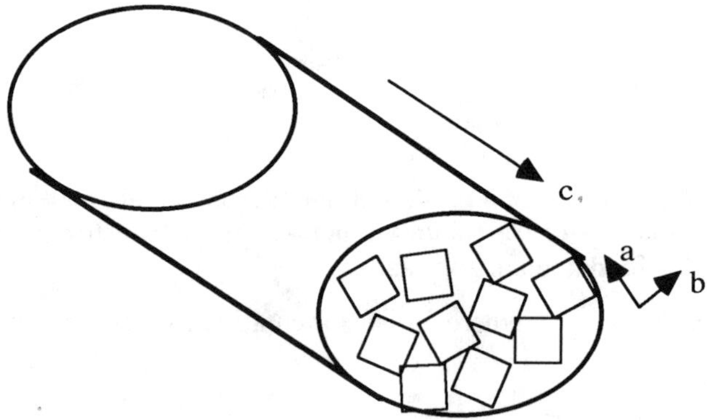

Figure 2 - Schematic of polycrystalline aggregate orientations

The basic principles involved in establishing the determination of an ODF in R-F space are as follows:

i) R-F polyhedron is subdivided into a set of points evenly distributed to simulate total random texture. For the purposes of this computation, we used 8897 points which is feasible with the capabilities of personal computer. Each point is represented by a Dirac δ function.

ii) Each of these points in the R-F fundamental zone for random texture is described by a "screening potential"(S) ,which effectively defines a "sphere of influence" of a given texture component in R-F space. This gives a texture or sampling strength component for each point. This screening behavior is defined by a Gaussian type function.

iii) An analytic form of the ODF is described by a three dimensional convolution of the sampling function (S) over the input misorientation data (for N "experimental" data or simulated points) described by the summation of the Dirac δ function:

$$\Re = \iiint S \sum_{1}^{N} \delta\, dxdydz$$

iv) The convolution of the sampling function over the fundamental zone results in a set of values which is effectively the probability distribution for the misorientation space. The ODF ,$\Re$, is then expressed as a value which compares by ratio, the value of the probability density of the actual (or in this case simulated) data with the value expected at random.

v)   The actual simulation involves inputting a number of orientations and the corresponding misorientations of neighboring grains.   The ODF calculation then simply involves the value of number density of the input data on misorientations. A separate algorithm is written for generating the misorientation data needed to be used for the R-F , ODF calculations.

vi) It should be pointed out that all the above calculations account not only for the crystal structure symmetry but also the sample geometry. For the present study the long axis of the sample (as would be applicable to wire or tape forms for long lengths of superconductors) was aligned with respect to the R-F fundamental zone as to make the use of cubic crystal symmetry applicable to the pseudo-cubic nature of the a/b plane.

Below is a 2-dimensional cut along the z- axis of the ODF contours in the R-F polyhedra for c - axis texture. For the purposes of the simulation a variation of approximately $10°$ was allowed to accommodate possible misorientations within the a/b plane.

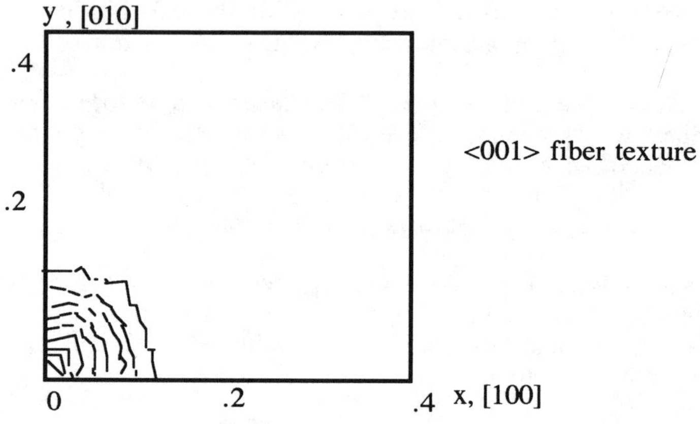

Figure 3 - ODF contours in R-F space for simulated c - axis texturing accommodating slight grain misorientations within the a/b plane

We can also simulate the degradation of this macroscopic texture with the appearance of other texture components (say for example [111]) as shown below in Figure 4.

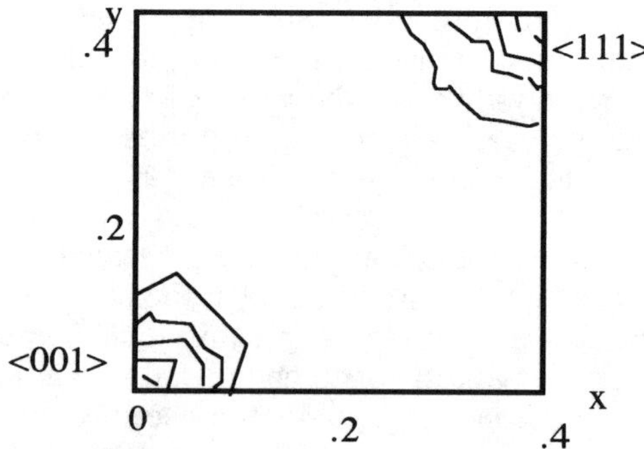

Figure 4 - ODF showing appearance of <111> texture components

Discussion

The most significant aspect of the calculations is that we do not need to input a large amount of grain orientation data. In fact for the purposes of this study we used less than one hundred grains. Conventional analysis of orientation measurements, on the other hand, involves thousands of grains in order to discern significant changes in the orientation distribution functions. The application of R-F polyhedra representations however provides a means of effectively condensing the misorientation data allowing for discerning qualitative changes in texture components more easily. The R-F space analysis also is very sensitive to grain specific orientation information. Hence the application of experimental techniques which allow for such information to be detected , for example Electron Backscattered Patterns in the scanning electron microscope or Kikuchi lines in the transmission electron microscope are very useful experimental techniques from which data can be processed in the manner outlined here in these simulation studies.

The present simulations can also be used to provide an insight into the relative contributions of individual grains to achieving an overall macroscopic texture. For instance *Wei and Vander Sande* [4] have shown that significant changes in the J_c values for BSCCO superconductor tapes occur for varying degrees of macroscopic texture as measured by X-ray diffraction. Figure 5 shows a schematic diagram summarizing their earlier observations.

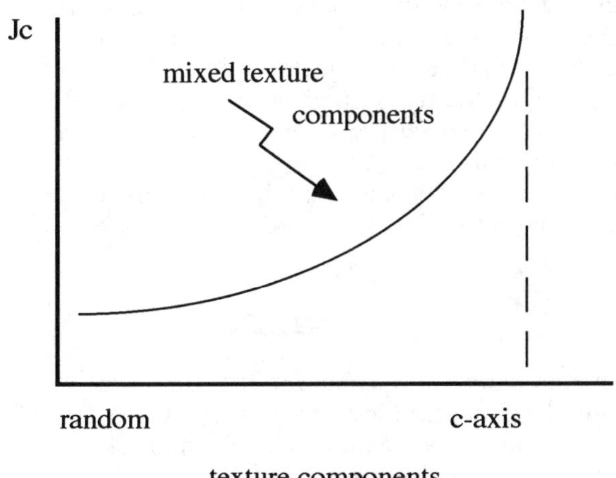

Figure 5 - Variations in J_c associated with deformation texture in BSCCO

Our simulations in Figures 3 and 4 indicate the possible variations in the ODF components (and hence the corresponding J_c values) associated with the strong c - axis texture and mixed texture regimes respectively as indicated in Figure 5. As noted earlier, these simulation studies are applicable not only to variations in texture associated with critical current variations but also to local variations in microtexture associated with microstructure.

The identification of the textures is also important for optimising the deformation processes or parameters through a correlation of desired textures and the nature of the deformation. As has been pointed out by *Wright et.al.* [5] the deformation processing can develop or enhance textures; however, an understanding of the interrelationship is desirable to achieve the necessary deformation effects.

The success of any mathematical modeling of a metal forming process warrants a proper constitutive relationship depicting the material property within the deforming zone. It is also well known [6,7] that the properties may vary within the deformation zone and the extent of modeling error would depend on the ability of the constitutive relationship to take account of the property variation [8]. It is envisaged that with the increasing use of superconductive materials, there will be a growing need to fabricate these materials using conventional and non-conventional metal forming processes [9,10]. At present for superconductive materials, there is , however, little understanding of how the deformation modes/characteristics influence the eventual microstructure and other related properties of the products. Nevertheless to produce superconductors with optimized electronic transport properties, it is critically important to understand the deformation mechanism and the effects of processing parameters on the texture structure. The present study raises the possibility of developing a semi-empirical method of predicting textures and hence improving the constitutive relationships of the deforming materials.

Conclusions

We have in this paper proposed for the first time the application of Rodrigues-Frank representations to analyze grain specific information associated with macroscopic texture in long lengths of superconductors. The development of a simulation procedure to calculate in orientation distribution functions in R-F space is also presented. It is shown that one of the primary advantages of this analytical approach is the ability to utilize relatively small sampling volumes in order to discern gradients in texture in superconductors. Experimental techniques for microtexture characterization to supplement these analytical descriptions is also discussed.

References

1. F.C.Frank, "Orientation Mapping", <u>Eighth Intern. Conf. on Textures of Materials</u>, ed. J.S.Kallend and G.Gottstein (Warrendale,PA: The Metals, Materials and Minerals Society,1988) 3-13.

2. R.Becker and S.Panchandeeswaran, "Crystal Rotations Represented as Rodrigues Vectors", <u>Textures and Microstructures</u>, **10** (1989), 167-194.

3. R.N.Petkie, K.N.Tu and K.Rajan, "Crystallographic Evolution of Thin Film Microstructures: II- Grain Boundary Structure", <u>Journal of Electronic Materials</u> (in press).

4. W.Gao and J.B.Vander Sande, "Textured BSCCO/Ag Superconducting Microcomposites with Improved Critical Current Density Through Mechanical Deformation" , <u>Supercond.Sci Technol.</u>, **5** (1992), 318-326.

5. R.N.Wright, R.H.Doremus, R.M.German, D.B.Knorr, R.K.MacCrone and K.Rajan, "Deformation Processing of High T_c Superconducting Wire" ,<u>Processing and Applications of Superconductors</u>, ed.W.E.Mayo, (Warrendale,PA.: TMS,1988) ,139.

6. B.Avitzur, <u>Handbook of Metal Forming Processess</u>, Wiley-Interscience,N.Y. (1983).

7. E.M.Mielnik, <u>Metalworking Science and Engineering</u>, McGraw-Hill,N.Y. (1991).

8. D.Bhattacharyya, P.J.Richards and A.A. Somashekar, "Modelling of Metal Extrusion Using the PHOENICS Package" , <u>J.Matls. Processing Technol.</u>, **35** (1992) 93.

9. W.Gao and J.B.Vander Sande, "Synthesis of High-T_c Superconducting Materials by Oxidation and Press Coating of Metallic Precursor Alloys", <u>Materials Letters</u> **10** (1991) 444.

10. S.K.Samanta, 'Hydrostatic Extrusion of High T_c Superconducting Ceramics" (Paper presented at the NSF Grantees Meeting on High Temperature Superconducting Materials, Washington D.C. 26 April 1988).

Acknowledgements

The authors wish to acknowledge R.Petkie for his assistance in developing the computer programs used in this study.

LOW-TEMPERATURE SUPERCONDUCTORS

SOME EXPERIENCES WITH THE FACTORS AFFECTING THE PRODUCTION OF

LONG LENGTHS OF NIOBIUM-BASED SUPERCONDUCTORS

Eric Gregory

IGC Advanced Superconductors Inc.
1875 Thomaston Avenue,
Waterbury, CT 06704.

Abstract

A problem which has thwarted the development of practical superconductors throughout the last three decades has been the production of these materials in long lengths with consistent properties throughout. Each time a new material or new configuration has been developed the problem reoccurs and attempts are made to apply similar solutions. The paper traces the history of these attempts from the early work on NbZr and Nb_3Sn monofilaments through multifilamentary NbTi-MRI materials to complete conductor designs for accelerator magnets. Some factors which are believed to influence the "piece length" are discussed as is its importance on the economics of practical superconductor manufacture.

Processing of Long Lengths of Superconductors
Edited by U. Balachandran, E.W. Collings and A. Goyal
The Minerals, Metals & Materials Society, 1994

<u>Introduction</u>

It is obvious to anyone who has attempted to draw wire of any design or composition, that the most essential set of circumstances for economical production is that the wire does not break. Whether the drawing machines are multidie, highly automated or simply single pass, setting them up is a much more time consuming and manpower intensive operation that the actual drawing operation itself.

In superconducting wires the problem of wire breakage is particularly important as raw material costs are high and wires cannot be joined by spot welding without markedly degrading the mechanical and electrical properties of the product.

For economic and performance reasons, the tendency is to specify higher and higher current densities, (J_c's), at constantly increasing fields. As the J_c increases, fabrication problems become more serious, often resulting in wire breakage.

The belief exists, not without some justification, that, this breakage during manufacture, is an indication of inherent defects. It is difficult to check the integrity of the wire along its length until a coil is fabricated from it, and so, in an attempt to protect himself, the magnet designer frequently specifies long piece lengths. He has been burnt in the past. Excellent data has been presented on short lengths of material made under some idealized conditions. The designer has used this data to propose a device, only to find that, when he went out to order large quantities of the material in usable lengths, it was either not available , exceedingly costly or its delivery subject to intolerable delays.

To meet constantly increasing requirements, superconductors have become more complex and consequently more difficult to fabricate. At each stage in the development, feasibility has first been proved on short lengths. After this, considerable time and effort has been devoted to the production of long lengths of reliable and reproducible material. It is hoped that by recounting some of these past experiences and describing how the problems were overcome, information will be forthcoming that will help in meeting the challenges of the future applications.

<u>Monofilamenttary conductors</u>

<u>Uniformity, workability and ductility of the superconductor itself.</u>

In 1961 the first "high field" [8T] laboratory superconducting magnet was made [1] and this is generally considered to be the start of the commercial manufacture of superconducting wires and tapes. The conductor used in these early magnets was monofilamentary and made by assembling pressed powder pellets of Nb and Sn in a Nb tube sheathed by a monel tube, the idea being to maintain the ductility of Nb and Sn down to the final size and then to react the two materials to form the brittle intermetallic. While no significant piece length problems were encountered in the manufacture of these early lengths, the requirements were not very highly developed and the coils were very small. The statement made by Kunzler in 1986 [2] when reviewing the 1961 activity, was that "if there were no breaks" a coil having between 10 and 20 thousand feet of wire resulted. He did not elaborate on how frequently the material broke but, since it was monofilamentary and all the constituents were relatively ductile, breakage may have been infrequent.

It was realized early in the development of ductile superconducting materials, that low work hardening characteristics of the basic superconductor were desirable. It is primarily for this reason that Nb 46.5 wt. %Ti has become established as the most widely used alloy and why it displaced NbZr, which was the first ductile alloy used as a commercial superconductor.

The uniformity of the NbTi alloy has been a problem that has existed over the years and, while it has been improved recently, problems connected with it are still experienced on occasions. One of the manufacturers made a special grade of high homogeneity material and introduced radiographic inspection techniques to ensure that the specifications were met. Despite this, hard areas are sometimes encountered, as are inclusions and high niobium areas. These problems are exceedingly difficult to detect until fracture actually occurs.

Introduction of the stabilizer into monofilaments.

It was discovered very early in the manufacture of superconducting coils that a Cu stabilizer was necessary if reliable performance was to be achieved [3]. Initially for the fine wires of NbZr , a thin layer of copper was plated onto the wire. Even though the wire was drawn to very fine sizes in some cases, no particular piece length problems were reported. The amount of copper in the strands was often quite small but if larger amounts were required, they were supplied by cabling with pure Cu strands [4].

As copper plating lost its popularity due to poor adhesion and thus poor thermal conduction across the interface, copper cladding by various metallurgical processes took its place. The first method was simply to take a large diameter NbTi rod, clean it and insert it into a cleaned copper tube and cold draw the two materials together after simply swaging the front end. While monofilaments are not now used extensively as such, they are assembled into multifilamentary arrays and on these occasions, a coextrusion process is used to develop them. The precautions that have to be employed are outlined below.

Multifilamentary stabilized conductors.

The early approach for making conductors capable of carrying larger currents, was to cable them from these small diameter monofilaments [4]. It soon became evident however, that cabled conductors were less desirable for large magnets than multifilamentary laminates and so, since then, multifilamentary conductors in copper matrices have been used almost universally. To complicate matters further, shortly after the introduction of multifilamentary material, it was determined that twisting was necessary to reduce losses [5]. This exaggerated the piece length problems, as the twisting operation put a stress on poorly bonded material. Most of the conductors made today are twisted multifilamentary designs and these will be the topic of the rest of the discussion.

Design Aspects

Three different approaches are used primarily for the production of multifilamentary NbTi conductors, depending principally on the number of filaments required.

These are shown schematically in Figure 1. They consist of 1. Drilled billets, 2. Rod and tube assemblies and 3. Clad monofilamentary assemblies. Generally long piece lengths are easier to maintain the fewer the number of filaments and the larger their diameter.

Magnetic Resonance Imaging (MRI)

For MRI applications, the drilled billet approach is frequently used as the number of filaments is small, their diameter large and a high percentage of copper often employed Since the number of surfaces is relatively small and the copper is essentially monolithic, bonding, the lack of which is a cause of breakage, is not as big a problem as it is in the case of fine filamentary materials.

It is still important that the extrusion billet is designed correctly, made out of forged copper parts, evacuated and welded accurately Any leakage of air into the billet prior to or during extrusion can lead to gaseous contamination and piece length problems. A good check on the integrity of the billet is to hot isostatically press, (HIP), it, before extrusion. Shrinkage during this HIP'ing operation is an indication of its vacuum tightness. HIP'ing is, however, a costly operation and manufacturers tend to try to avoid it wherever possible.

While MRI conductors are not the most difficult to produce in long lengths, their application usually demands specific piece lengths. This is because these conductors operate in a persistent mode. Any joints must be resistanceless and carefully located in the magnet in order to minimize decay effects. This requirement for specific piece lengths is, perhaps, the most demanding from a manufacturing cost point of view. It permits very few breaks indeed, accurate yield predictions and carefully designed billets. It demands excellent engineering, flawless manufacturing and good quality assurance. It is also important that the design is such as to minimize "sausaging" i.e. uneven reduction along the length. It is desirable to arrange the array so that it is as uniform and circular as possible and that no filaments are left in unsupported locations, see Figures 2a & b This is particularly important in drilled billets as it is not possible for the gun driller to drill holes very close to

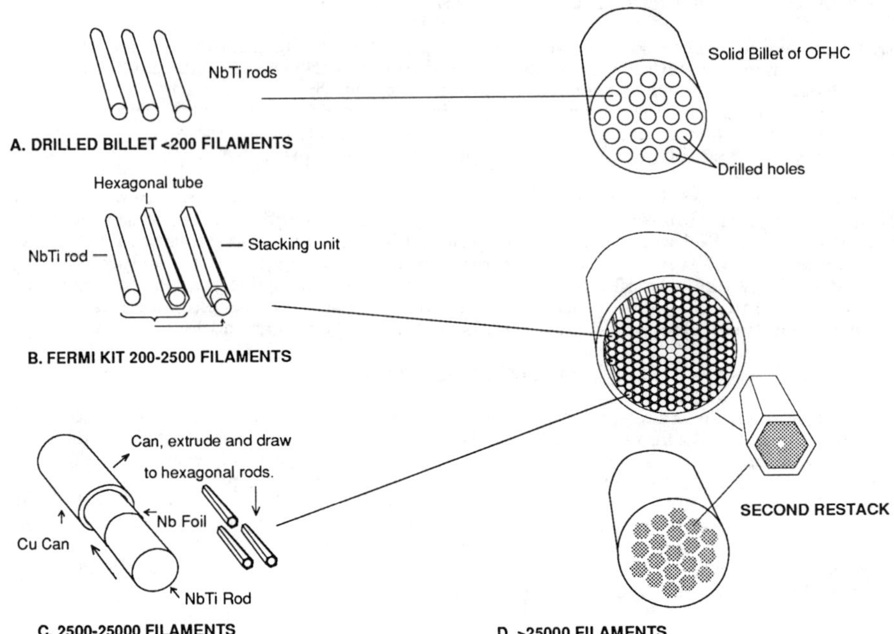

Figure 1. Schematic representation of various methods of making multifilamentary superconductors

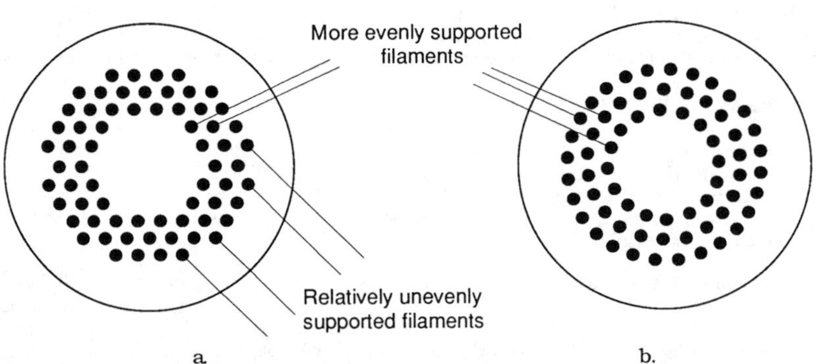

Figures 2a & b. Schematic representation of a, poorly designed drilled billet, and, b. an improved design giving a more uniformly supported filament array.

one another in relatively long billets. The spacing between the filaments is therefore relatively high and, this causes filament "sausaging".

<u>Fermilab accelerator magnet conductor</u>

This was a multifilamentary conductor, containing ~ 2000 filaments and one of the first made in large quantity by the rod and tube approach, shown in Figure 1. The idea of making a monofilament by drawing copper onto a NbTi rod and cutting the rod, discussed above, seemed to be a cumbersome and wasteful process. The rod and tube approach appeared to be superior, it eliminated one step at the wire manufacturer, as Nb rod was purchased at the size required for the multifilamentary assembly. The piece length problems that resulted from the approach, were primarily ones resulting from inclusions and problems of cleaning both rods and tubes. The subcontractor was not aware of the seriousness of the surface condition, steel inclusions were frequently found in the tubes and abrasive particles on the surfaces of the rods, see below.

Part of the piece length problems encountered in these rod and tube materials was due to the "upsetting" of the billets during the extrusion process. This was caused by the fact that, as the number units increased and little attention was paid to tolerances, extent of etching of the components and the extrusion liner condition, a large volume of unoccupied space had to be filled before extrusion took place. This caused the filaments to buckle before extrusion, producing areas with greatly varying filament cross sections. This was a condition that frequently led to wire breakage. More careful control of the extrusion liner condition, all the tolerances of the components, and the introduction of cold isostatic pressing (CIP'ing) and remachining of the billets before extrusion, tended to reduce upsetting and its deleterious effects. Bonding was believed to take place during extrusion provided the conditions were correctly determined. This was subsequently shown to be an over simplification of the true state of affairs.

In addition the critical current densities, (J_c's) and "n" values of the materials were far from ideal. The "n" value is defined by the equation

$$\rho = \text{constant} \times I^n \qquad\qquad (1)$$

where ρ is resistivity and I is the current determined in the sample, and "n" is frequently regarded as a measure of the quality of the wire , i.e. "sausaged" wires have low "n" values. Due to the fact that, in an effort to increase J_c, the number of heat treatments had been increased as had the duration of each, a significant amount of interaction occurred between the Ti in the filaments and the Cu matrix. Figure 3a shows an extracted filament exhibiting the interaction and Figure 3b shows an electron microprobe trace across the the filament - matrix interface. It was originally thought that this interaction occurred primarily during extrusion, but it was later found to occur mainly during the multiple heat treatments required to develop J_c [6].

At this stage specified piece lengths were around 0.3 km. and the J_c's were between 1800 A/mm^2 and 2000 A/mm^2 at 5T. The values for the conductors for the Isabelle accelerator, {a machine which was in the past projected to be built at the Brookhaven National Laboratory (BNL)}, were better than those for the Fermilab conductor. The reason for this was that, although the Isabelle wire was smaller in diameter, the filaments were not but the spacing to diameter ratio, (s/d), was also smaller. The significance of this on J_c and piece length will be explained later.

<u>Conductor for the Superconducting Supercollider (SSC) dipoles and quadrupoles.</u>

The SSC required smaller filaments than Fermilab, at least initially, and much higher J_c's. When an attempt was made to develop this combination of properties, piece length problems became even more acute. The following illustrates the magnitude of the piece length problem: In the early work on the SSC conductor the design was essentially established and the desired electrical and thermal properties proved in one year's time [7], because of the experience gained from the manufacture of the other fine filamentary NbTi conductors. It took another 3 to 4 years to produce this material reproducibly and reliably in the long lengths required for the machine [8]. Much of the problem was in achieving piece length <u>in conjunction with the other properties already established in short lengths.</u> It is important to point out that, when the development work started on the SSC conductor we had 20 years of experience with the basic material which is a highly ductile alloy with low work hardening characteristics. i.e. few of the ductility problems were due to poor ductility in the superconductor itself.

Several possible factors leading to these ductility problems are given below: one is filament <u>"sausaging"</u> which, in the extreme, can lead to strand breakage, a second one is inclusions but a more serious one is lack of <u>bonding</u>.

<u>"Sausaging"</u> In order to develop high J_c' s in NbTi, it is necessary to introduce a large amount of cold work and give multiple heat treatments as mentioned above. These

327

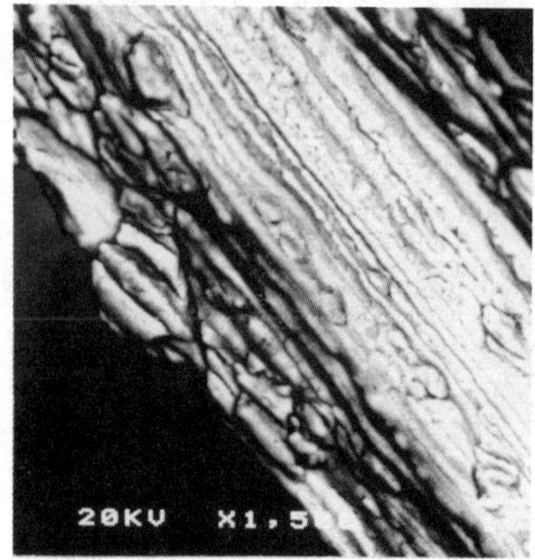

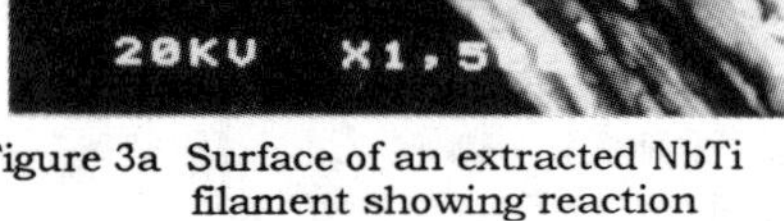

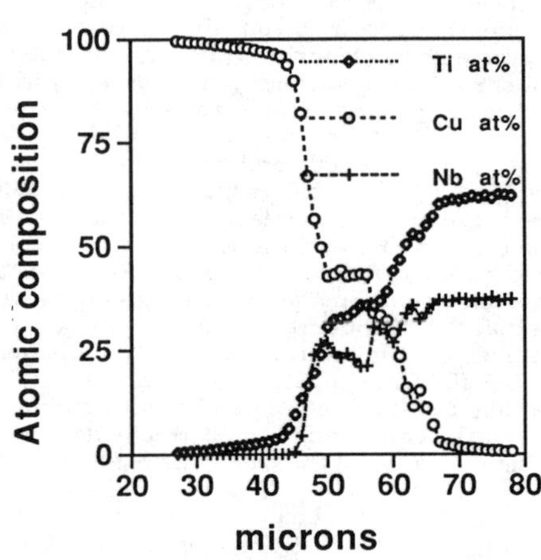

Figure 3a Surface of an extracted NbTi filament showing reaction between Ti and Cu.

Figure 3b Electron Microprobe trace across reacted interface.

operations cause the precipitation of α-Ti particles which pin magnetic flux and thus give the high J_c's, [9]. Unfortunately the α–Ti also pins dislocations, vacancies, etc. and causes strengthening and hardening of the material. At the same time the heat treatments soften the copper and this difference in strength can lead to "sausaging" of filaments. There are various ways to reduce this wire "sausaging", four of which are mentioned below:

1. Try to maintain as good a match of mechanical properties between the filaments and the matrix as possible. This may mean changing the matrix around the filaments from Cu to a Cu-based alloy.

2. Space the filaments as closely as possible in order that they can support one another. "Sausaging" is very sensitive to the s/d. Figure 4. This Figure shows area analysis plots of two conductor designs with different s/d's, 0.15 (standard SSC) and 0.06.

3. Improve the uniformity of the filament arrays in an SSC type billet by stacking as many small diameter rods as possible into the first restack and avoiding double restacking, Figures 5a & b.

4. Reduce any interaction between the NbTi filaments and the copper matrix as mentioned above. There are two ways of reducing such interaction.

a. Introduction of a barrier of a material such as Nb on the surface of the filament [7] .

b. Introduction of a material such as silicon into the matrix between the filaments which will concentrate in the interface during the processing and heat treatment [10],[11].

Figs. 6 a & b, show the surfaces of filaments after these two approaches have been used. These should be compared with an unprotected Cu-NbTi interface shown in Figure 3a..

All the above efforts to reduce "sausaging" and compound formation, certainly improve the J_c's obtained in these multifilamentary materials. Wire breakage which can be due to extreme cases of these two phenomena, is also reduced. It is important to realize, however, that, under certain conditions and designs of billets, it is possible to draw the wire to finished size and still have all the filaments broken!! In one instance NbTi rods were ground with an Al_2O_3 abrasive and this was not removed before the rod was supplied for assembly in the rod and tube approach. The resulting material was drawn down to finished wire size without any wire breakage. The resulting product was then etched to dissolve the Cu , a

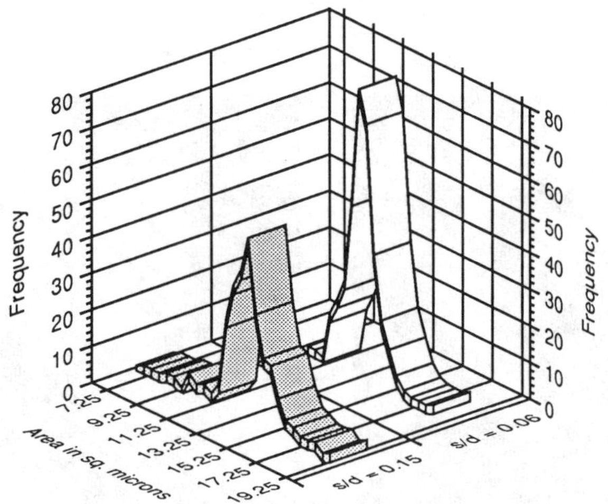

Figure 4. Histograms of filament area distributions in 0.15 & 0.06 s/d Cu 0.5 wt. % Mn matrix materials.

Figure 5a A cross section of a doubly restacked
type of array.

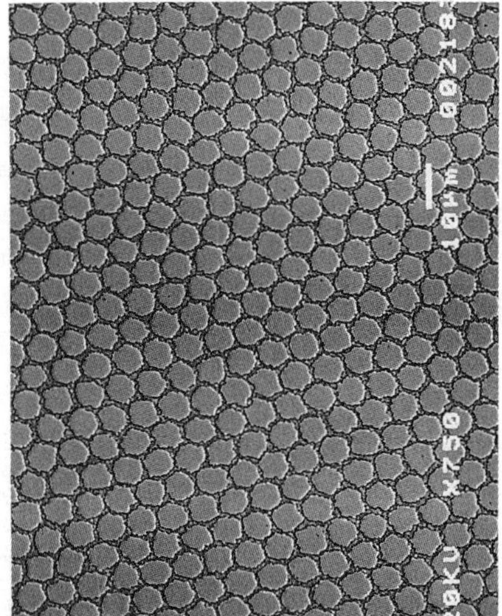

Figure 5b. A cross section of an SSC
single restack

procedure frequently used to enable the determination of the Cu / superconductor ratio. The filaments were revealed to be in short lengths, each broken by micron sized Al_2O_3 particles. This shows that filament breakage may or may not be evidenced by wire breakage. This depends on a number of factors such as matrix integrity (this material was a drilled billet), quantity of superconductor relative to that of the matrix. In the above example, the Cu / superconductor ratio was relatively high, 7/1.

In the development of the SSC conductor, even when all the above precautions were taken, it was still difficult to meet the piece length specifications (85% of the material > 1.5 km) The reason is believed to be due to a combination of factors. Inclusions and bonding between the various surfaces in both the monofilament and the multifilamentary billets,

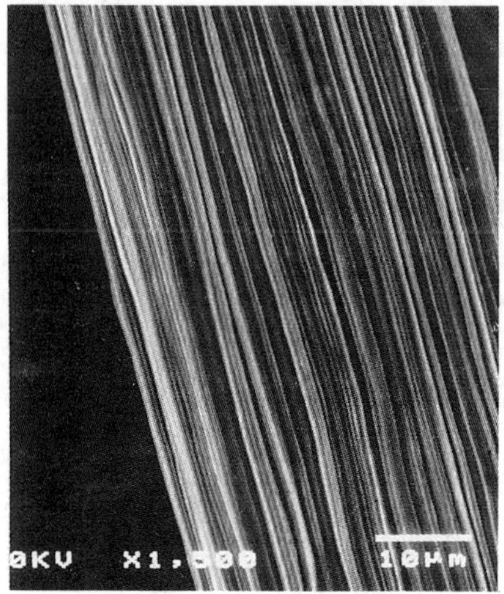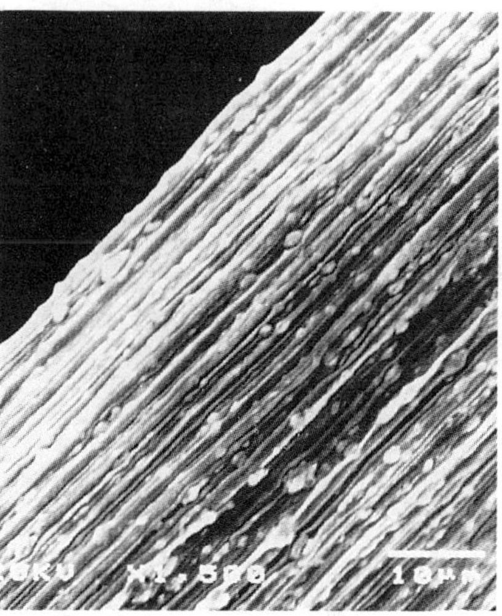

Figure 6a. Surface of an extracted NbTi filament filament with a Nb barrier.

Figure 6b Surface of an extracted NbTi from material with a Cu 2.5 wt. % Cu matrix.

being the most obvious cnes. The former are relatively straight forward and will only be discussed briefly below, the latter is in the author's opinion one of the most important factors affecting piece length.

<u>Inclusions.</u> Inclusions are picked up at all stages of the processing and complete cleanliness is obviously required throughout all operations, particularly during billet loading which is generally done in a cleanroom. Provided that inclusions are relatively few in number and not localized in one monofilament, they are less deleterious than poorly bonded material. One problem with inclusions occurs when failure analysis is carried out by an examination of fracture surfaces under the SEM. Frequently an inclusion of some type is found on the fracture surface and this is deemed to be the cause of breakage. It is possible however, that, had other conditions been more ideal, breakage would not have occurred. For instance, if dies are the wrong size, wrong angle, not cleaned, or drawing tensions are not controlled correctly, wire may break and this will frequently occur at a weak point i.e. where there is an imperfection such as an inclusion.

<u>Bonding</u>. As mentioned above in recent designs [7], a coextrusion process has been used in order to develop a Nb clad monofilament for further processing into the multifilamentary conductor for the SSC dipole and quadrupole magnets. This caused the introduction of an increased, number of different materials and surfaces and, as a result, contamination increased as did the bonding problems between the various constituents. The effect of these on drawability and piece length are not immediately obvious as monofilaments are not usually drawn to a small diameter. It only takes one monofilament that is not in excellent condition however, to cause an entire multifilamentary billet to exhibit poor piece length. The importance of high quality of the monofilament cannot be over emphasized if it is to be used in a multifilamentary array. Cleanliness and the integrity of the extrusion billet are of the greatest importance.

The monofilament is sometimes checked by bend tests to detect poor bonding and always checked by eddy currents for surface inclusions, when found these were removed. The same monofilament is shaved shortly before it is restacked to remove any surface inclusions as they would become internal inclusions in the multifilamentary billet.

Bonding in the multifilamentary billet was also a serious factor affecting piece length. In some early work done on a conductor which was of similar design to that of the SSC dipole conductor, cracks appeared across the filament array in the early stages of drawing, indicating that the hexagonal monofilamentary rods were poorly bonded, Figure 7a.

Three changes were made at this stage:

1. The array was changed so that more copper appeared outside the filaments and less in the central core, Figure 7b. .

2. Hot isostatic pressing, HIP'ing, was employed to bond the multifilamentary array. As an additional safety step, the same operation was introduced into the monofilament fabrication process.

3. The monofilament was annealed before assembly to ensure that it was as soft as possible to assist in bonding during HIP'ing and extrusion.

The effect of these changes was such that much improved piece lengths were achieved. It is important to realize that the changes introduced to improve piece length, as outlined above, did reduce somewhat the J_c's obtained and that a balance has always to be struck between piece length and other properties. The material as presently made for the SSC meets the piece length specification of 85% of the material greater than 1.5 km. Occasionally, however, even now, the odd billet exhibits poor piece length, which underlines the importance of quality assurance and careful adherence to all the procedures which have been established to ensure the best combination of piece length and J_c.

In Nb3Sn made by the internal tin process, shown schematically in Figure 8,. bonding problems are particularly severe. First a subelement is fabricated and the same design aspects apply as explained above for NbTi. Somewhat larger s/d's can be tolerated since no intermediate heat treatments are given and therefore the filaments and the matrix increase in hardness together. A well bonded subelement is essential for successful drawing of the final assembly, and HIP'ing is frequently used to ensure this. The bonding of these subelements to each other is however, where the worst problems arise. Once the tin has been introduced, all further processing has to be done cold. Another problem is that the components involved have widely differing work hardening characteristics. It is particularly difficult to cold bond these subelements when they are relatively hard. For this reason Nb 7.5 wt.%Ta filaments have been preferred to Nb 1.3 wt.%Ti for high field applications.

Summary

The economic importance of reducing breakage has been stressed. The history of multifilamentary superconductor development has been reviewed and an attempt made to illustrate the problems affecting piece length at each stage of conductor development. Some of the methods by which these problems have been overcome are outlined in the hope that the information will help in meeting the challenges of future applications.

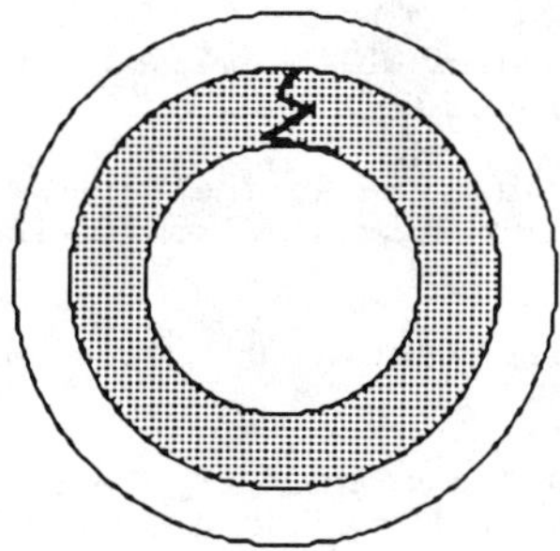

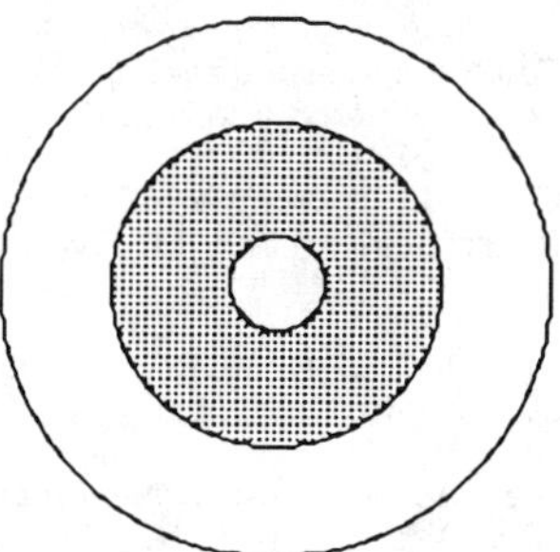

Figure 7a. Schematic representation of a conductor with a poorly designed array with cracks across the filament area

Figure 7b. Schematic presentation of a the conductor with improved design

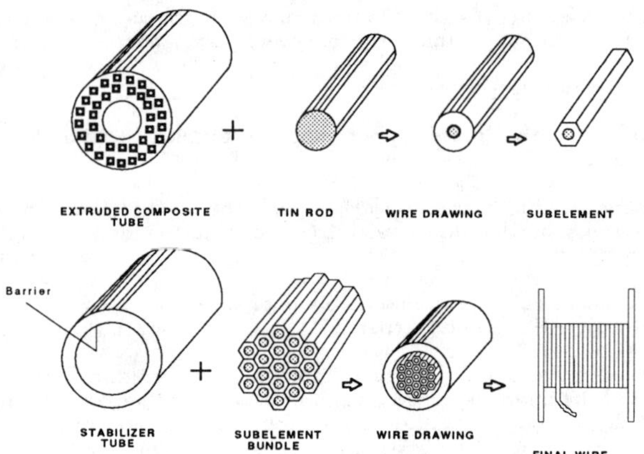

Figure 8 Schematic representation of the internal - tin Nb3Sn tin-core process.

References

1. J.E. Kunzler et al., "Superconductivity in Nb3Sn at High Current Density in a Magnetic Field of 88 kgauss", Phys Rev. Lett. 6, (1961),. 89-91.

2. J.E. Kunzler, "Recollections of Events Associated with the Discovery of High Field-High Current Superconductors", Trans. IEEE, MAG-23, (2,) (1987), 396-402.

3. Z.J.J Stekly and J.L. Zar, "Stable Superconducting Coils", Trans. IEEE on Nucl. Sci., NS-12, (3), (1965) 367.

4. C. Laverick, "Experiments with Superconducting Cables", MT-1 SLAC, Stanford Univ., eds. H. Brechna and H.S. Gordon, (1965), 560-567.

5. P.F. Smith, in Proc. 1968 Brookhaven Summer Study on Superconducting Devices and Accelerators, BNL Publ.. No. BNL-50155 (C-55), (1968).

6. D.C Larbalestier, P.J. Lee and R.W. Samuels, 'The Growth of Intermetallic Compounds at a Copper-Niobium-Titaanium Interface", Adv. Cryo Eng., eds. R.P. Reed and A.F. Clark, 32 (1986), 715-722.

7. T.S Kreilick, E. Gregory and J. Wong, "Fine Filamentary NbTi Superconducting Wires", Adv. Cryo. Eng. eds. R.P. Reed and A.F. Clark, 32, (1985), 739-745.

8. T.S. Kreilick et al., "Superconducting Wire and Cable for the Superconducting Supercollider", Supercollider 1, ed. M. McAshan, (Plenum Press, New York NY, 1989), 235-242.

9. Li Chengren and D.C. Larbalestier, Development of High Critical Current Densities in Niobium 46.5 wt.% Titanium, Cryogenics, 27, (4), (1987), 171.

10. K. Tachikawa et al., "Recent Studies on Composite Superconductors", (Paper presented at the 7th U.S-Japan Workshop on High Field Superconductors, Fukuoka, Japan, Oct. 22-24, 1991).

11. H. Liu, K.J. Faase, E. Gregory and B.A. Zeitlin, 'The addition of silicon to the copper between NbTi filaments to reduce interdiffusion" (Paper IV-F-6, presented at IISSC 5 San Francisco May 1993).

FLUX PINNING IN NBTI SUPERCONDUCTORS WITH

ENGINEERED PINNING CENTERS

L. R. Motowidlo and B. A. Zeitlin
IGC Advanced Superconductors Inc., Waterbury, CT 06704

X. S. Ling, N. D. Rizzo, J. D. McCambridge, and D. E. Prober
Department of Applied Physics, Yale University
New Haven, CT 06520-2824

Abstract

In previously reported NbTi superconductors with artificially engineered pinning centers (APC), dramatic improvements in the critical current density (J_c) over conventionally processed commercial NbTi superconductors have been demonstrated in magnetic fields below 4 Tesla. At higher fields however, the J_c and H_{c2} are reduced markedly. This is found to be a consequence of the pin material and basic design of the artificially engineered superconducting filament. In previous APC designs the volume percentage and choice of the pin material have been shown to be important for the proximity effect. Measurements also suggest 20% reduction in J_c from the proximity of the copper surrounding the first assembly of subfilaments and pinning centers. Based on these findings, new APC designs have been implemented. We present preliminary results on the new APC designs for NbTi over extended magnetic field ranges.

Work at Yale University and part of the work at IGC Advanced Superconductors Inc. is supported by the Connecticut University/Industry Partnerships Programs, Connecticut Department of Economic Development, Grant # 92G036 and the National Science Foundation through the Small Business Innovative Research Programs, Contract # III-9203072.

Processing of Long Lengths of Superconductors
Edited by U. Balachandran, E.W. Collings and A. Goyal
The Minerals, Metals & Materials Society, 1994

Introduction

Over the last ten years, researchers in superconducting wire development have been studying a new approach for improving the critical current density (J_c) of commercial NbTi.[1-16] In contrast to the conventional state-of-the-art of metallurgically precipitated normal flux pinning defects[17], engineered pinning centers or so called artificial pinning centers (APC) seem to offer an exciting new approach of controlling better the volume percent, spacing, and elementary pinning force through material choice and conductor design.[1,2] Thus, it may be possible to enhance critical current densities at desired operating magnetic fields and reduce conductor costs.

Calculated flux pinning strengths in NbTi at about 5 Tesla give theoretical J_c's approaching 20,000 A/mm^2.[18] The best J_c results for optimized commercial NbTi wire is slightly over 3,000 A/mm^2 at 5 Tesla.[17] NbTi conductors recently fabricated with the APC approach have already shown a factor of two improvement in J_c compared to commercial wires at low fields (< 3 Tesla)[13], and a substantial increase (3800 A/mm^2) at 5 Tesla.[15] Furthermore, by adjusting the design parameters, even the high field (> 5 Tesla) properties of APC conductors have shown to be comparable to the best optimized commercial NbTi.[19]

In this paper, we compare the bulk pinning force (F_p) curves of similarly designed APC conductors using different pinning materials, as a function of magnetic field and pin spacing. In addition, we will present preliminary results on a new APC design that is partially completed. The critical current density (J_c) and the specific pinning force (Q_p) of this new conductor at a pin spacing which is not yet fully optimized is presented and compared to previous results.

Experimental

An example of IGC's patented artificial pinning centers (APC) consists of the following: a copper tube having a diameter of 1.9 cm is utilized to assemble 61 elements of Nb47wt%Ti rods clad with with different percentage by area of either copper or niobium. In this configuration or design, the NbTi clad with a barrier of copper or niobium are the subfilaments and the copper or niobium are considered to be the engineered pinning centers. In this design, the assembly of 61 subfilaments constitutes the artificially engineered superconducting filament containing the pinning centers.[1] An SEM cross-section of this design is shown in Figure 1. The assembly or billet is then cold drawn to a final diameter and cut into short lengths for a second assembly or restack. The process is repeated with two more restacks in order to obtain dimensions or pin spacing of the order of the spacing of the flux lattice. Ideally, the pin spacing is designed to match the flux lattice spacing for a particular operating magnetic field. For example, at 5 Tesla this would mean a pin repeat distance on the order of 22 nm. The design and fabrication details have been previously described in earlier communications.[1,7,8]

A new APC approach is under development jointly by the groups at Yale University and IGC Advanced Superconductors Inc. The details of the processing and the choices of materials will be presented elsewhere. We will present here some preliminary results on this new APC approach.

The critical current of samples approximately 200 cm long were measured at 4.2K in external magnetic fields from 1 through 8 Tesla. A standard four probe technique was used on helically wound samples. Voltage taps were spaced 75 cm apart. The critical current was defined as occurring at resistivity onset of 10^{-12} Ohm-cm. The current density and resistivity were calculated based on the sum of the areas of NbTi and superconducting or normal pinning material matrix of the subfilaments shown in Figure 1, treating the composite filament area as a single superconducting phase.

FIGURE 1. SEM showing actual subfilament structure within the pinning matrix at final wire diameter. Each bundle is an engineered superconducting filament, ~ 1μm diam.

<u>Results and Discussion</u>

A comparison of the bulk pinning force for APC conductors are shown in Figures 2a and 2b. A proximity effect reduction of H_{c2} in NbTi is observed as the overall dimension of the pin spacing is reduced for both designs including copper and niobium pinning centers. Furthermore, it is observed that the proximity effect is more pronounced for the design which uses copper pinners. This is clearly observed in Figure 2a where we find that the H_{c2} decreases with decreasing pin spacing. In both designs the optimum pinning force is observed to be at a pin center-to-center spacing of roughly 33.5 nm with a maximum occurring at about 3 Tesla. However, the bulk pinning force for the APC with niobium is considerably larger as compared to the APC with copper pinning centers. For example, the F_p is about 17.5 GN/m^3 for APC with niobium versus about 10.5 GN/m^3 for APC with copper at the optimum pin spacing. The results shown in Figures 2a and 2b suggest that the properties of the APC conductors can be dramatically affected by the choice of the pin material through the proximity effect.

More recently, it was also shown that the APC conductor can be further optimized if particular attention is given to the design.[20] For example, it was observed that the proximity effect from the surrounding copper matrix on the NbTi filament, can reduce the measured J_c by about 20%. Thus the effective cross-sectional area of each artificial superconducting filament is reduced by a fraction $(4\xi_{sc}/d) \sim 0.15$, with ξ_{sc} the coherence length in NbTi and d the filament diameter.

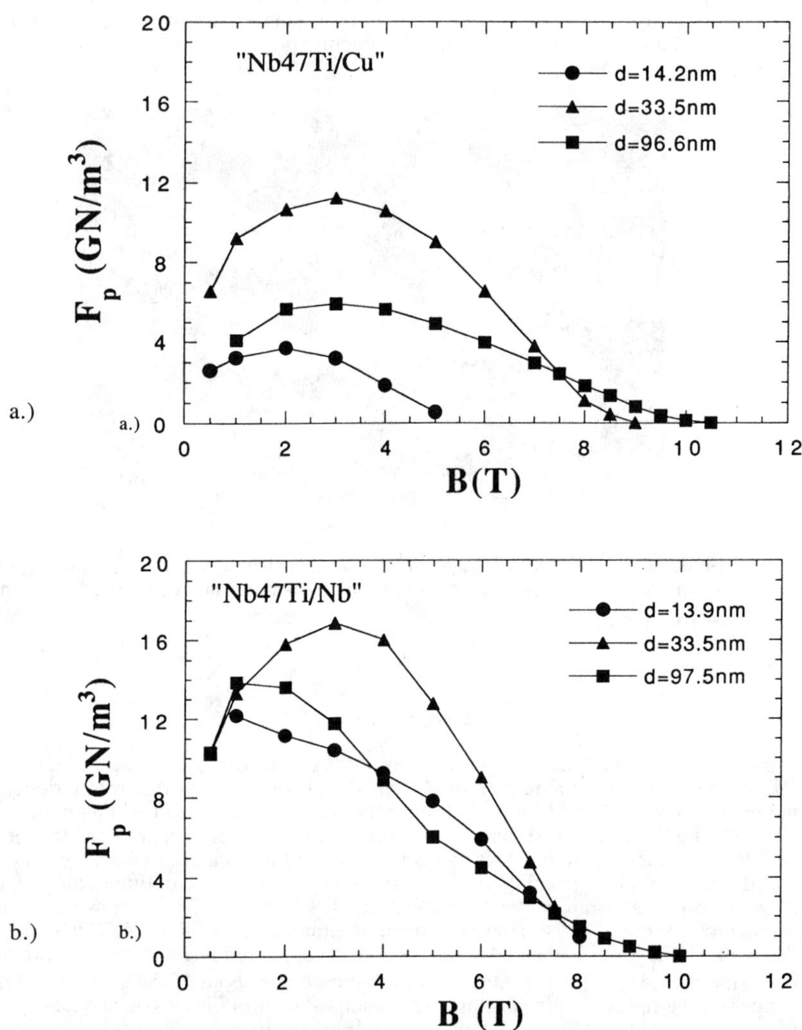

FIGURE 2. A measure of the bulk pinning force as a function of applied magnetic field and pin spacing for a.) Nb47wt%Ti with copper pinning centers and b.) Nb47wt%Ti with niobium pinning centers.

The importance of optimizing the pin spacing for a given applied magnetic field is emphasized in Figure 3. At low fields the optimum pin spacing is found to be above 60 nm for the APC design with Nb pins. However, as the magnetic field is increased the flux lattice spacing decreases and we find a maximum critical current density at smaller pin spacing. For example, at 5 Tesla the most effective interaction between Nb pins and the flux lattice is obtained at about 37 nm.

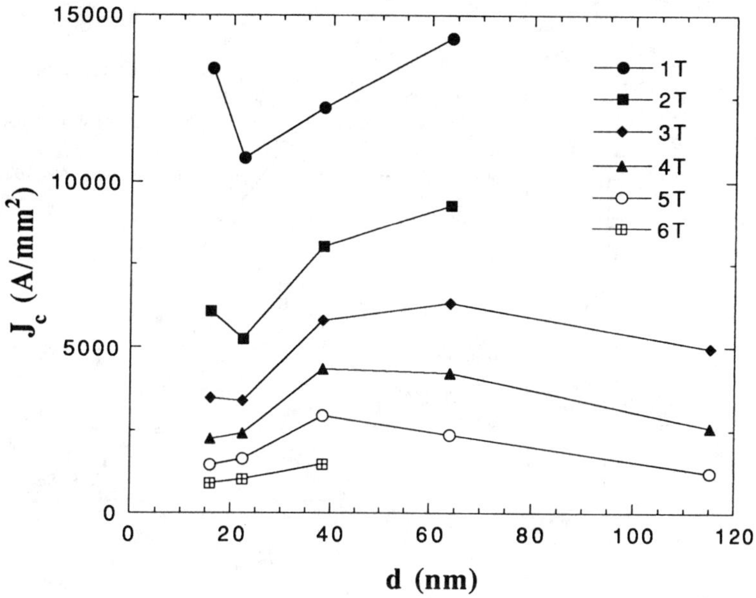

FIGURE 3. The critical current density as a function of the pin center-to-center repeat distance and applied magnetic field for APC Nb52wt%Ti with Nb pins.

Based on these results new APC designs were initiated. Shown in Figure 4 is a comparison of the J_c versus B for an old APC design at 100 nm and 33.5 nm, an optimized SSC conductor[22], and a new APC design including a new pinning material which has yet to be fully optimized with regard to the spacing of the pinning centers. In order to achieve a comparable pin spacing in the new APC design, the conductor was rolled. In this case the pin repeat distance, which is perpendicular to the wire length and the applied magnetic field, is estimated to be about 122 nm. Processing and material specifics will be given in a separate communication. The high field properties of this new, unoptimized APC conductor exhibit significantly larger critical currents at 7 and 8 Tesla as compared to the optimized old APC conductor. The new APC conductor has significantly larger J_c at 5

Tesla and above compared to the old unoptimized APC conductor with pin spacing of 100 nm. In addition, at 8 Tesla, J_c of the new APC approaches the SSC optimized conductor.

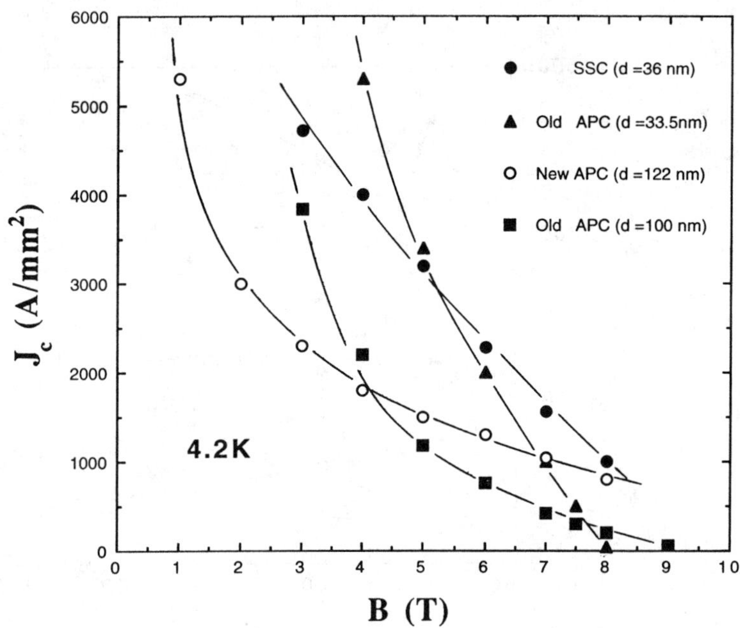

FIGURE 4. The critical current density as a function of the applied magnetic field for an optiimized SSC conductor, an old APC design and a new APC design.

It is believed that some of the enhancement of the J_c may be due to alignment of the APC structure as well as the design and pin material. In order to estimate the possible pinning strength that may be reached by the APC method or by the pin material, we plotted the specific pinning force as a function of applied magnetic field, shown in Figure 5. The specific pinning force ($Q = F_p/\rho$) with ρ being the density of pins, was first introduced by Kramer[23] as a relative measure of the effectiveness of the pinning center. In this paper, we take the pin repeat distance as a measure of the linear pin density and thus obtain a conservative estimate of Q for each APC design.

In Figure 5 the specific pinning force of each APC design is compared to a rolled optimized NbTi conductor[21] where the α-Ti ribbon like precipitates have been highly aligned by the rolling process. The magnetic field (B) is parallel to the wide face of the rolled wire but perpendicular to the current direction. B is thus parallel to the wide direction of most α-Ti precipitates for the rolled NbTi conductor. In an optimized NbTi round conductor the flux pinning is considered to be from clusters of α-Ti ribbons with an estimated spacing of about 36 nm.[22] In the highly aligned rolled NbTi material the clusters are also expected to be the pinners but with enhanced alignment and with estimated cluster spacing of about 30 nm.[24] This enhanced alignment of α-Ti ribbons in the optimized rolled NbTi has resulted in a J_c of about 5200 A/mm^2 at 5 Tesla. Thus, the specific pinning force is estimated for the rolled optimized NbTi using a pin spacing of 30 nm, and the specific pinning force of the new APC design is estimated using a pin spacing of 122 nm. Also, shown for comparison is the old APC conductor. This comparison in Figure 5, suggests that the new APC design with the better pin material has the possibility of achieving improved J_c in the mid to high magnetic field range.

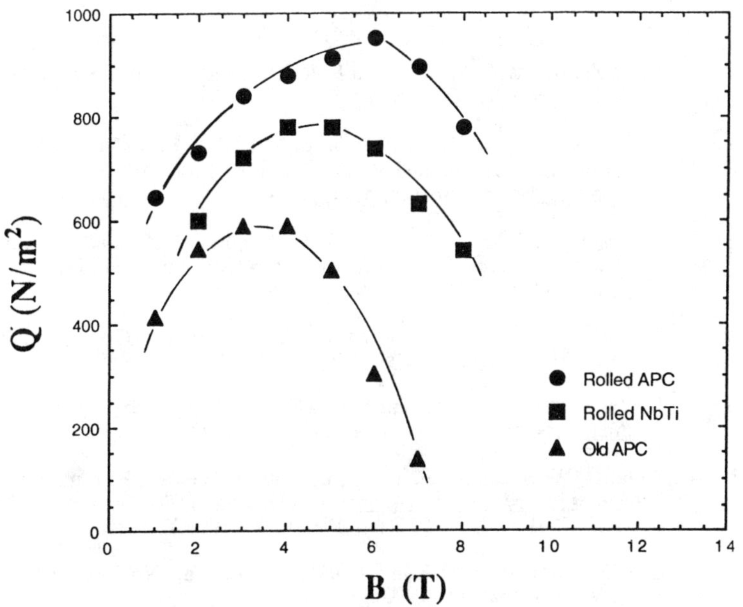

FIGURE 5. The specific pinning force (F_p*d) as a function of the applied magnetic field for the new APC design compared to the old APC design and the rolled optimized SSC conductor(Ref. 21).

<u>Concluding Remarks</u>

In our study of NbTi conductors engineered with artificial pinning centers, we have found that the pin material is an important factor in developing high pinning strength. If proper consideration is given to the choice of the pin material as well as to the design and pin spacing, then it is possible to achieve very high pinning capability and therefore, very high critical current density in NbTi. Thus, critical currents may be obtainable with NbTi conductors having artificially engineered pinning centers that are a factor of two larger than obtained with conventional conductors.

<u>Acknowledgements</u>

Work at Yale University and part of the work at IGC Advanced Superconductors Inc. is supported by the Connecticut University/Industry Partnerships Programs, Connecticut Department of Economic Development, Grant # 92G036 and the National Science Foundation through the Small Business Innovative Research Programs, Contract # III-9203072.

<u>References</u>

1.   B. A. Zeitlin, M. S. Walker, and L. R. Motowidlo, United States Patent No. 4,803,310, Feb. 7, 1989.

2.   G. L. Dorofejev, E. Yu. Klimenko, S. V. Frolov, E. V. Nikulenkov, E. I. Plashkin, N. I. Salunin, and V. Ya. Filkin, In Proc., 9th Int'l Conf. Magnet Technology, p. 564, Boston, Mass., 1985.

3.   I. Hlasnik, S. Takacs, V. P. Burjak, M. Majoros, J. Krajcik, L. Krempasky, M. Polak, M. Jergel, T. A. Korneeve, O. N. Mironova and I. Ivan, <u>Cryogenics</u>, Vol 25, pp. 558-565, October,1985.

4.   P. Dubots, A. Fevrier, J C. Renard, J P. Tavergnier, J. Goyer, and Hoang Gia Ky, <u>IEEE Trans. on Magnetics</u>, MAG-21, No.2, 1985.

5.   K. Yasohama, K. Morita, and T. Ogasawara, <u>IEEE Transactions on Magnetics</u>, MAG-23, No.2, p. 1728, March, 1987.

6.   S Gauss, W. Specking, F. Weiss, E. Seibt, J. Wu, and R. Flukiger, <u>Advances in Cryogenics Engineering</u>, (Materials), Vol. 34, p. 843, Editied by A. F. Clark and, R. P. Reed, Plenum Press, NY, 1988.

7.   L. R. Motowidlo, H. C. Kanithi, and B. A. Zeitlin, "NbTi   in <u>Adv. Cryo. Eng.</u>, V. 36A, pp. 311, 1990.

8.   L. R. Motowidlo, P. Valaris, H. C. Kanithi, M. S. Walker and B. A. Zeitlin,<u>NbTi In Supercollider 2</u>, Edited by Michael McAshan, Plenum Press, New York, pp. 341, 1990.

9.   L. D. Cooley, P. J. Lee, and D. C. Larbalestier, in <u>IEEE MAG</u>, Vol. 27, No.2, p. 1096, March 1991.

10. K. Yamafuji, N. Harada, Y. Mawatari, K. Furaki, T. Matsushita, K. Matsumoto, O. Miura, and Y. Tanaka, <u>Cryogenics</u> 31, 431, 1991.

11. L. R. Motowidlo, B. A. Zeitlin, M. S. Walker, and P. Haldar, <u>Applied Physics Letters</u>, Vol 61, No. 8, pp. 991, 24 August, 1992.

12. D. R. Dietderich, S. Eylon, and R. Scanlon, <u>Adv.Cryog. Eng.</u> 38B, p. 685, 1002, 1992.

13. H. C. Kanithi, P. Valaris, L. R. Motowidlo, B. A. Zeitlin, and R. Scanlon, in <u>Adv. Cryo. Eng.</u>, 38B, 675, 1992.

14. L. R. Motowidlo, B. A. Zeitlin, M. S. Walker, P. Haldar, J. D. McCambridge, N. D. Rizzo, X. S. Ling and D. E. Prober, <u>IEEE Transactions on Applied Superconductivity</u>, Vol. 3, p. 1366, 1993.

15. K. Matsumoto, Y. Tanaka, K. Yamafuji, K. Funuki, M. Iwakuma, and T. Matsushita, <u>IEEE Transactions on Applied Superconductivity</u>, Vol. 3, p. 1362, 1993.

16. R. Zhou, S. Hong, W. Marancik, and B. Kear, <u>IEEE Transactions on Applied Superconductivity</u>, Vol. 3, p. 986, 1993.

17. C. Meingast, P. J. Lee, and D. C. Larbalestier, <u>J. Appl. Phys.</u>, 66, p. 5962, 1989.

18. E. H. Brandt, <u>Phys. Lett.</u>, A 77, p. 484, 1980.

19. R. M. Scanlan, A. Lietzke, J. Royet, A. Wandesfonde, C. E. Taylor, J. Wong, and M. K. Rudziak, Presented at Thirteenth International Conference on Magnet Technology, 20-24 September, 1993.

20. X. S. Ling, J. D. McCambridge, D. E. Prober, L. R. Motowidlo, B. A. Zeitlin, and M. S. Walker, Presented at the XX International Conference on Low Temperature Physics, Aug. 4-11, 1993; to be published in J. Low. Temp. Phys.

21. L. D. Cooley, P. D. Jablonski, P. J. Lee, and D. C. Larbalestier, <u>Appl. Phys. Lett.</u>, 58 (25), p. 2984, 1991.

22. C. Meingast and D. C. Larbalestier, <u>J. Appl. Phys.</u>, 66. p. 5971, 1989.

23. E. J. Kramer, <u>J. Nucl. Materials</u>, 72, p. 5-33, 1978.

24. L. D. Cooley, private communication with author L. Motowidlo, NIST, 15 October 1993.

EFFECT OF COMPRESSIVE AND TENSILE STRAINS ON THE CRITICAL CURRENT DENSITY OF LIQUID-PHASE DIFFUSION PROCESSED Nb3Sn

Melissa L. Murray, Mark G. Benz, Robert J. Zabala and Tom R. Raber

General Electric Research and Development Center
1 River Road
Schenectady, New York 12301

The influence of strain on the critical current density of liquid-phase diffusion processed Nb_3Sn foil has been examined. Eight different levels of plane stress were applied to the Nb_3Sn by using differential thermal contraction. This differential thermal contraction was achieved by soldering Nb_3Sn foil samples between two plates of eight different materials, and cooling the composite structures from the soldering temperature to the test temperature. Brass, copper, stainless steel, Hastelloy X, Inconel 600, nickel, niobium and tungsten plates were used. In addition, bending strains were applied to the Nb_3Sn foil using a non-symmetric bi-metal structure with a brass or niobium plate placed in a bending fixture. In a transverse field of 5T and at a test temperature of 4.2 K, a thermally applied axial strain of 0.4% increased the critical current by 60%. An axial tensile bending strain of the same magnitude resulted in a critical current increase of only about 9%. These results will be discussed and compared to observations in the literature for solid-phase diffusion processed Nb_3Sn.

Processing of Long Lengths of Superconductors
Edited by U. Balachandran, E.W. Collings and A. Goyal
The Minerals, Metals & Materials Society, 1994

Introduction

The effect of strain on the superconducting properties of Nb_3Sn has been studied for nearly thirty years. It is well known that compressive strains in Nb_3Sn protect the brittle material from damage which may be encountered during handling, coiling, or from electromagnetic field forces in an energized magnet[1]. These compressive strains are usually imparted by the materials that encase the superconductor, such as bronze in bronze-processed multifilamentary wire or Cu foil in laminated liquid-phase diffusion processed Nb_3Sn foil. The difference in thermal expansion coefficients of the Cu or bronze and the Nb_3Sn leads to compression of the Nb_3Sn when the composite is cooled from the processing temperature. A strain scaling relation[2] has been developed to predict the effect of strain on the critical current, I_c, based on the axial strain dependence of the critical field, B_{c2} for multifilamentary superconductors.

Most Nb_3Sn strain-effect research has been performed at high fields on mono- and multifilamentary bronze-processed wire which is formed by a solid-phase diffusion process. The complicated geometry of these filamentary composites leads to an uncertain residual thermal strain state in the Nb_3Sn. In this study, liquid-phase diffusion processed Nb_3Sn foil was used to investigate the effect of strain on I_c. The plate geometry of this material allows the measurement and prediction of a three-dimensional strain state in the Nb_3Sn. Two methods of applying strain to the superconductor were investigated. First, tensile and compressive strains were induced in the Nb_3Sn as a result of a thermal contraction mismatch between the Nb_3Sn and a testing substrate. Second, the strain state was mechanically altered by applying bending strains at the cryogenic test temperature through the use of a three-point bend tester. The results provide an accurate, measurable relation between strain in the Nb_3Sn and critical current. The experiments show that the axial thermally-induced strains produce a much larger change in the I_c than the axial bending strains. A three-dimensional representation of the strain state is thus necessary for understanding the effect of strain on the superconducting properties of Nb_3Sn.

Experimental

<u>Superconducting Foil</u>

The Nb_3Sn superconductor used in this study is in the form of flat, laminated tape. The superconducting laminate was made in our laboratory using a liquid-phase diffusion technique[3]. In this method, 25.4 μm-thick Nb alloy foil (Nb - 1at. % Zr - 2 at. % O) is dipped in a tin alloy melt (Sn - 17 at. % Cu). This tin-coated Nb alloy foil is then reacted at 1050°C for a short time (about 200 sec.). At this temperature, the tin alloy coating is liquid; tin alloy diffuses through the forming Nb_3Sn layer to react with the solid Nb alloy core. The process forms a layer of fine-grained, superconducting Nb_3Sn about 7μm thick on either side of the remaining Nb foil (Figure 1). For thermal and mechanical stability, strips of copper foil are then soldered to the reacted foil to form a Nb_3Sn-Cu laminate.

<u>Thermally-Induced Strain Method</u>

Strips of superconducting Nb_3Sn-Cu laminate were soldered in between two flat, significantly thicker, metal substrates. The thicknesses of the Nb-Nb_3Sn foil, the Cu foil, and the metal substrates were 25.4 μm, 76.2 μm, and 1520 μm, respectively. The length of the foil-substrate laminate was 7.6 cm, the substrates were 1.1 cm wide, and the Nb_3Sn-Cu laminate was 3 mm wide. The metals for the substrates were chosen to include a wide range of linear thermal expansion coefficients. The difference in thermal expansion properties of the substrates and the Nb_3Sn induces uniform tensile or compressive strains in the superconducting layer during cooling from the processing temperature to the test temperature. The calculation of these strains is outlined in Appendix A. Two substrates, one on each side, were used to provide a symmetric laminate which prevents bending that would occur during the cooling of a bi-metal strip.

344

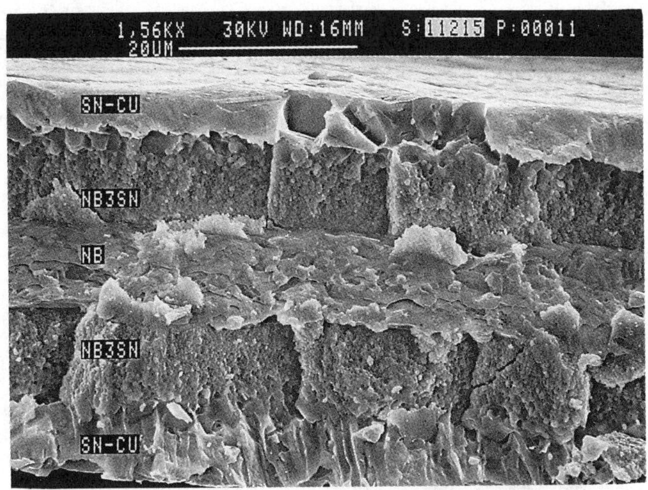

Figure 1. Fracture section of liquid diffusion-processed Nb3Sn foil.

Application of Bend Strain

Another method of applying strain to the superconductor involved soldering Nb$_3$Sn foil samples to a single thicker substrate of either Nb or brass and straining in three-point bending. It was necessary to use a single substrate so that the Nb$_3$Sn was removed from the neutral axis of the beam. The effect of the un-symmetric configuration is discussed later, and the calculation of the bending strains in the Nb$_3$Sn layer is outlined in Appendix B. The apparatus used to apply the bend force is shown in Figure 2. The mechanism consists of two rings of non-metallic material which hold the beam at the ends, and a threaded rod which applies the deflection in the center of the beam. Both rings have milled slots to align the beam. The depth of the slot is slightly greater than the thickness of the beam to allow unrestrained contraction of the beam during cooling. This also allows the beam ends to deflect upward, as is necessary for three-point bending. The radius of the lower ring at the point of contact with the beam was also milled away to provide linear contact along the ends of the beam surface. Electrical continuity between the beam and the threaded rod indicated when the straining of the sample began. The strains were determined from a calibrated relation between the rotation of the threaded rod and strain measured on the bend bar.

Critical Current Testing

The composite metal-Nb$_3$Sn laminate samples were tested for critical current in liquid helium in a field of 5T. The temperature and magnetic field strength were not varied. For the thermal strain measurements, the field was oriented along the z-axis (at 0° to the wide side of the foil), while in the bending experiments, the field was directed along the y-axis (at 90° to the wide side of the foil). Previous critical current measurements[4] of Nb$_3$Sn foil on brass substrates have established that changing the field orientation from 0° to 90° produces a 14% increase in the critical current. In all of the experiments, current flows in the x-direction. Voltage probes were placed on the middle 1 cm of the samples and the critical current was defined by a voltage difference of 0.2μV.

Strain Measurement

The bending strains of cold-worked substrates were measured with strain gages. The gages, Micromeasurements type WK with a gage length equal to 0.32 cm, are self-temperature compensating and are designed for use at cryogenic temperatures. The axial (x-direction) strains were measured at room temperature and at 4.2 K, and the transverse or width (z-direction) strains were measured only at room temperature. The strain was not measured simultaneously with the critical current measurements since applied magnetic field (5T) would affect the voltage output of the gages, resulting in an error in the strain recorded. Dummy strain gages were attached to another un-strained brass substrates to compensate for thermal output. Thermal output is caused by the temperature-dependence of resistivity of the gage material.

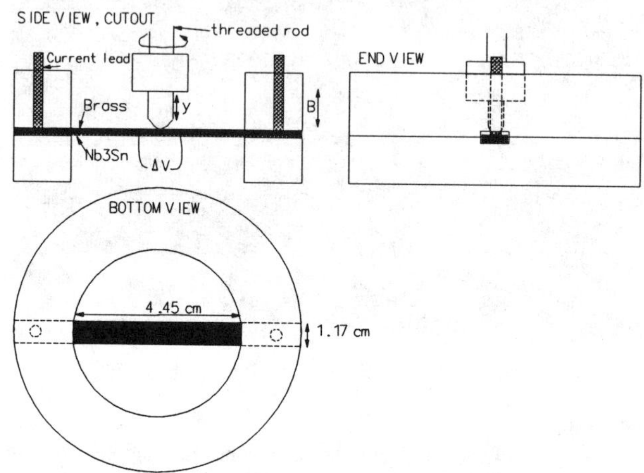

Figure 2. Schematic of apparatus for applying bending strains to the superconductor.

<u>Strain Calculation</u>

Thermally Induced Strain

Residual strains in the superconducting foil result from cooling the laminate from the solder solidification temperature 456 K (183 °C) to liquid helium temperature, 4.2 K. The calculated strains are shown in Table I. An additional residual strain is induced in the Nb_3Sn during cooling the foil from the reaction temperature, 1323 K (1050 °C), to the soldering temperature. At the reaction temperature, there is a liquid intergranular phase[3] which provides no resistance to shear deformation, and thus no straining occurs in the Nb_3Sn. This liquid phase, which is approximately 65at.% Sn - 30at.% Nb - 5at.% Cu, is assumed to solidify at 1223 K (950 °C). The resulting residual thermal strain in the Nb_3Sn is approximately +0.03%. This was found assuming constant moduli for the Nb_3Sn and Nb in the temperature range, and using a Nb_3Sn thickness of 14 μm and a Nb thickness of 18 μm.

From Table I, it can be seen that the contribution of the Cu- and Nb-foils to the reduction of the thermal contraction of the thicker substrates is small, ranging from 11% for Nb to 0.2% for brass. This means that the total contraction or strain of the laminate due to the temperature drop is approximately equal to the thermal contraction of the substrate material.

Bend Strain

Bending strain measurements of ε_x and ε_z are shown in Figure 3. The thickness strain, ε_y, could not be measured with gages. At large axial strains, greater than about 0.32%, the strain-deflection relation deviates from elementary beam theory calculations. The total curve can be represented by two lines of different slopes: a linear elastic region and a linear plastic region. Bending deflections greater than those measured with the gages (>0.6%) become increasingly less accurate.

Table I. Properties of substrate materials and calculated residual thermal strains in Nb$_3$Sn.

Substrate material	Young's modulus (GPa)[a]	Poisson's ratio[a]	$\Delta L/L_0$ for ΔT=4.2K-456K (%)[b]	ε_x $(=\varepsilon_z)$ (%)	ε_y (%)	Ic 4.2K, 5T (A)
Brass	105	0.35	0.676	-0.393	0.336	393
Copper	112	0.36	0.596	-0.312	0.267	424
Stainless steel	190	0.3	0.590	-0.306	0.263	451
Hastelloy	200	0.3	0.471	-0.185	0.158	500
Inconel	200	0.3	0.463	-0.177	0.152	511
Nickel	207	0.31	0.456	-0.169	0.145	528
Niobium	103	0.38	0.263	0.002	-0.002	628
Tungsten	407	0.32	0.161	0.116	-0.099	565
Nb$_3$Sn	165[c]	0.3	0.402	- -	- -	- -

[a] From [5].

[b] From [6], except for brass, which is from [7].

[c] From [15].

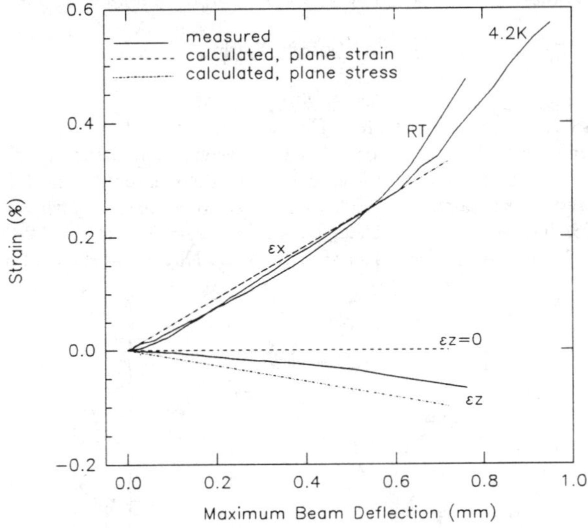

Figure 3. Measured bending strain versus the deflection at the midpoint of the beam at both room temperature and 4.2 K compared to the strains calculated assuming either a state of plane strain or plane stress.

Figure 3 also shows a comparison of the data with the calculations, assuming either plane stress or plane strain states. The calculated ε_x, which is the same for plane stress and plane strain, agrees well with the measured data in the elastic region. However, from the transverse strain data, the beam appears to be in neither a state of plane stress or plane strain. The figure shows that the measured transverse strain is half-way between the plane stress and plane strain states. This is probably due to the beam geometry, for the beam is not wide and thin enough to be in a state of plane strain, nor narrow enough for plane stress to apply. The strain relations which are used to plot elastic strain versus critical current in the remainder of the study are the measured values, ε_x and ε_z. The thickness strain, ε_y, is approximated as being half-way between the plane stress and plane strain states, since the measured transverse strain, ε_z, follows a similar relation.

The thermal residual strain for laminated Nb$_3$Sn on a single substrate, as in the bending tests, is not significantly different from that on double substrates. For a single brass substrate, the total thickness of the brass is halved, and this produces only 0.2% change in the calculated strain value, or an axial strain decrease of 0.001%. Bending due to the un-symmetric layup of the laminate reduces the Nb$_3$Sn axial strain by an additional 0.01%[9], for a total strain change of 0.011%.

The triaxial strain state, ε_x, ε_y, and ε_z, for the residual thermal strain and the bending strains can now be compared. The thermal residual strains are $\varepsilon_x = \varepsilon_z$ and $\varepsilon_y = -0.86\varepsilon_x$ and the bend strains are $\varepsilon_x = 0.47y$, $\varepsilon_y = -0.36\varepsilon_x$, and $\varepsilon_z = -0.15\varepsilon_x$, where y is the maximum deflection in mm. By the principle of strain superposition, the total strains are

$$\varepsilon_x = \varepsilon_{x(\text{thermal})} + \varepsilon_{x(\text{bend})}$$
$$\varepsilon_y = -0.86\varepsilon_{x(\text{thermal})} - 0.36\varepsilon_{x(\text{bend})} \tag{1}$$
$$\varepsilon_z = \varepsilon_{x(\text{thermal})} - 0.15\varepsilon_{x(\text{bend})}$$

For axial bending strains greater than about 0.3%, the slope of the strain-deflection curve increases, and the axial and transverse strain relations become $\varepsilon_x = 0.87(y - 0.28)$ and $\varepsilon_z = -0.10(y - 0.20)$.

Critical Current Testing

Critical Current Variation with Residual Thermal Strain

The metal substrates which were tested on include yellow brass, copper, 304L stainless steel, Hastelloy X, Inconel 600, nickel, niobium, and tungsten. The thermally induced strains, ε_x, which are shown in Table 1 are plotted in Figure 4 versus the measured critical current, I_c. Each point represents at least six measurements made on at least three separate superconducting samples. The data shows a 60% increase in I_c for a change in strain of $\varepsilon_x = -0.39$ to $\varepsilon_x = 0.0$. This represents a difference in I_c of 235 A for Nb$_3$Sn foil soldered to brass and Nb$_3$Sn foil soldered to niobium.

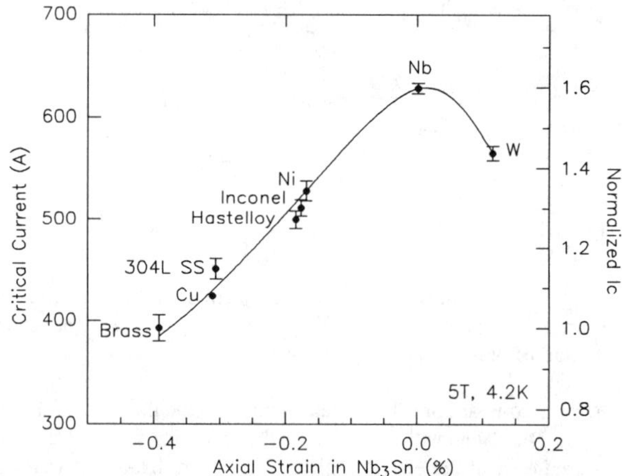

Figure 4. Critical current of Nb$_3$Sn versus the axial thermal residual strain in the Nb$_3$Sn caused by the different substrate materials.

Critical Current Variation with Bending Strain

While incrementally increasing the bending deflection, the critical current of the Nb$_3$Sn foil soldered to a single brass substrate was measured. The measurements were performed on nine different samples and the results for axial strain on all samples are shown in Figure 5. The critical current data in this figure was adjusted to account for the effect of field orientation as discussed earlier, so that the bending results can be compared to the residual thermal strain results.

The measurements show that as the bending strain was increased, the critical current increased a small amount, 7% to 12%, to a maximum value. The strain at the maximum I_c varied from -0.31% to +0.25% with an average of -0.10% ±0.17. When the bending load was removed in this region, the effect of strain on I_c was reversible and the critical current returned to the original value at zero applied strain. At large applied strain values, the I_c rapidly dropped off, and this effect was irreversible. The axial strain value where the I_c decreased rapidly varied widely among the samples, from -0.13% to +0.40%, or an average of +0.05% ±0.16. The dramatic fall-off in I_c was probably due to damage or cracking in the Nb_3Sn. Kroeger et al[10] and Sekine[11] have observed evidence for microstructural damage at high strains in bronze-processed Nb_3Sn wire. Since the failure strain of a brittle material such as Nb_3Sn is a statistical quantity, this variation in the point of measurable damage or irreversible decrease in I_c should be expected. In addition, the region of maximum strain in a 3-point bend test is small, so that a small section of the Nb_3Sn material is experiencing the maximum strain. This small sample size adds to the statistical variation in the point of failure. A 4-point bend test, which can have a larger region of constant strain, may reduce this variation in failure strain.

Similar bend tests were preformed on Nb substrates for four different samples. In this case, the initial strain state, due to thermally applied strains, is nearly zero. The hysterisis of applied strain and Ic, shown in Figure 6, reveals a much smaller region of reversible strain. In one case, there was no region of reversible strain (open circles in Figure 6).

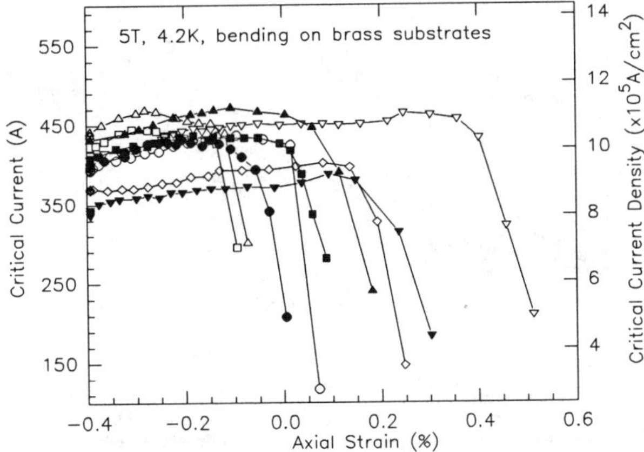

Figure 5. Total axial strain (bending strain + thermal residual strain imposed by substrate) versus the critical current for Nb_3Sn laminate soldered to a single brass substrate for bending tests on nine different samples.

Discussion

A comparison of the effect of thermally applied strains and the bending strains on I_c show that the magnitude of the critical current variation with strain is strongly dependent on the conditions of the applied strain. The change in I_c for axial strains caused by thermal mismatch is much larger than the change in I_c caused by axial bending strains. One notable difference between the two methods of applied strain is the temperature where the loading is applied. While the bend strains are all applied at 4.2 K, the thermal-mismatch strain is applied continuously from 456 K to 4.2 K. Kroeger[10] suggests that microstructural features which act as flux pinning centers may undergo stress-induced phase transformations, thus affecting the I_c. This effect was investigated by applying the bend strain at room temperature and comparing this to the bend strain applied at 4.2 K. Figure 7 shows that the I_c is insensitive to the temperature where the strain was applied since the critical currents for a deflection applied at room temperature and the same deflection applied at 4.2 K are nominally the same.

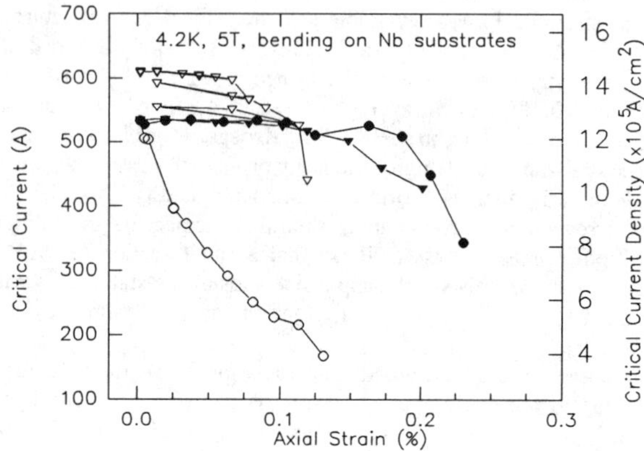

Figure 6. Strain versus I_c for Nb$_3$Sn on Nb substrates with applied bending loads.

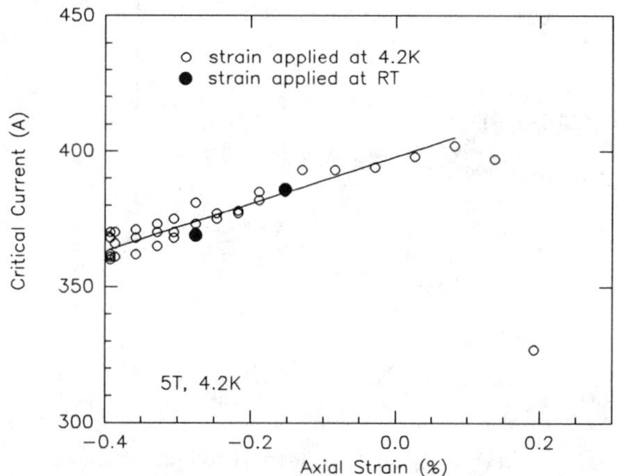

Figure 7. Comparison of the I_c measurements for strains applied at room temperature and strains applied at 4.2 K.

Another difference between the methods of applied strain is the total strain state. For the thermally applied strains, the axial and transverse strains are equal, while the thickness strain is lower in magnitude and oppositely directed. The axial bending strain in this study was always tensile, and thus the transverse and thickness strains were compressive and lower in magnitude. The variation in I_c with the average principle strains, ε_x, ε_y and ε_z, for bending on brass is shown in Figure 8. The figure shows that while the axial strain, ε_x, passes through zero when bending strains are applied, the transverse, ε_z, and thickness strains, ε_y, are never equal to zero for bending on brass. However, all of the principal strains were simultaneously zero, $\varepsilon_x \approx \varepsilon_y \approx \varepsilon_z \approx 0$, when the Nb$_3$Sn was soldered to Nb substrates, and the critical current was thus a maximum (I_c=628 A, J_c=15.0x10^5 A/cm^2) .

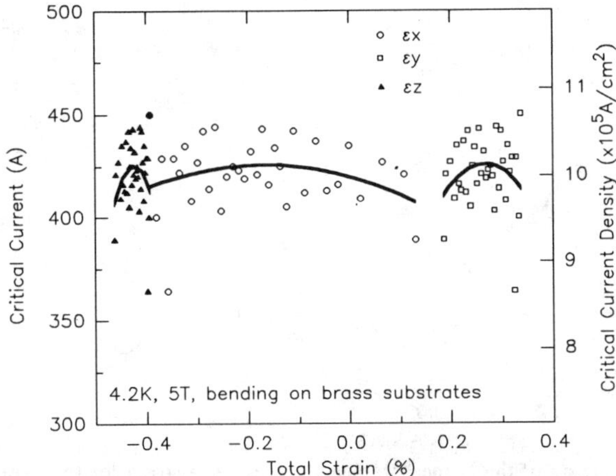

Figure 8. Principal strains, ε_x, ε_y, and ε_z, versus critical current of Nb$_3$Sn on brass substrates with applied bending loads. The curves represent a parabolic regression fit of the average strains in the reversible strain region of the nine bend samples.

The axial strain has historically been used to relate the I_c, B_{c2} and strain[1,2]. Figure 9 compares the data from this study, normalized by $I_{c(max)}$, with other data on Nb$_3$Sn foil[13] and to the strain scaling law from Ekin[2] for bronze-processed wire calculated at 5T. A ten-point running average was applied to the bend data to reduce the scatter. The figure shows a similar discrepancy between the conditions of strain and the effect on I_c. These observations suggest that it is not sufficient to consider only the axial strain when determining the strain effect on I_c for Nb$_3$Sn foil; another representation of strain which includes all of the principal strains is probably necessary.

Testardi[8] represents the strain dependence of the transition temperature, $T_c(\varepsilon)$, by a Taylor expansion around $T_c(0)$ as

$$T_c(\varepsilon) = T_c(0) + \Gamma \cdot \varepsilon + 1/2 \, \varepsilon \cdot \Delta \cdot \varepsilon \qquad (2)$$

where Γ and Δ are the tensor forms of first and second derivatives of T_c by strain, respectively, and ε is the 6x6 strain tensor. Welch[12] has simplified this equation for a random polycrystalline aggregate of cubic crystals and expressed the strain dependence in terms of two strain invariants: the deviatoric (non-hydrostatic) strains and the dilatation,

$$T_c(\varepsilon) = T_c(0) + \Gamma_1 e + 1/2 \, \Delta_v e^2 + 2/5 \, \Delta_t(\hat{\varepsilon}_x^2 + \hat{\varepsilon}_y^2 + \hat{\varepsilon}_z^2) \qquad (3)$$

where Γ_1, Δ_v, and Δ_t are constants, the dilatation is $e = \varepsilon_x + \varepsilon_y + \varepsilon_z$ and the deviatoric strain in the i-direction is $\hat{\varepsilon}_i = \varepsilon_i - e/3$. The strain dependence of the critical current density can be represented by the Kramer scaling law[12], assuming that the flux pinning characteristics of the material are constant. Then, if the critical temperature is strongly dependent on either the hydrostatic or non-hydrostatic strains, e or $\hat{\varepsilon}_i$, then the I_c (or J_c) will show a similar relation. The data for the thermally applied strains and the bending experiments are represented in terms of e and $(\hat{\varepsilon}_x^2 + \hat{\varepsilon}_y^2 + \hat{\varepsilon}_z^2)$ in Figure 10. The data show that the deviatoric strain representation most closely relates both conditions of strain, bending and thermal-mismatch, to the critical current. This suggests that the critical current may be dependent on the non-hydrostatic strain, and not the axial strain or the hydrostatic strain.

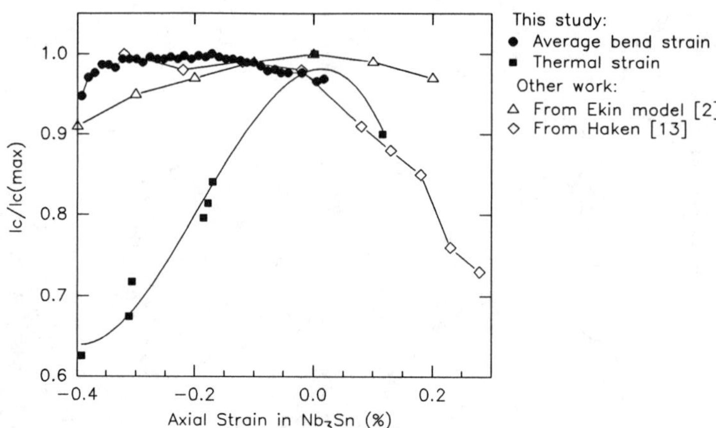

Figure 9. Comparison of the I_c measurements versus axial strain for this work and published relations for Nb3Sn tape (Haken) and bronze processed wire (Ekin).

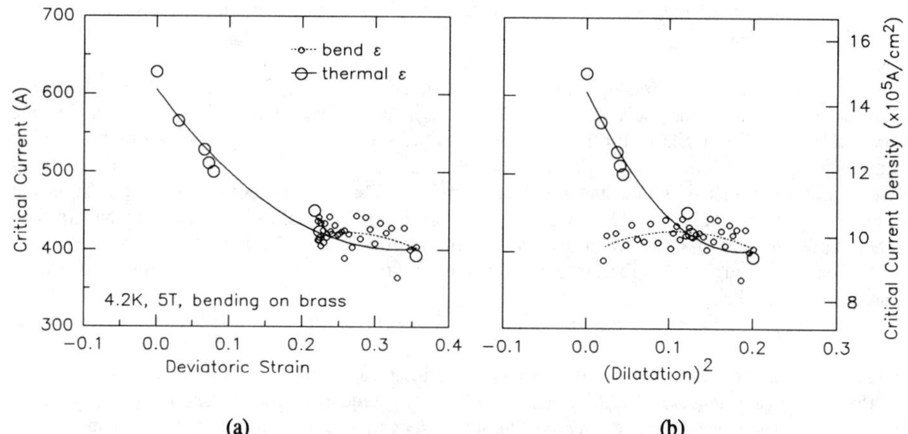

Figure 10. The measured critical current plotted versus a) sum of the squares of the deviatoric strain components and b) dilatation squared.

Conclusions

1) Strains were applied to superconducting Nb3Sn foil using two methods: differential thermal contraction and three-point bending. The critical current was measured at various strain states.

2) A triaxial strain state was determined in the Nb3Sn layer for both methods of strain application.

3) Results show that the axial strain state in the Nb3Sn is not a sufficient parameter for understanding the strain dependence of I_c for liquid-pahse diffusion processed Nb3Sn foil. Rather, the three-dimensional strain state should be used to relate strain and critical current.

4) The application of a relationship developed by Testardi and Welch to this data suggests that the critical current can be more accurately related to the deviatoric strain components than to the axial strain.

Acknowledgements

The authors are grateful to Howard R. Hart, Jr., James W. Bray, Bruce Knudsen, and Neil A. Johnson for valuable discussions, advice, and experimental assistance.

Appendix A. Calculation of Thermally Induced Strains in Nb$_3$Sn

The model for the thermal strain calculation is an elastic, symmetric, 5-layer laminate of superconducting foil bounded by Cu foil and then by thicker metal substrates. The laminate consists of a-b-c-b-a stacking, where a is the substrate material, b is the Cu foil, and c is the Nb-Nb$_3$Sn foil. The solder connecting the laminae is a small fraction of the total plate thickness. Thus the shear of this layer is small and can be neglected, and the interface between the foil and substrate is considered ideal. The Nb and Cu foils are also thin, and thus their shear strains are negligible. The shear force acting on the thicker substrate is small, resulting in small shear strains. Thus, the shear in the laminae are also neglected. Similarly, plasticity of the substrate material is neglected since the strains in this layer are small.

The xyz axes are aligned with the principal directions, where x, y, and z are the axial, thickness, and width directions, respectively. For a homogeneous, isotropic, and elastic medium, then only two elastic constants are necessary to describe the stress state: Poisson's ratio, ν, and Young's modulus, E. Since there is no constraint in the y-direction, a state of plane stress exists and σ_y is zero. Hooke's law for the stress and strain reduces to

$$\begin{aligned}
\varepsilon_x &= (\sigma_x - \nu\sigma_z)/E \\
\varepsilon_z &= (\sigma_z - \nu\sigma_x)/E \\
\varepsilon_y &= -(\nu\sigma_x + \nu\sigma_z)/E
\end{aligned} \tag{A1}$$

Ideal interfaces mean that the total strains are compatible, in both the x- and z-directions, or $\varepsilon_{tot(a)} = \varepsilon_{tot(b)} = \varepsilon_{tot(c)}$ where $\varepsilon_{tot} = \varepsilon_{thermal} + \varepsilon_{mechanical}$. If $\alpha_{(a)} > \alpha_{(b)} > \alpha_{(c)}$, where α is the thermal expansion coefficient, then there will be compressive stresses in layers a and b and tensile stress in layer c with a decrease in temperature. In the x-direction, compatibility of strains gives

$$\alpha_{(a)}\Delta T + \frac{\sigma_{x(a)} - \nu\sigma_{z(a)}}{E_{(a)}} = \alpha_{(b)}\Delta T + \frac{\sigma_{x(b)} - \nu\sigma_{z(b)}}{E_{(b)}} \tag{A2}$$

$$\alpha_{(a)}\Delta T + \frac{\sigma_{x(a)} - \nu\sigma_{z(a)}}{E_{(a)}} = \alpha_{(c)}\Delta T + \frac{\sigma_{x(c)} - \nu\sigma_{z(c)}}{E_{(c)}}$$

A force balance requires that $A_{(a)}\sigma_{x\ (a)} - A_{(b)}\sigma_{x\ (b)} - A_{(c)}\sigma_{x\ (c)} = 0$ where A is the cross-sectional area in the y-z plane. Isotropic thermal expansion coefficients mean that the thermal strains in the x- and z- directions are equal, or $\alpha_{x\ (c)}\Delta T = \alpha_{z(c)}\Delta T$. Since the area ratios, ie. $A_{(a)}/A_{(b)}$, are the same in the x- and z-directions, namely $A_{(a)}/A_{(b)} = t_{(a)}/t_{(b)}$, where t is the layer thickness, then $\sigma_{x\ (a)} = \sigma_{z\ (a)} = \sigma_{(a)}$ and similarly for layers b and c. Combining equations (2) and (3) yields the stress in the Nb-Nb$_3$Sn foil layer, or layer c,

$$\sigma_{(c)} = \frac{E_{(c)}E_{(a)}t_{(a)}\Delta T \left[\alpha_{(a)} - \alpha_{(c)} - \dfrac{E_{(b)}t_{(b)}}{E_{(a)}t_{(a)}} (\alpha_{(c)} - \alpha_{(b)}) \right]}{(1-\nu)[E_{(c)}t_{(c)} - E_{(b)}t_{(b)} - E_{(a)}t_{(a)}]} \tag{A3}$$

The residual strains in the Nb$_3$Sn are then

$$\varepsilon_{x\ (c)} = \varepsilon_{z(c)} = \left[\frac{(1-\nu)}{E_{(c)}} \right] \sigma_{(c)} \tag{A4}$$

$$\varepsilon_{y(c)} = \left[\frac{-2\nu}{(1-\nu)} \right] \varepsilon_{x\ (c)}$$

353

This calculation is valid for a brass substrate only, since $\alpha_{(brass)} > \alpha_{(Cu)} > \alpha_{(NbsSn)}$ (see Table I). A similar calculation can be done for $\alpha_{(c)} < \alpha_{(a)} < \alpha_{(b)}$, which applies to Cu, stainless steel, stelloy, Inconel, and Ni substrates. For the substrates Nb, W, and Mo, the coefficients of thermal expansion are related by $\alpha_{(a)} < \alpha_{(c)} < \alpha_{(b)}$.

Appendix B. Strain Calculation for Applied Bending

For calculating the bending strains in the Nb$_3$Sn, the composite beam is again assumed to be elastic and isotropic and shearing strain in the solder layer is neglected. Elementary beam theory is used to calculate to calculate the bending strains in three-point bending. The geometry of the beam, shown in Figure B1, suggests a state of plane strain, since the beam is wide and long. In this case, the strains are

$$\varepsilon_x = (1-\nu^2)\left(\sigma_x - \frac{\nu}{(1-\nu)}\sigma_y\right)/E \tag{A5}$$

$$\varepsilon_y = (1-\nu^2)\left(\sigma_y - \frac{\nu}{(1-\nu)}\sigma_{x'}\right)/E$$

$$\varepsilon_z = 0.$$

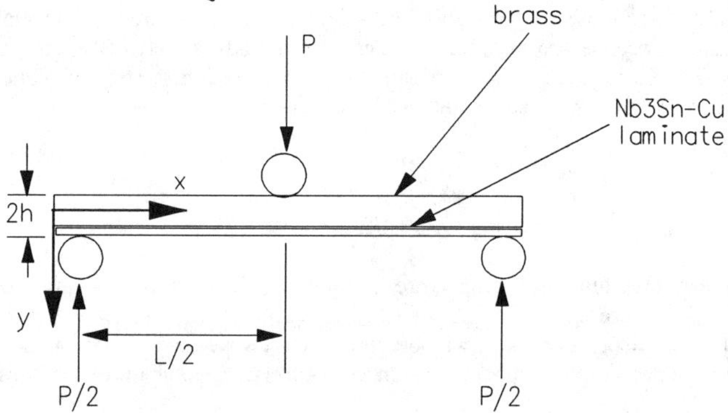

Figure B1. Geometry of three-point bending test on Nb-Nb3Sn laminate soldered to a brass substrate. A similar geometry is used for testing on Nb substrates.

The stresses are $\sigma_x = M_z y / I_z$ and $\sigma_y = 0$, where y is the distance from the neutral axis, M_z is the moment about the z-axis and I_z is the moment of inertia about the z-axis. From Timoshenko[14], in three-point bending, the thickness stress, σ_y, due to the lateral loading is zero at the outer fibers of a beam near the point of application of the load. This is the location that we are interested in determining the strains, since this is the location of the critical current measurement.

For a wide beam, the deflection in the y-direction is given by

$$\frac{d^2y}{dx^2} = \frac{M_z(1-\nu^2)}{EI_z} \tag{A6}$$

Integration of this equation gives the deflection, and the maximum stress is found

$$\sigma_{x(max)} = \frac{12Eh}{L^2(1-\nu^2)}y_{max} \tag{A7}$$

where h is the distance from the neutral axis, or one-half the beam thickness plus the Cu foil thickness. Hooke's law then gives the maximum strains,

$$\varepsilon_{x(max)} = \frac{12h}{L^2}y_{max}, \qquad \varepsilon_{y(max)} = \frac{-\nu}{(1-\nu)}\varepsilon_x, \qquad \varepsilon_z = 0 \tag{A8}$$

If the beam is considered to be narrow, then a state of plane stress exists. The only stresses on the outer fibers of the beam are $\sigma_x = M_z y/I_z$, and $\sigma_y = \sigma_z = 0$. Hooke's law for plane stress is then $\varepsilon_x = \sigma_x/E$, $\varepsilon_y = \varepsilon_z = -\nu\varepsilon_x$, and the deflection equation is

$$\frac{d^2y}{dx^2} = \frac{M_z}{EI_z} \tag{A9}$$

and the resulting maximum stress and strains are

$$\sigma_{x(max)} = \frac{12Eh}{L^2}\, y_{max} \tag{A10}$$

$$\varepsilon_{x(max)} = \frac{12h}{L^2}\, y_{max}\,, \qquad \varepsilon_{y(max)} = \varepsilon_{z(max)} = -\nu\varepsilon_x$$

Plane stress and plane strain states thus have similar axial strains, ε_x, but different transverse and thickness strains, ε_z and ε_y.

References

1. G. Rupp, "The Importance of Being Prestressed," in Multifilamentary A15 Superconductors, ed. M. Suenaga and A.F. Clark (Plenum Press, New York, 1980) 155-170.
2. J.W. Ekin, "Strain Scaling Law and the Prediction of Uniaxial and Bending Strain Effects in Multifilamentary Superconductors," in Multifilamentary A15 Superconductors, ed. M. Suenaga and A.F. Clark (Plenum Press, New York, 1980) 187-203.
3. E.L. Hall, M.G. Benz, L.E. Rumaner, and K.D. Jones, "Interface Structure, Grain Morphology, an Kinetics of Growth of the Superconducting Intermetallic Compound Nb_3Sn Doped With ZrO_2 and Copper," Mater. Res. Soc. Symp. Proc., 205 (1992), 263-268.
4. Mark G. Benz, unpublished research, 1992.
5. J.R. Davis et al, eds., Metals Handbook, vol. 2 (Metals Park, OH: American Society for Metals, 1990)
6. Y.S. Touloukian et al, eds., Thermophysical Properties of Matter, vol. 12 (New York: IFI/Plenum).
7. R.J. Corruccini and J.J. Gniewek, "Thermal Expansion of Technical Solids and Low Temperatures," NBS Monograph 29 (1961).
8. L.R. Testardi, "Elastic Behavior and Structural Instability of High Temperature A-15 Structure Superconductors," Physical Acoustics, vol 10, ed. W.P. Mason and R.N. Thurston (Academic Press, New York, 1973) 193-292.
9. S. Timoshenko, "Analysis of Bi-Metal Thermostats," J.O.S.A. & R.S.I, 11 (1925), 233-255.
10. D.M. Kroeger, D.S. Easton, C.C. Koch, and A. DasGupta, "Evidence for Microstructural Effects Under Strain in Bronze Process Nb_3Sn," in Multifilamentary A15 Superconductors, ed. M. Suenaga and A.F. Clark (Plenum Press, New York, 1980) 205-220.
11. H. Sekine, K. Inoue, T. Kuroda and K. Tachikawa, "Strain Dependence of the Critical Current of Nb_3Sn Superconductors Doped with Hf and Ga," Cryogenics, 29 (1989) 96-100.
12. D.O. Welch, "Alteration of the Superconducting Properties of A15 Compounds and Elementary Composite Superconductors by Nonhydrostatic Elastic Strain," Advances in Cryo. Eng., 26 (1980) 48-65.
13. B.T. Haken, A. Godeke, and H.J. ten Kate, "The Influence of Various Strain Components in the Critical Parameters of Layer Shaped Nb_3Sn ," to be published in ICMC Proc. (1994).
14. S. Timoshenko and J.N. Goodier, Theory of Elasticity, (New York: McGraw Hill Book Co., 1951), 99.
15. D.S. Easton, D.M. Kroeger, W. Specking and C.C. Koch, "A Prediction of the Stress State in Nb_3Sn Superconducting Composites," J. Appl. Phys., 5 (1980) 2748-2757.

Subject Index

critical current density (Jc),
14-18, 20
four-point probe dc
measurement, 15
Electron microscopy
electron backscattered patterns,
317
kikuchi lines, 317

F
Flux pinning, 333-335

G
Grain boundaries, 279-288
Grain orientation, 186, 190, 191

H
High-T_c cuprate superconductor
Tl-2223, 52-58
powder-in-tube (PTI), 53
pancake coil, 58
solid-state reaction, 53
superconducting magnet, 58
High-T_c superconductors
Ag-clad Bi-2223 conductor, 14, 15,
17, 19, 20
powder-in-tube process (PIT),
14, 19, 20
rolling, 15, 17, 20
texturing, 16, 17
thermomechanical treatment, 14,
16
uniaxial pressing, 15-17
Ag-sheathed Bi-2223, 32
bend strain, 33, 37, 38
bend test, 37
critical current density, 32-34,
36-38
multifilament tapes, 32-34, 37, 38
powder-in-tube, 32
brick-wall model, 293, 294, 296
dynamic magnetic compaction,
254
I-V curves, 255
monofilamentary, 270

silver sheathed tape, 270, 275
spray-pyrolyzed, 290
thallium powder-in-tube, 244
thick film, 245
Hot isostatic pressing (HIP)
processing step, 132-133

J
J_c-strain, 153-155

L
Long conductors
high temperature superconductor,
23, 25, 26
monofilament tapes, 23-25
pancake coils, 23, 26, 27
powder-in-tube (PIT) process, 23
silver-clad Bi-2223, 23
test magnets, 23, 26, 27
wind and react, 23
pancake shaped coils, 14, 15, 20
test magnet, 14, 20
two-step rolling, 14, 20
wind and react technique, 14, 15,
20
Low temperature superconductors
multifilamentary Nb-based
lengths, 324

M
Magnetic fields
the effect on critical current
densities of Bi-2212 tapes at 4.2K,
91-93, 95, 97, 98
of Bi-2212/Ag pancake coil
magnets, 91, 93-95, 97, 98
Mechanical testing
bending of Nb_3Sn foil, 345
compression of Nb_3Sn foil, 344
Microstructure of Bi-2212/Ag
composites)
alkaline earth cuprates, 75, 76, 79
imperfect grain alignment, 71, 76,
77
voids, 74, 76, 79

Author Index